ISNM
International Series of Numerical Mathematics
Vol. 139

International Series of Numerical Mathematics, Vol. 139

Supplement

Optimal Control of Complex Structures

International Conference in Oberwolfach, June 4–10, 2000

K.-H. Hoffmann
I. Lasiecka
G. Leugering
J. Sprekels
F. Tröltzsch
Editors

Springer Basel AG

Corresponding Editor:

Günter Leugering
Fachbereich Mathematik, Arbeitsgruppe 10
Technische Universität Darmstadt
Schlossgartenstrasse 7
64289 Darmstadt
Germany
e-mail: leugering@mathematik.tu-darmstadt.de

2000 Mathematics Subject Classification 49-06; 49J20

A CIP catalogue record for this book is available from the Library of Congress, Washington D.C., USA

Deutsche Bibliothek Cataloging-in-Publication Data

Optimal control of complex structures : international conference in
Oberwolfach, June 4 - 10, 2000 / K.-H. Hoffmann ... ed.. - Basel ; Boston ; Berlin : Birkhäuser, 2002
(International series of numerical mathematics ; Vol. 139)

ISBN 978-3-0348-9456-2 ISBN 978-3-0348-8148-7 (eBook)
DOI 10.1007/978-3-0348-8148-7

Originally published by Birkhäuser Verlag, Basel, Switzerland in 2002

Printed on acid-free paper produced of chlorine-free pulp. TCF ∞

9 8 7 6 5 4 3 2 1

www.birkhäuser-science.com

Contents

Preface

The international **Conference On Optimal Control of Complex Structures** was held at the **Mathematisches Forschungsinstitut Oberwolfach** (www.mfo.de, Germany) from June 4 to 10, 2000. The conference was attended by 44 scientists from eight countries. The scientific program included 35 invited talks in the areas of optimization, optimal control, shape optimization, controllability and stabilizability, as well as numerical simulation and mathematical modelling related to applications in control and optimization of complex structures.

Interest in the area of control of systems defined by partial differential equations has increased strongly in recent years. A major reason has been the requirement of these systems for sensible continuum mechanical modelling and optimization or control techniques which account for typical physical phenomena. Particular examples of problems on which substantial progress has been made are the control and stabilization of mechatronic structures, the control of growth of thin films and crystals, the control of Laser and semi-conductor devices, and shape optimization problems for turbomachine blades, shells, smart materials and microdiffractive optics.

The intention of the organizers of the conference was to bring together top scientists from the international community in order to establish a common understanding of the state-of-the-art in this expanding and exciting field of applied mathematics and to both enhance and encourage interdisciplinary research. Therefore, a majority of the 21 selected and refereed papers presented in this volume concentrate precisely on modelling analysis and optimal control or shape design in the context of the above mentioned applications.

The demand for more sophisticated mathematical models, optimization-, control-, identification- and simulation techniques, has enhanced analytic research such as that found in controllability problems, asymptotic models of piezoelectric shells, and shape-derivative calculus. On the other hand, numerical simulations, scientific computing and, in particular, the relation between discretization and control in the context of partial differential equations have also proven to be useful techniques in the solution of real manifestations of these problems.

This volume highlights some of the most fruitful directions of research being currently pursued and offers a wealth of open problems which might stimulate even the inexperienced reader to take part in their solutions. These problems offer opportunities for exploration of deep mathematical ideas in many clear-cut mathematical fields of research. At the same time, they invite development of the mathematical tools needed for modelling, analysis of control/optimization and simulation of complex industrial processes in a broad spectrum of mechanical-, chemical- and civil engineering.

The editors express their gratitude not only to the contributors to this volume but also to the Oberwolfach Institute which again proved to be a unique location with an inspiring atmosphere enhancing research and discussions.

K.-H. Hoffmann
I. Lasiecka
G. Leugering
J. Sprekels
F. Troeltzsch

International Series of Numerical Mathematics, Vol. 139, 1–17

Reduced Order Based Compensator Control of Thin Film Growth in a CVD Reactor

H.T. Banks and H.T. Tran

Abstract. This paper reports on an interdisciplinary effort, which involves applied mathematicians, material scientists and physicists at North Carolina State University, to integrate new intelligent processing approaches with advanced mathematical modeling, optimization, and control theory to guide the construction and experimental implementation of a series of high pressure (up to 100 atm) organometallic chemical vapor deposition (CVD) reactors. An integral component of this research program is the design of the reactor so that control and sensing are a basic component of the optimal design efforts for the reactor. We report here on the successful use of mathematics in a fundamental role in the development of linear and nonlinear feedback control methods for real-time implementation on the reactor. This is achieved in the required context of gas dynamics coupled with nonlinear surface deposition processes. The problems are optimal tracking problems (for the chemical component fluxes over the substrate) that employ state-dependent Riccati gains with nonlinear observations and the resulting dual state dependent Riccati equations for the compensator gains. This control methodology is successfully combined with reduced order model methods based on proper orthogonal decomposition techniques. Computational results to support the efficacy of our approach and methods are also included.

1. Introduction

Chemical vapor deposition (CVD) is an important industrial technique used to grow thin films with certain desired properties. This process involves the deposition of precursor vapor sources onto a heated substrate where they react to form the desired material. CVD is a key element in a wide variety of advanced industrial applications, ranging from the development of short wavelength light sources to detectors and integrated sensors, in particular the integration of III-V optoelectronics and silicon technology. In addition, wide bandgap materials are of particular interest for advanced silicon ULSI technology in the context of dielectric isolation, vertical integration, optically interconnected common memory, integrated sensors and microwave applications. These materials are also of considerable interest in the context of optoelectronics for applications in displays, optical recording, signal processing, printing and medicine [1]. In these processes, the control of the concentrations and distributions of both point defects and extended defects and the related control of the surface morphology and interfacial chemistry for compound semiconductor heterostructures are essential because of their important role in the control of the electrical/optical properties and the reliability of wide bandgap semiconductor devices and circuits.

For the past five years, an interdisciplinary effort at North Carolina State University has been carried out to explore new intelligent processing approaches that access conditions outside the capabilities of conventional methods. Generally the control of the stoichiometry, and the related issue of the control of the point defect chemistry of the heteroepitaxial layers of mixed III-V compounds requires, under the conditions of thermally activated growth, high ratios in the fluxes of the group V to the group III source materials. In this context we have explored organometallic CVD (OMCVD) under superatmospheric conditions (up to 100 atm), where high partial pressure ratios can be established without compromise with respect to the growth rate. Because the process is operating at high flows/vapor densities, control of the fluid dynamics becomes essential for optimal growth conditions. Therefore advanced methods of mathematical modeling, optimization, and control theory have been applied to guide the development and experimental implementation of these processes. In particular, advanced simulations for flow processes in a computer-aided design (CAD) mode resulted in several generations of reactor geometries before a suitable configuration that promised desirable flow characteristics near the substrate was obtained. In this presentation we discuss a third generation reactor design (see Fig. 1) resulting from this developmental process. For a fourth generation reactor design see [23].

The mathematical models for OMCVD (under superatmospheric pressure) possess one of the most complex fluid dynamics systems imaginable. Some of the complex issues in computing chemically reacting flows include the simulation of three-dimensional flows governed by Navier-Stokes equation coupled with equations for energy and species and strong temperature dependence of the physical properties of gases. Chemical reactions are taking place in the gas-phase as well as on the substrate. However, since only a trace amount of reactants mixed with carrier gas is used, a dilute approximation is assumed. This leads to a quasi-steady gas-phase model with steady-state nonlinear coupled system of equations for the continuity, momentum and energy that is decoupled from the time dependent species equations. This gas phase model is coupled with a reduced order model of the surface reactions involved in the decomposition of source vapors from the gas phase and the growing film on the substrate.

The resulting mathematical model is a system of partial differential equations coupled with nonlinear ordinary differential equations describing the surface deposition process. Numerical simulations and control designs and syntheses of such systems are faced with considerable challenges regarding dimensionality and nonlinearity. This paper describes our efforts during the last five years to overcome these difficulties. More specifically, §2 describes the proper orthogonal decomposition (POD) technique, also known as the Karhunen-Loève procedure, that is used to obtain low dimensional dynamic models of distributed parameter systems. The POD method, which is well known in statistical and pattern recognition fields [2], has been shown to be an effective tool for the analysis of complex systems such as turbulence flows, shear flows, and weather prediction (see e.g., [3] and the references therein). Roughly speaking, POD is an optimal technique of finding a basis that spans an ensemble of data, collected from an experiment or a numerical simulation of a dynamical system, in the sense that when these basis functions are used in a Galerkin procedure, they will yield a finite dimensional system with the smallest possible degrees of freedom. Thus this method may well be suited to treat optimal control and parameter estimation of distributed parameter systems. In §3, we describe our successful use of POD techniques as a reduced basis method for computation of feedback controls and compensators in a high pressure CVD reactor. More specifically, we present a proof-of-concept computational implementation of this method with a simplified growth example of group III-V compounds that includes multiple species and controls, gas phase reactions (no surface reactions), and time dependent tracking signals that are consistent with pulsed vapor reactant inputs. In §4 state estimation and feedback tracking control methods for nonlinear systems are presented. The methods, which are based on the "state-dependent Riccati equations", allow the construction of nonlinear estimators and nonlinear feedback tracking controls for a wide class of systems including high pressure CVD systems considered here. The performance of the nonlinear estimator and tracking control will be presented on a flight dynamics simulation example. Finally, §5 contains our overall conclusions.

2. Proper Orthogonal Decomposition

In general, the discretization of linear/nonlinear partial differential equations using finite element, finite volume, or finite different methods involves basis functions that have little to do with the differential equation. For example, piecewise polynomials are used in the finite element method, grid functions are used in the finite difference method, and Legendre or Chebyshev polynomials are used in some spectral methods. POD, on the other hand, uses basis functions that span a data set, collected from an experiment or numerical simulation of a dynamical system, in a certain "optimal" fashion. Because POD basis elements are optimal in the sense that they are the extractions of characteristic features of the data set, frequently only a small number of POD basis functions are needed to describe the solution. POD based approximation methods have been applied to numerous applications including turbulent coherent flows [4], shear flows [5], characterization of human faces [6], and image recognition [7]. More recently, the possibility of POD based control design and parameter estimation has been proposed. In particular, applications of POD to optimization or open loop control were developed in [8, 9, 10], to feedback control design were reported in [11, 12, 13, 14, 15], and to parameter estimation or inverse problems were discussed in [16]. References to recent work of other authors can be found in [11]–[16] and [23].

We now outline an algorithm to obtain the POD basis element. The mathematical basis for the algorithm has been described in numerous articles (see e.g., [8]). Let $\{\mathbf{U}_i(\vec{x}) : 1 \leq i \leq N; \vec{x} \in \Omega\}$ denote the set of N observations (also called *snapshots*) of some physical processes over a domain Ω. In the context of CVD process, these observations could be experimental measurements or numerical solutions of velocity fields, temperatures, species etc. taken at different physical parameters (Reynolds number, input flow rates etc.) or time steps.

Step 1. *Compute the covariant matrix* $\mathbf{C}$. The matrix elements of $\mathbf{C}$ are given by

$$\mathbf{C}_{ik} = \frac{1}{N} \int_{\Omega} \mathbf{U}_i(\vec{x}) \mathbf{U}_k(\vec{x}) d\vec{x},$$

for $i, k = 1, 2, \ldots, N$.

Step 2. *Solve the eigenvalue problem* $\mathbf{CV} = \lambda \mathbf{V}$. Since $\mathbf{C}$ is a nonnegative, Hermitian matrix, it has a complete set of orthogonal eigenvectors

$$\mathbf{V}^1 = \begin{bmatrix} a_1^1 \\ a_2^1 \\ \vdots \\ a_N^1 \end{bmatrix}, \mathbf{V}^2 = \begin{bmatrix} a_1^2 \\ a_2^2 \\ \vdots \\ a_N^2 \end{bmatrix}, \ldots, \mathbf{V}^N = \begin{bmatrix} a_1^N \\ a_2^N \\ \vdots \\ a_N^N \end{bmatrix}$$

with the corresponding eigenvalues arranged in ascending order as $\lambda_1 \geq \lambda_2 \geq \cdots \geq \lambda_N \geq 0$.

Step 3. *Compute the POD basis vectors.* The POD basis elements $\Phi_i(\vec{x})$ such that $X^{\text{POD}} = \text{span}\{\Phi_1, \Phi_2, \ldots, \Phi_N\}$, where X^{POD} is the finite-dimensional POD space, are given as

$$\Phi_k = \sum_{i=1}^{N} a_i^k \mathbf{U}_i,$$

where $1 \leq k \leq N$ and a_i^k are the elements of the eigenvector $\mathbf{V}^k$ corresponding to the eigenvalue λ_k.

To approximate a distributed parameter system by a finite-dimensional problem one uses a combination of Galerkin procedures and POD basis elements (see [11, 12] for details). However, to this point we have not discussed any model reduction features associated with using POD basis elements in approximation schemes. In the algorithm described above, the number N may be large, $100 - 1000$ or even more, depending on the complexity of the dynamics represented in the "snapshots" $\mathbf{U}_i$. In general, one should take N sufficiently large so that the snapshots $\mathbf{U}_i$ contain all salient features of the dynamics being investigated. Thus, the POD basis functions Φ_i, used with the original dynamics in a Galerkin procedure, offers the possibilities of achieving a high fidelity model, albeit with perhaps a large dimension N.

To achieve model reduction, one chooses $M \ll N$ and carries out a Galerkin procedure with the set of elements $\{\Phi_1, \Phi_2, \ldots, \Phi_M\}$. The crucial question is how to choose M. As discussed elsewhere (see e.g., [11, 12]) the percentage of the total snapshots set data variability contained in a certain POD mode Φ_k is given by the ratio of the eigenvalue λ_k to the total of all eigenvalues, $\lambda_k / \sum_{j=1}^{N} \lambda_j$. The reason for ordering the POD modes from highest to lowest eigenvalues is to include as much of the variability of the system into the first few modes as possible. Therefore to capture most of the data variability of the system contained in the N POD elements, it suffices to choose M, where M is sufficiently smaller than N, so that $\sum_{i=1}^{M} \lambda_i \approx \sum_{i=1}^{N} \lambda_i$. For the CVD examples studied in [8, 11, 12], the POD system was constructed for $N = 100 - 200$ and a reduced order model with $M = 2 - 10$ resulted in a truly significant computational savings.

While the above comments suggest the proper choice for the dimension of the reduced order model to be used in simulations, there are additional order questions related to the linear control system to be used in determining reduced order gains and compensators. We have found that the ranks of the controllability and observability matrices have sometimes been useful criteria (see the discussion in [13, 11, 12]) to help in the choice of the number of modes to use in control design applications for the reduced basis representation.

In the next section, we present a proof-of-concept computational implementation of this method with a simplified growth example for III-V layers. In this example we implement Dirichlet boundary control of dilute reactants transported by convection and diffusion to an absorbing substrate after they undergo gas phase reactions.

3. Application of POD to Compensator Control of CVD

The particular geometry of the differentially pressure controlled (DPC) reactor system under consideration here features horizontal flow of the process gases and source vapor/carrier gas mixtures into an expansion section leading into a rectangular channel that contains the substrate (see Fig. 1). The substrate wafer is mounted on a rotating induction heated SiC coated graphite susceptor. The exhaust gases are vented through a vertical exhaust tube. Loading and unloading of substrate wafers is accomplished through a load-lock chamber beneath the radio frequency (rf) section of the reactor that can be evacuated by a turbomolecular pump. After purging with ultra-pure nitrogen, sample transfer can be executed using a magnetic transfer rod. Gas is purging through the gap between the susceptor and the reactor's base to avoid flow of gas mixtures to the mechanical workings behind the susceptor. The quartz glass reactor is connected at the inlet to a source vapor/process gas flow control and switching panel that directs individual streams of source vapor saturated carrier gas either to a vent line or to the reactor. Thus, pulsed operation separating plugs of source vapor saturated carrier gas by plugs of high purity carrier gas, flow rate modulated flow or continuous flow can be implemented for all source vapors without change in reactor pressure or total flow. Two optical windows at the Brewster angle of the substrate are attached to the sides of the reactor. They allow for the real-time process monitoring utilizing p-polarized reflectance spectroscopy (PRS) (see e.g., [17] and the references therein).

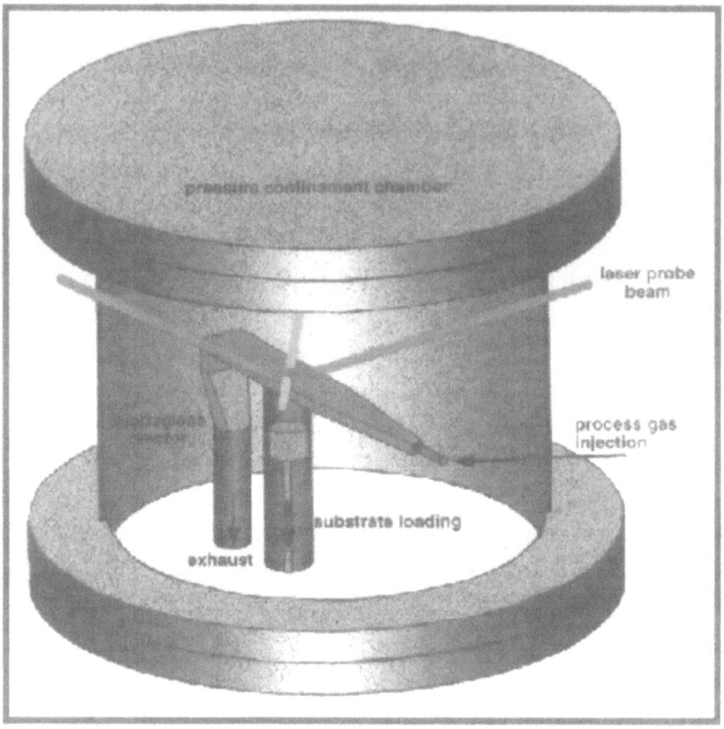

FIGURE 1. Schematic representation of a horizontal, quartz reactor in a steel confinement shell

To demonstrate the feasibility of using POD technique as a reduced basis method for computation of feedback controls and compensators in a high pressure CVD reactor, we will restrict our study to a two-dimensional rectangular domain (Fig. 2) representing the longitudinal cross section through the center of the reactor. We consider the deposition of InP using pulsed trimethyl-indium (TMI) and phosphine (PH_3) as source vapors and hydrogen as carrier gas. In particular, at first only carrier gas flows through the reactor. After the flow reaches steady state, a pulse of reactant (e.g., TMI) diluted with carrier gas enters the reactor. After the pulse, the reactor is then flushed with carrier gas. This process is then repeated for another reactant. Pulsing of the III-V source materials prevents nucleation of the film in the gas phase and make PRS observation and analysis possible.

We consider only trace amounts of reactants mixed with the carrier gas. Under this dilute approximation, we can classify CVD processes as a quasi-transient

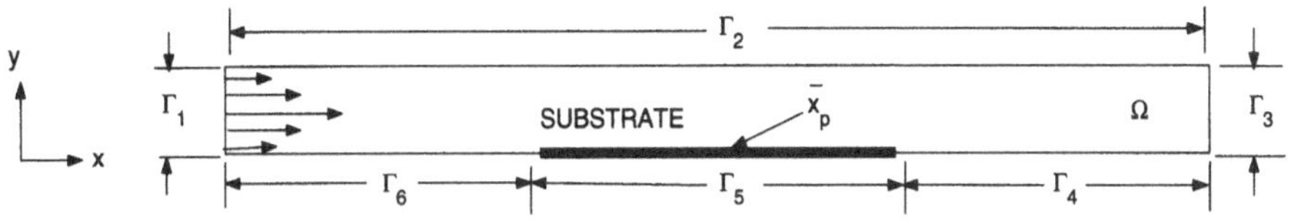

FIGURE 2. Two-dimensional cross section of the CVD reactor (height=0.011m, length=0.156m, substrate length=0.048m)

flow (steady-state flow with transient species). The steady-state flow is described by the following set of equations [12]
(continuity)

$$\vec{\nabla}\cdot(\rho\vec{v})=0, \tag{1}$$

(momentum)

$$\rho\vec{v}\cdot\vec{\nabla}\vec{v}=-\vec{\nabla}P+\vec{\nabla}\cdot\vec{\tau}-\rho\vec{g}, \tag{2}$$

where the viscous stress tensor is of the form

$$\vec{\tau}=-\frac{2}{3}\mu(\vec{\nabla}\cdot\vec{v})\vec{I}+\mu(\vec{\nabla}\vec{v}+\vec{\nabla}\vec{v}^{T}), \tag{3}$$

(energy)

$$\rho c_p\vec{v}\cdot\vec{\nabla}T=\vec{\nabla}\cdot(k\vec{\nabla}T), \tag{4}$$

where $\vec{g}$ is the gravitational acceleration, $\vec{u}$, T, and P are the velocity, temperature, and pressure, μ, c_p, and k are the viscosity, specific heat, and thermal conductivity of the carrier gas. The density variations are modeled as [12]

$$\rho=\rho_0[1-\beta(T-T_0)], \tag{5}$$

where T_0 is a reference temperature, ρ_0 is a reference density calculated from the ideal gas law at the reference temperature and reactor pressure, and β is the volume coefficient of expansion ($\beta = 1/T$). In addition, we consider a hydrogen carrier gas at atmospheric pressure. Temperature dependent values for μ, k, and c_p are linearly interpolated from measurements taken from the available literature [12]. A parabolic velocity flow profile is specified at the inlet ($\Gamma 1$), with an average inlet velocity of 0.1147 m/s. No slip (zero velocity) boundary conditions are imposed on those portions of the model corresponding to the reactor walls ($\Gamma 2, \Gamma 4, \Gamma 5$, and $\Gamma 6$). Room temperature boundary conditions are imposed at the inlet and along the upper wall ($\Gamma 2$). Along the bottom wall, the substrate ($\Gamma 5$) temperature is fixed at 800°K , with a non-linear temperature decrease from the substrate edge to the inlet ($\Gamma 6$) and, similarly, from the substrate edge to the outlet ($\Gamma 4$) (see Fig. 2).

Steady-state solutions for v, T, and ρ obtained from equations (1–4) are then used in the time-dependent species equations for the precursor mass fractions [12],

$$\frac{\partial Y_n}{\partial t}+\vec{v}\cdot\vec{\nabla}Y_n=\frac{1}{\rho}\vec{\nabla}\cdot\left(\rho D_n\vec{\nabla}Y_n\right)+\sum_{i=1}^{N_R}r_{ni}\,, \tag{6}$$

where D_n is the diffusivity of the species, Y_n is the mass fraction of the nth species, N_R is the number of gas phase reactions, and r_{ni} is the rate of production of species n in the ith chemical reaction.

Under the reactor conditions considered here (H_2 carrier gas, 800°K substrate temperature, and 1 atm pressure), there are no effective gas phase reaction mechanisms for phosphine, and the only significant gas phase reaction for TMI is the decomposition of TMI to MonoMethylIndium (MMI) and two methyl molecules, $In(CH_3)_3 \rightarrow InCH_3 + 2CH_3$. This reaction can be described as a first-order Arrhenius reaction [12]

$$r_n = \nu_n \frac{W_n}{W_{\mathrm{TMI}}} k_0 e^{(-E/RT)} Y_{\mathrm{TMI}}, \tag{7}$$

where ν_n refers to the stoichiometry of species n in the reaction, W_n and W_{TMI} refer to the molecular weight of species n and TMI, respectively, $k_0 = 5.25 \times 10^{15}$ s^{-1} is the rate constant, and $E = 47.2$ kcal/mol is the activation energy [18].

The tracking control problem that we formulate is to find the mass fractions of TMI and phosphine at the inlet (Γ_1) of the reactor in order to obtain a desired flux $q_T(t)$ of reactants at point $\vec{x}_p$ on the susceptor (Γ_5). That is, we consider to minimize a cost functional of the form

$$J(u) = \int_0^\infty [u'Ru + (q - q_T)'Q(q - q_T)]\, dt, \tag{8}$$

subject to

$$\begin{array}{rcll}
\frac{\partial Y_n}{\partial t} + \vec{v} \cdot \vec{\nabla} Y_n & = & \frac{1}{\rho} \vec{\nabla} \cdot \left(\rho D_n \vec{\nabla} Y_n \right) + \lambda_n Y_1 & \\
Y_n(0, \vec{x}) & = & y_{n0}(\vec{x}) & \\
Y_1(t, \vec{x}) & = & u_1(t) & \text{on } \Gamma 1 \\
Y_2(t, \vec{x}) & = & u_2(t) & \text{on } \Gamma 1 \\
Y_3(t, \vec{x}) & = & 0 & \text{on } \Gamma 1 \\
Y_n(t, \vec{x}) & = & 0 & \text{on } \Gamma 5 \\
\frac{\partial Y_n(t,\vec{x})}{\partial n} & = & 0 & \text{on } \Gamma 2 \cup \Gamma 3 \cup \Gamma 4 \cup \Gamma 6 \\
n & = & 1, 2, 3, &
\end{array} \tag{9}$$

where Y_1, Y_2, and Y_3 refer to the mass fractions of TMI, phosphine, and MMI, respectively; $u_1(t), u_2(t)$ are the controls corresponding to TMI and phosphine; $\lambda_1(T) = -k_0 e^{(-E/RT)}$, $\lambda_2(T) = 0$, and $\lambda_3(T) = (W_3/W_1) k_0 e^{(-E/RT)}$; $\vec{v}$, ρ, T are the steady state solutions to equations (1–4); and $\frac{\partial}{\partial n}$ denotes the outward normal derivative. Finally, the general flux vector, $q(t)$, at the point $\vec{x}_p$ is given by

$$q(t) = \begin{bmatrix} q_{\mathrm{In}}(t) \\ q_P(t) \end{bmatrix} = -\rho \begin{bmatrix} D_1 \frac{W_{\mathrm{In}}}{W_1} \frac{\partial Y_1}{\partial n}\Big|_{\vec{x}_p} + D_3 \frac{W_{\mathrm{In}}}{W_3} \frac{\partial Y_3}{\partial n}\Big|_{\vec{x}_p} \\ D_2 \frac{W_P}{W_2} \frac{\partial Y_2}{\partial n}\Big|_{\vec{x}_p} \end{bmatrix}, \tag{10}$$

where W_{In} and W_P are the molecular weights of indium (a component of Y_1 and Y_3) and phosphorus (Y_2), respectively.

We note that since the methyl (CH_3) molecules do not participate in film growth or otherwise affect the transport properties (under the dilute approximation), we do not include them in the state equations (9). The reactor walls ($\Gamma 2$, $\Gamma 4$, and $\Gamma 6$) are assumed non-absorbing, and the substrate ($\Gamma 5$) is assumed to be perfectly absorbing (concentration of zero). Temperature dependent values for the diffusivities D_n in hydrogen are linearly interpolated from values taken from the available literature [18].

We next use a penalty boundary formulation on the species state equations (9) to change all Dirichlet boundary conditions to Neumann conditions. For example, the Dirichlet condition $Y_1(t,\vec{x}) = u_1(t)$ is reformulated as $\frac{\partial Y_1(t,\vec{x})}{\partial n} = \frac{1}{\epsilon}(Y_1(t,\vec{x}) - u_1(t))$, where ϵ is a small parameter (for most of our calculations we used $\epsilon = 10^{-3}$). This boundary conditions reformulation provides a natural setting for the Galerkin procedure as well as for the control formulation. More specifically, writing the state equation in weak form using test function w_j, integrate by parts, and applying the modified Newmann conditions, we obtain

$$\begin{aligned} \int_\Omega \frac{\partial Y_n}{\partial t} w_j \, d\Omega &= -\int_\Omega \left(\vec{v} \cdot \vec{\nabla} Y_n\right) w_j \, d\Omega - \int_\Omega D_n \vec{\nabla} Y_n \cdot \vec{\nabla} w_j \, d\Omega \\ &\quad + \int_\Omega \frac{1}{\rho} w_j D_n \vec{\nabla} Y_n \cdot \vec{\nabla} \rho \, d\Omega + \int_\Omega \lambda_n Y_n w_j \, d\Omega \\ &\quad + \frac{1}{\epsilon} \int_{\Gamma 1, \Gamma 5} w_j D_n Y_n \, ds - \frac{1}{\epsilon} \int_{\Gamma 1} w_j D_n u_n \, ds \,, \end{aligned} \tag{11}$$

where $n = 1, 2, 3$ and $u_3 \equiv 0$. The mass fraction of the nth species is approximated as a linear combination of the M_n most significant POD basis elements as

$$Y_n^{M_n}(t,\vec{x}) = \sum_{i=1}^{M_n} y_{n,i}(t) \Phi_{n,i}(\vec{x}), \tag{12}$$

where $M_n \ll N$ and $\Phi_{n,i}$ is the ith POD basis element corresponding to the nth species. Using the representation (12) and the orthonormality of the $\{\Phi\}'s$, we apply a Galerkin procedure to the weak form (11) to obtain a system of M ordinary differential equations for the coefficients $y_{n,i}$

$$\dot{y}^M(t) = A^M y^M(t) + B^M u(t), \quad u(t) = \begin{bmatrix} u_1(t) \\ u_2(t) \end{bmatrix}, \tag{13}$$

where $M = \sum_{n=1}^{3} M_n$, A^M is an $M \times M$ matrix, and B^M is an $M \times 2$ matrix.

We remark that the POD modes for each species are constructed from 150 snapshots ($N = 150$) taken during the three second cycle (2 s pulsing, 1 s clearing) of each source species. Each solution vector represents the species mass fraction at the 453 nodal points and corresponds to a time increment of 0.03-s in the time range from 0 to 3 seconds. However, for the reduced order model, we used only 19 POD modes ($M = 19$: $M_{\text{TMI}} = 8$, $M_{\text{Phospine}} = 8$, $M_{\text{MMI}} = 3$), which yield a worst captured variability of 99.995% for MMI ($\sum_{j=1}^{3} \lambda_{\text{MMI}j} / \sum_{k=1}^{150} \lambda_{\text{MMI}k} = 99.995$). While the captured variability suggests the proper order for accurate reduced order model simulations, there are additional order questions related to the control system to be used in determining reduced order gains and compensators.

For example, the ranks of the controllability and observability matrices have sometimes been found to be useful criteria to help in the choice of the number of modes to use in control design applications for the reduced basis representation [11, 12].

The reduced order species state model (13) is linear in both the state and control variables and, therefore, linear control methodologies can be applied to obtain the optimal feedback tracking control. However, the application of this feedback control to the reduced order model requires a linear state estimator since only partial state observations of the fluxes of In and P at the substrate center are available. These partial state observations are compatible with current PRS sensing technology. For detailed calculations of the POD basis elements, the reduced order model, and the compensator-based optimal feedback tracking control see the recent article [12].

To show how well the control/compensator system designed above via the reduced order model performs when used in the actual physical experiments, we computationally test the reduced order control/compensator design on the full system, which is approximated by 453×3 quadratic finite elements. We emphasize that the reduced order model is formulated using only 19 POD basis elements, a substantial order reduction from 453×3. Fig. 3 depicts plots of the observed fluxes as functions of time. It clearly shows that the system is able to closely track the time dependence of the desired flux profile (shown in dotted line) without significant delays. It also confirms that the ability of the system to match the target flux is sensitive to the design parameter Q.

4. Nonlinear Tracking Control and State Estimator

4.1. Tracking control for nonlinear systems

In general, the mathematical model developed in §3 has to be linked with a surface kinetics model describing the decomposition kinetics of the organometallic precursors involved and their incorporation into the film deposition. For the growth of epitaxial GaP heterostructures on Si(001) substrates, a reduced order surface kinetics (ROSK) model is proposed in [19]. This surface kinetics model is a nonlinear system of ordinary differential equations and when combining with the gas phase model will yield a system of nonlinear differential equations. In this section, we summarize our development of nonlinear estimators and nonlinear feedback tracking controls that are applicable to a wide class of systems including the high pressure CVD systems as considered here [20].

Consider the following nonlinear control system

$$\begin{cases} \dot{x}(t) &= f(x(t)) + Bu(x(t), t) \\ x(0) &= x_0 \\ y(t) &= Hx(t), \end{cases}$$

where, for this presentation, the tracking variable y is taking to be a linear function of the state variables (see [20] for the nonlinear tracking variable case). In addition

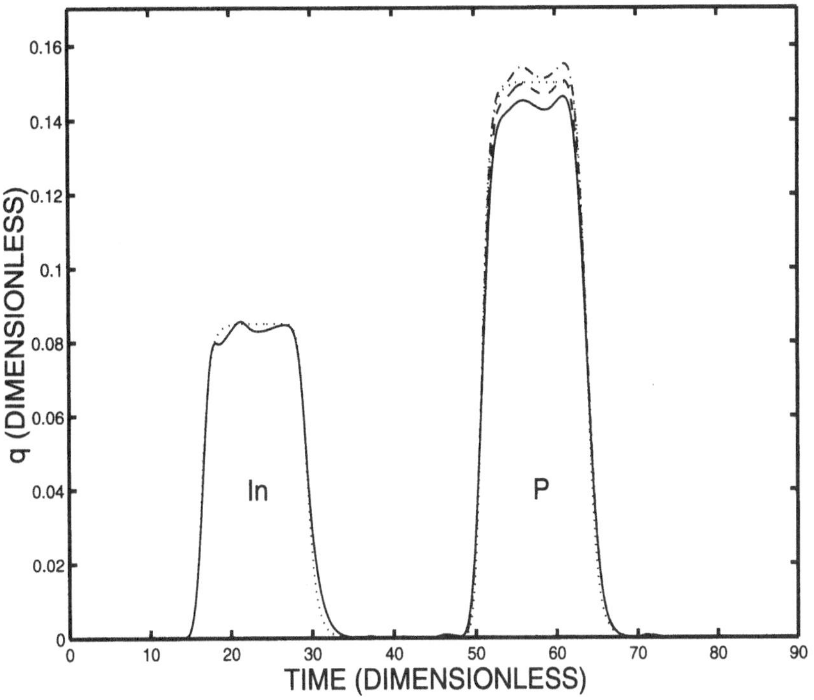

FIGURE 3. Observed fluxes as a function of time for different values of the control parameter Q: small (solid line), medium (dashed line), and large (dash-dot line). The target flux profile is also shown (dotted line) for reference. The system is able to closely track the desired flux profile.

to the nonlinear state equation, the cost function for the tracking problem, with a desired trajectory $r(t)$, is given by

$$J(x_0, u) = \frac{1}{2} \int_0^\infty \left((y-r)^T Q (y-r) + u^T R u \right) dt.$$

Now rewriting the nonlinear function as $f(x) = A(x)x$ and solving the necessary optimality conditions, we obtain the optimal feedback control

$$u(x,t) = -R^{-1} B^T \left[\Pi \left(x(t) \right) x(t) + s(t,x) \right], \tag{14}$$

where $\Pi(x)$ is the solution to the so-called state dependent Riccati equation (SDRE)

$$\Pi \left(x \right) A(x) + A^T(x) \Pi \left(x \right) - \Pi \left(x \right) B R^{-1} B^T \Pi \left(x \right) + H^T Q H = 0 \tag{15}$$

and $s(t,x)$ is the solution to the following two-point boundary value problem

$$\left\{\begin{array}{rcl}\dot{s} & = & -A^T(x_{nom})s+\Pi\left(x_{nom}\right)BR^{-1}B^Ts+H^TQr \\ & & -\sum_{i=1}^m\left(x_{nom}\right)_i\left(\frac{\partial A_{1\to m,i}}{\partial x_{nom}}(x_{nom})\right)^T\left(\Pi\left(x_{nom}\right)x_{nom}+s\right) \\ & & -D_t\Pi\left(x_{nom}\right)x_{nom} \\ \dot{x}_{nom} & = & A(x_{nom})x_{nom}-BR^{-1}B^T\left(\Pi\left(x_{nom}\right)x_{nom}+s\right)\end{array}\right. \tag{16}$$

with $x_{nom}(0)=x_0$ and $s(T_f,x_{nom}(T_f))=0$, (see [20] for details).

The SDRE (15) is solved using a power series approximation. We begin by splitting A into a constant part and a state-dependent part as $A(x)=A_0+\varepsilon g(x)\Delta A_C$, where ε is a temporary variable used for the expansion that will be set to 1 later. We next write Π as a power series in ε, as

$$\Pi(x,\varepsilon)=\sum_{n=0}^{\infty}\varepsilon^n g^n(x)(L_n)_C, \tag{17}$$

where Π as well as each $(L_n)_C$ is symmetric. Substituting these expansions into the state-dependent Riccati equation (15) and matching terms with the same powers of ε we obtain the following set of equations for determining the constant-valued matrices $(L_n)_C$:

$$(L_0)_CA_0+A_0^T(L_0)_C-(L_0)_CBR^{-1}B^T(L_0)_C+Q = 0 \tag{18}$$

$$(L_1)_C\left(A_0-BR^{-1}B^T(L_0)_C\right)+\left(A_0^T-(L_0)_CBR^{-1}B^T\right)(L_1)_C +(L_0)_C\Delta A_C+\Delta A_C^T(L_0)_C = 0 \tag{19}$$

$$(L_n)_C\left(A_0-BR^{-1}B^T(L_0)_C\right)+\left(A_0^T-(L_0)_CBR^{-1}B^T\right)(L_n)_C +(L_{n-1})_C\Delta A_C+\Delta A_C^T(L_{n-1})_C-\sum_{k=1}^{n-1}\left((L_k)_CBR^{-1}B^T(L_{n-k})_C\right) = 0. \tag{20}$$

Equation (18) is the standard Riccati equation for the linear part of the system, A_0, which can be solved easily. Equations (19) and (20) are constant-valued matrix Lyapunov equations, for which stable and efficient algorithms also exist in the literature.

4.2. State estimation for nonlinear systems

We consider a nonlinear control system with a nonlinear measurement of the form

$$\left\{\begin{array}{rcl}\dot{x}(t) & = & f(x(t))+Bu(x_e(t),t) \\ z(t) & = & c(x(t)).\end{array}\right.$$

The control for a tracking problem is given by

$$u(x_e,t)=-R^{-1}B^T\left(\Pi\left(x_e\right)x_e+s(t,x_{nom})\right)$$

as discussed in the last section except now in terms of the estimated state x_e.

The estimated state will be formulated by an ordinary differential equation similar to the state equation, with a gain matrix (found using a dual state-dependent Riccati equation) applied to the difference between the measurements of the actual and estimated states. The coupled actual and estimated states are given by

$$\begin{cases} \dot{x} &= A(x)x - BR^{-1}B^T\left[\Pi(x_e)x_e + s(t, x_{nom})\right] \\ \dot{x}_e &= A(x_e)x_e - BR^{-1}B^T\left[\Pi(x_e)x_e + s(t, x_{nom})\right] \\ & \quad + L(x_e)\left[z - c(x_e)\right], \end{cases} \tag{21}$$

where the state estimation gain is found by

$$L(x_e) = \Sigma(x_e)(C_0)^T V^{-1} \tag{22}$$

with $\Sigma(x_e)$ is the solution to the dual state-dependent Riccati equation

$$\Sigma(x_e)A^T(x_e) + A(x_e)\Sigma(x_e) - \Sigma(x_e)(C_0)^T V^{-1} C_0 \Sigma(x_e) + U = 0. \tag{23}$$

For the purposes of finding the estimator gain in equations (22)–(23) the nonlinear measurement function $z(t) = c(x(t))$ is rewritten as matrix function multiplication $c(x) = C(x)x$ and to choose $C_0 = C(0)$. We note that the nonlinearity of the measurement function does remain in the estimator system (21) itself, and the nonlinearity of the system dynamics remains in (21)–(23). The estimator gain SDRE (23) is solved using the power series approximation in an analogous fashion as described in previous section.

4.3. A flight dynamics example

We apply the nonlinear estimator and feedback tracking control presented in previous sections to the flight dynamics example from [21]. The control system is given by

$$\dot{x} = (A_0 + x_2 A_{NL})\,x + Bu, \qquad z(t) = [x_1, x_2, x_5]^T$$

where the matrices A_0, A_{NL} and B are given by:

$$A_0 = \begin{bmatrix} -0.0443 & 1.1280 & 0.0 & -0.0981 & 0.0 \\ -0.0490 & -2.5390 & 1.0 & 0.0 & -0.0854 \\ -0.0730 & 19.3200 & -2.2700 & 0.0 & 22.6834 \\ 0.0490 & 2.5390 & 0.0 & 0.0 & 0.0854 \\ 0.0 & 0.0 & 0.0 & 0.0 & 20.0 \end{bmatrix}$$

$$A_{NL} = \begin{bmatrix} -0.2317 & 0.0 & 0.0 & 0.0 & 0.0 \\ -1.2760 & -0.7922 & 0.0 & 0.0 & 0.0206 \\ 0.1020 & 64.2940 & -13.9710 & 0.0 & -5.4167 \\ 1.2760 & 0.7922 & 0.0 & 0.0 & -0.0206 \\ 0.0 & 0.0 & 0.0 & 0.0 & 0.0 \end{bmatrix}$$

$$B = \begin{bmatrix} 0.0 & 0.0 & 0.0 & 0.0 & 20.0 \end{bmatrix}^T.$$

The cost functional to be minimized is

$$J(x_0, u) = \frac{1}{2}\int_0^\infty \left((y - r)^T Q(y - r) + u^T Ru\right) dt,$$

The state variables in this model represent the flight conditions of the aircraft: x_1 is the deviation of the velocity from the level flight trim value of 1(100m/s) (given in units of (100m/s)), x_2 is the deviation of the angle of attack from the trim value of $4.2(\pi/180)$ radians, x_3 is the pitch rate in rad/s, x_4 is the flight path angle in radians, and x_5 is the deviation of the canard deflection angle in radians from the trim value, which is not given. The control u is the input canard deflection in radians. The canards are control flaps which can deflect downward by up to $90(\pi/180)$ radians. Finally, $y(t) = x_4(t)$ is the tracking flight path angle and $r(t)$ is the desired flight path angle.

The weights are $Q = 1$, $R = 1$, $U = 100I_5$ and $V = I_3$. The actual state starts at the origin, but the estimated state starts slightly off the actual, at $(x_e)_0 = (0, 0, 0, 5(\pi/180), 0)^T$. Figure 4 depicts the estimated state almost converging to the actual state by the time of the desired x_4 increase, and remaining close to the actual state for the rest of the time period. In Figure 5 we plot the actual

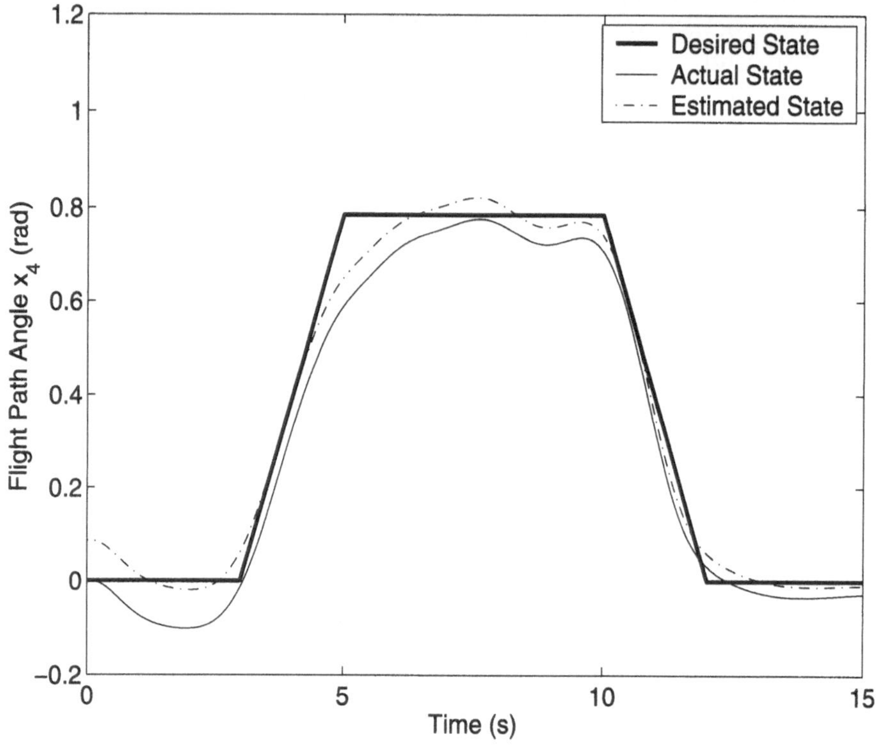

FIGURE 4. Actual and estimated states for nonlinear tracking control/state estimator

state when controlled using our fully nonlinear algorithm, as well as when using the linear SE gain control (as proposed in [22]), and the fully linearized control. It can be seen that the linear control overshoots significantly at the top of the ascent and is very slow to return to 0. The other two methods produce virtually identical results (the difference is indiscernable in the plots in Figure 5).

In the recent manuscript [23], we have successfully applied these methodologies to feedback tracking control of the GaP film thickness in a high-pressure CVD

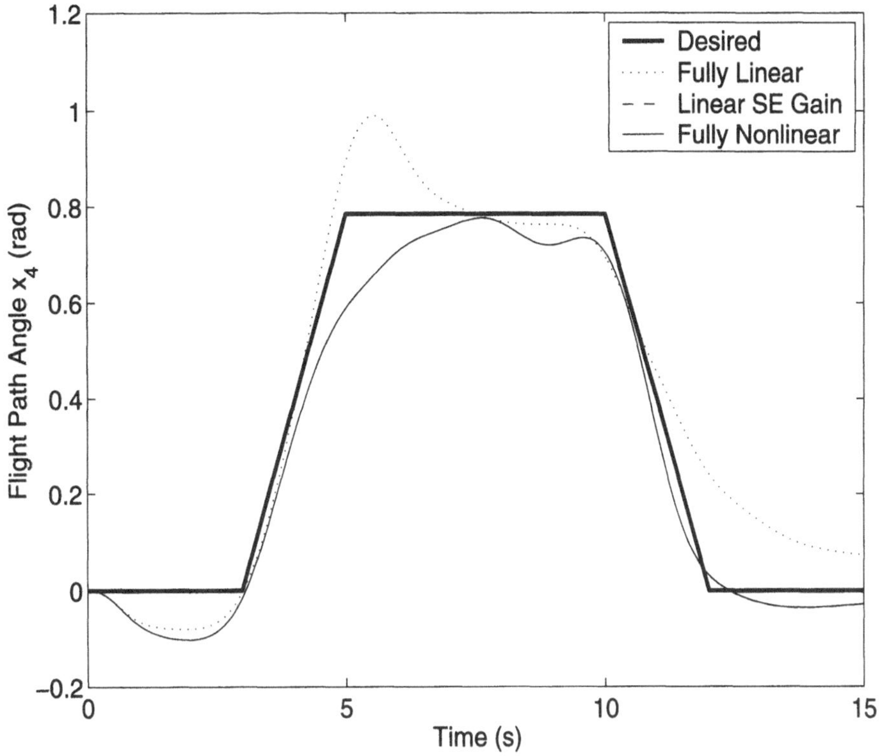

FIGURE 5. Comparison of tracking controls/state estimators with inaccurate $(x_e)_0$.

reactor that is more complex and realistic than the one presented here. Some of the complexities that were considered in [23] include 3-dimensional flow region, 10 atm operating condition, multiple species that include gas-phase kinetics as well as nonlinear surface kinetics, and nonlinear partial state observation.

5. Conclusions

In this paper we report on the development of nonlinear compensators and nonlinear feedback tracking control methodologies that can be applied to high pressure CVD systems. We also present successful computational implementation of reduced order feedback control of pulsed high pressure CVD III-V film growth involving the transport of multiple species with linear gas phase reactions. The combination of reduced order model methods based on proper orthogonal decomposition techniques with nonlinear compensator-based feedback tracking control can provide a powerful tool for treating more complex situations that may be encountered when nonlinear gas phase and/or nonlinear surface phase reactions are present.

6. Acknowledgements

This research was supported in part by a DOD/AFOSR MURI Grant AFOSR F49620-95-1-0447, in part by the AFOSR under Grant F49620-98-1-0180, and in part by the AFOSR under Grant F49620-96-1-0292 (AASERT).

References

[1] D. Olego, *Status and projections of blue light emitters,* Plenary Lecture, TMS Electronic Materials Conference, June 22, 1994.

[2] K. Fukunaga, *Introduction to Statistical Pattern Recognition,* (1972), Academic Press, New York.

[3] G. Berkooz, P. Holmes and J.L. Lumley, *The proper orthogonal decomposition in the analysis of turbulent flows,* Annual Review of Fluids Mechanics, **25** (1993), N5:539–575.

[4] G. Berkooz, P. Holmes, J.L. Lumley and J.C. Mattingly, *Low-dimensional models of coherent structures in turbulence,* Physics Reports-Review Section of Physics Letters, **287** (1997), N4:338–384.

[5] M. Rajaee, S.K.F. Karlson and L. Sirovich, *Low-dimensional description of free-shear-flow coherent structures and their dynamical behavior,* J. of Fluid Mechanics, **258** (1994), 1–29.

[6] M. Kirby and L. Sirovich, *Application of the Karhunen-Loève procedure for the characterization of human faces,* IEEE Trans. on Pattern Analysis and Machine Intelligence, **12** (1990), N1:103–108.

[7] R. Hilai and J. Rubinstein, *Recognition of rotated images by invariant Karhunen-Loève expansion,* J. of the Optical Society of America A-Optics Image Science and Vision, **11** (1994), N5:1610–1618.

[8] H.V. Ly and H.T. Tran, *Proper orthogonal decomposition for flow calculations and optimal control in a horizontal CVD reactor,* CRSC Tech. Report **98-13** (1998), North Carolina State University, Raleigh, North Carolina; Quarterly of Applied Mathematics, to appear.

[9] H.V. Ly and H.T. Tran, *Modeling and control of physical processes using proper orthogonal decomposition,* Computers and Mathematics with Applications, to appear.

[10] C. Theodoropolous, R.A. Adomaitis and E. Zafiriou, *Model reduction for optimization of rapid thermal chemical vapor deposition systems,* IEEE Trans. Semiconductor Manufacturing, **11** (1998), 85–98.

[11] G.M. Kepler, H.T. Tran and H.T. Banks, *Reduced order model compensator control of species transport in a CVD reactor,* CRSC Tech. Report **99-15** (1999), North Carolina State University, Raleigh, North Carolina; Optimal Control Applications and Methods, to appear.

[12] G.M Kepler, H.T. Tran and H.T. Banks, *Compensator control for chemical vapor deposition film growth using reduced order design models,* CRSC Tech. Report **99-41** (1999), North Carolina State University, Raleigh, North Carolina; IEEE Trans. on Semiconductors, to appear.

[13] H.T. Banks, R.C.H. del Rosario and R.C. Smith, *Reduced order model feedback control design: Numerical implementation in a thin shell model*, CRSC Tech. Report **98-27** (1999), North Carolina State University, Raleigh, North Carolina; IEEE Trans. on Aut. Control, to appear.

[14] K. Kunisch and S. Volkwein, *Control of Burgers' equation by a reduced order approach using proper orthogonal decomposition*, Optimierung und Kontrolle Bericht **138** (1998), Universität Graz, Austria; J. Opt. Theory Applic., to appear.

[15] J.A. Atwell and B. King, *Proper orthogonal decomposition for reduced basis feedback controllers for parabolic equation*, ICAM Report **99-01-01** (1999), VPISU, Blacksburg, Virginia; Math. and Comp. Modeling, to appear.

[16] H.T. Banks, M.L. Joyner, B. Wincheski and W.P. Winfree, *Nondestructive Evaluation using reduced-order computational methodology*, Inverse Problems, **16** (2000), 929–945.

[17] K.J. Bachmann, N. Sukidi, C. Hopfner, C. Harris, N. Dietz, H.T. Tran, S. Beeler, K. Ito and H.T. Banks, *Real-time monitoring of steady-state pulsed chemical beam epitaxy by p-polarized reflectance,* J. of Crystal Growth, **183** (1998), 323–337.

[18] C. Theodoropolous, N.K. Ingle, T.J. Mountziaris, Z.Y. Chen, P.L. Liu, G. Kioseoglou and A. Petrou, *Kinetic and transport modeling of the metallorganic chemical vapor deposition of InP from trimethylindium,* J. Electrochem. Soc., **142** (1995), 2086–2094.

[19] S. Beeler, H.T. Tran and N. Dietz, *Representation of GaP formation by a reduced order surface kinetics model using p-polarized reflectance measurements,* J. of Applied Physics, **86** (1999), 674–682.

[20] S. Beeler, H.T. Tran and H.T. Banks, *State estimation and tracking control of nonlinear dynamical systems*, CRSC Tech. Report **00-19** (2000), North Carolina State University, Raleigh, North Carolina; Automatica, submitted.

[21] W.L. Garrard, D.F. Enns and S.A. Snell, *Nonlinear feedback control of a highly maneuvrable aircraft,* Int. J. of Control, **56** (1992), 799–812.

[22] F.E. Thau, *Observing the state of non-linear dynamic systems,* Int. J. of Control, **17** (1973), 471–479.

[23] S.C. Beeler, G.M. Kepler, H.T. Tran and H.T. Banks, *Reduced order modeling and control of thin film growth in an HPCVD reactor*, CRSC Tech. Report **00-33** (2000), North Carolina State University, Raleigh, North Carolina.

Center for Research in Scientific Computation
Department of Mathematics
North Carolina State University
Raleigh, North Carolina 27695
E-mail address: `htbanks@eos.ncsu.edu; tran@control.math.ncsu.edu`

International Series of Numerical Mathematics, Vol. 139, 19–30

On Modeling and Design Problems in Micro Diffractive Optics

Gang Bao

Abstract. Consider a time-harmonic electromagnetic plane wave incident on a periodic surface. The optical medium may be linear, chiral, or nonlinear. The diffraction problem is to predict energy distributions of the propagating waves away from the structure. The process is governed by time-harmonic Maxwell's equations. In this paper, direct, inverse, and optimal design problems in the mathematical modeling of diffractive optics are studied. Results on mathematical and computational aspects of the model problems are presented.

1. Introduction

Micro diffractive optics is an emerging technology with many practical applications. Over the past two decades, significant technology developments have been made particularly in two areas. First, high precision micromachining techniques have permitted the creation of gratings (periodic structures) and other diffractive structures of extremely small scales. Second, rapid developments of laser technology and nonlinear optical materials have made important progress possible in nonlinear diffractive optics. The practical application of diffractive optics technology has driven the need for mathematical models and numerical algorithms.

This paper focuses on the diffraction or scattering by diffraction gratings (periodic structures). The scattering theory in periodic structures has many applications in micro diffractive optics, where doubly periodic structures are often called *crossed diffraction gratings* [35]. Consider scattering of electromagnetic waves by a biperiodic structure. The structure separates the whole space into three regions: Above and below the structure the medium is assumed to be homogeneous. However, inside the structure, the medium can be very general. In fact, the medium can be linear, chiral, or nonlinear. Given the structure and a time-harmonic electromagnetic plane wave incident on the structure, the scattering (diffraction) problem is to predict the field distributions away from the structure. The inverse or optimal design problem is to determine the structure from the measured or desirable scattered fields.

2. The Diffraction Problem

The electromagnetic fields are governed by the time harmonic Maxwell equations (time dependence $e^{-i\omega t}$):

$$\nabla \times E - i\omega\mu H = 0\,, \tag{1}$$
$$\nabla \times H + i\omega\varepsilon E = 0\,, \tag{2}$$

where E and H denote the electric and magnetic fields in $\mathbf{R}^3$, respectively. The magnetic permeability μ is assumed to be one everywhere. There are two constants Λ_1 and Λ_2, such that the dielectric coefficient ε satisfies, for any $n_1,\, n_2 \in Z = \{0, \pm 1, \pm 2, \ldots\}$,

$$\varepsilon(x_1 + n_1\Lambda_1, x_2 + n_2\Lambda_2, x_3) = \varepsilon(x_1, x_2, x_3)\,.$$

Further, it is assumed that, for some fixed positive constant b and sufficiently small $\delta > 0$, $\varepsilon(x_1, x_2, x_3) = \varepsilon_1$, for $x_3 > b - \delta$, $\varepsilon(x_1, x_2, x_3) = \varepsilon_2$, for $x_3 < -b + \delta$, where $\varepsilon(x) \in L^\infty$, $\mathrm{Re}(\varepsilon(x)) \geq \varepsilon_0$, $\mathrm{Im}(\varepsilon(x)) \geq 0$, ε_0, ε_1 and ε_2 are constants, ε_0, ε_1 are real and positive, and Re $\varepsilon_2 > 0$, Im $\varepsilon_2 \geq 0$. The case Im $\varepsilon_2 > 0$ accounts for materials which absorb energy.

Let $\Omega_0 = \{x \in \mathbf{R}^3 : -b < x_3 < b\}$, $\Omega_1 = \{x \in \mathbf{R}^3 : x_3 > b\}$, $\Omega_2 = \{x \in \mathbf{R}^3 : x_3 < -b\}$.

Consider a plane wave in Ω_1, $(E_I, H_I) = (se^{iq\cdot x}, pe^{iq\cdot x})$, incident on Ω_0. Here $q = (\alpha_1, \alpha_2, -\beta) = \omega\sqrt{\varepsilon_1}(\cos\theta_1 \cos\theta_2, \cos\theta_1 \sin\theta_2, -\sin\theta_1)$ is the incident wave vector whose direction is specified by θ_1 and θ_2, with $0 < \theta_1 < \pi$ and $0 < \theta_2 \leq 2\pi$. The vectors s and p satisfy $s = \frac{1}{\omega\varepsilon_1}(p \times q)$, $q \cdot q = \omega^2\varepsilon_1$, $p \cdot q = 0$.

We are interested in quasiperiodic solutions, *i.e.*, solutions E and H such that the fields E_α, H_α defined by, for $\alpha = (\alpha_1, \alpha_2, 0)$, $E_\alpha = e^{-i\alpha\cdot x}E(x_1, x_2, x_3)$, $H_\alpha = e^{-i\alpha\cdot x}H(x_1, x_2, x_3)$, are periodic in the x_1 direction of period Λ_1 and in the x_2 direction of period Λ_2.

Denote

$$\nabla_\alpha = \nabla + i\alpha = \nabla + i(\alpha_1, \alpha_2, 0)\,.$$

It is easy to see from (1) and (2) that E_α and H_α satisfy

$$\nabla_\alpha \times E_\alpha - i\omega H_\alpha = 0\,,$$
$$\nabla_\alpha \times H_\alpha + i\omega\varepsilon E_\alpha = 0\,.$$

Due to a consideration for coercivity, it turns out to be natural to solve the following problem:

$$\nabla_\alpha \times (\frac{1}{\varepsilon}\nabla_\alpha \times H_\alpha) - \nabla_\alpha(\frac{1}{\varepsilon_C}\nabla_\alpha \cdot H_\alpha) - \omega^2 H_\alpha = 0\,,$$

where ε_C is a fixed positive constant which satisfies $inf_{x\in\Omega_0}\,\mathrm{Re}\,\frac{1}{\varepsilon(x)} \geq \frac{3}{4\varepsilon_C}$.

We also need boundary conditions in the x_3 direction. These conditions may be derived by the radiation condition, the periodicity of the structure, and the Green functions.

Denote

$$\Gamma_1 = \{x \in \mathbf{R}^3 : x_3 = b\} \text{ and } \Gamma_2 = \{x_3 = -b\}.$$

Define for $j = 1, 2$ the coefficients

$$\beta_j^{(n)}(\alpha) = e^{i\gamma_j^n/2}|\omega^2\varepsilon_j - |\alpha_n + \alpha|^2|^{1/2}, \quad n \in Z,$$

where $\gamma_j^n = arg(\omega^2\varepsilon_j - |\alpha_n + \alpha|^2)$, $0 \leq \gamma_j^n < 2\pi$. We assume that $\omega^2\varepsilon_j \neq |\alpha_n + \alpha|^2$ for all $n \in Z$, $j = 1, 2$. This condition excludes "resonance".

For functions $f \in H^{\frac{1}{2}}(\Gamma_j)^3$, define the operator T_j^α by

$$(T_j^\alpha f)(x_1, x_2) = \sum_{n \in \Lambda} i\beta_j^{(n)} f^{(n)} e^{i\alpha_n \cdot x},$$

where $f^{(n)} = \frac{1}{\Lambda_1\Lambda_2}\int_0^{\Lambda_1}\int_0^{\Lambda_2} f(x)e^{-i\alpha_n \cdot x}$.

Denote the operator B_j by

$$B_j f = -i\sum_{n \in Z^2} \frac{1}{\beta_j^{(n)}}\{(\beta_j^{(n)})^2(f_1^{(n)}, f_2^{(n)}, 0) + ((\alpha + \alpha_n) \cdot f^{(n)})(\alpha + \alpha_n)\}e^{i\alpha_n \cdot x}.$$

Therefore, the scattering problem can be formulated as follows [13]:

$$\begin{aligned}
\nabla_\alpha \times (\frac{1}{\varepsilon}\nabla_\alpha \times H_\alpha) - \nabla_\alpha(\frac{1}{\varepsilon_C}\nabla_\alpha \cdot H_\alpha) - \omega^2 H_\alpha &= 0 \text{ in } \Omega_0\,, \\
\nu_1 \times (\nabla_\alpha \times (H_\alpha - H_{I,\alpha})) &= B_1(P(H_\alpha - H_{I,\alpha})) \text{ on } \Gamma_1\,, \\
\nu_2 \times (\nabla_\alpha \times H_\alpha) &= B_2(P(H_\alpha)) \text{ on } \Gamma_2\,, \\
(T_1^\alpha - \frac{\partial}{\partial\nu_1})H_{\alpha,3} &= 2i\beta_1 p_3 e^{-i\beta_1 b}\,, \text{ on } \Gamma_1\,, \\
(T_2^\alpha - \frac{\partial}{\partial\nu_2})H_{\alpha,3} &= 0\,, \text{ on } \Gamma_2\,,
\end{aligned}$$

where ν_j is the outward normal to the surface Γ_j and P is the projection onto the plane orthogonal to ν_1.

Theorem 2.1. [13] *For all but possibly a discrete set of ω, the above scattering problem attains a unique weak solution $H \in H^1(\Omega_0)^3$.*

Remark 2.1. In fact, one can further establish an equivalence of the current variational formulation and the original scattering problem.

Remark 2.2. Scattering of electromagnetic waves in a biperiodic structure has recently received considerable attention. We refer to [23], [24], [1], [21], [8], [13], [14], [36] for results and additional references on existence, uniqueness, and numerical approximations of solutions.

Computation. We have recently developed interface least-squares finite element methods for solving the diffraction problem [10], [19]. The idea is to formulate the problem as an interface problem with the grating surface as the interface. The model problem can then be solved by a new least-squares finite element method that incorporates the jump conditions at interfaces into the objective functional. The method allows the use of different finite element spaces on either side of the interface and the jump conditions are enforced through the least-squares functional. As with general least-squares finite element methods, the resulting discrete system is symmetric, positive definite, and so is easily treated by various existing preconditioning techniques, e.g., multigrid methods. Both electric and magnetic fields can be determined simultaneously, which avoids the unstable numerical differentiation process. With sufficiently smooth interfaces, significantly better estimates than that for the standard finite element methods can be expected.

3. Maxwell's Equations in a Perturbed Periodic Structure

Consider a time-harmonic electromagnetic plane wave incident on a an infinite periodic (grating) structure. An *inhomogeneous subwavelength* object is placed inside the periodic structure. The problem arises in the study of near-field optics and has many physical and biological applications. Recently, because of its capability of analyzing nano-scaled objects, near-field optics has generated great scientific interests. We refer to [27] for a survey of theoretical and experimental studies. A particular application of such phenomena is for testing the resolution limit of near-field optical microscopes.

The media are once again assumed to be nonmagnetic with a constant magnetic permeability $\mu = \mu_0$ everywhere. We focus on a two-dimensional geometry, *i.e.*, the media and materials are all assumed to be invariant in the x_3 direction. Let a plane wave be incident on an infinite periodic interface S which separates two domains D_1 and D_2. The incident wave is assumed to be TE (transverse electric) polarized, *i.e.*, the electric field is transverse to the incident plane or pointed to the x_3 direction. Underneath the interface S, a small inhomogeneous object is placed in Ω. Note that because of the presence of this small object in a single period, the scattering problem is no longer a standard grating problem. Above S, the medium is assumed to be homogeneous with a dielectric coefficient $\varepsilon_1 > 0$. Below S and outside of Ω, the medium is also assumed to be homogeneous with $\varepsilon(x) = \varepsilon_2 > 0$ and $\varepsilon_2 \neq \varepsilon_1$. However, the medium inside Ω can be general with $\varepsilon(x) \in L^\infty(\Omega)$.

Consider an incident plane wave $u_I(x) = e^{i\alpha x_1 - i\beta_1 x_2}$ where $\alpha = \omega\sqrt{\varepsilon_1\mu_0}\sin\theta$, $\beta_1 = \omega\sqrt{\varepsilon_1\mu_0}\cos\theta$, and $\theta \in (-\frac{\pi}{2}, \frac{\pi}{2})$ is the angle of incidence. For TE polarization, we deduce from the time harmonic Maxwell equations that $(\Delta + \omega^2\varepsilon\mu_0)u = 0$, where u is the total field. In addition, we require that the scattered field satisfy an appropriate radiation condition [4].

The scattering problem is as follows: For a given incoming plane wave, solve the Helmholtz equation for the scattered field. In the case $\varepsilon(x) = \varepsilon_2$ in Ω, the model problem becomes the well known grating problem. Let v be the quasiperiodic solution of the grating problem with quasiperiodicity α. In general, the model problem is a (non-periodic) scattering problem since there is only one object.

Denote $w = u - v$. We derive an integral representation of w.

Define

$$q_0^2(x) = \begin{cases} \varepsilon_1\mu_0 & \text{in } D_1\,, \\ \varepsilon_2\mu_0 & \text{in } D_2\,. \end{cases}$$

Obviously, by using the equations for u and v, w satisfies

$$(\Delta + \omega^2 q_0^2(x))w = \omega^2 \mu_0 (\varepsilon_2 - \varepsilon(x))(w + v)\chi_\Omega ,$$

where χ_Ω is the characteristic function of Ω. Let $\sum = \{\omega : \ \omega > 0$, the following periodic interface problem admits a unique solution when the incident plane wave is at the frequency $\omega.\}$

The periodic interface problem: For all $\xi = (\xi_1, \xi_2) \in U \times \mathbf{R}$, where U is a complex neighborhood of the real axis, find a function ϕ which is periodic in x_1 of the same periodic as the grating structure S, such that it satisfies

$$(\Delta + \omega^2 q_0^2(x))(e^{ix_1\xi_1}\phi(x,\xi)) = e^{ix\cdot\xi}$$

along with the periodic radiation condition with respect to x_2 for $|x_2| \to \infty$.

We are now ready to present an integral representation of w.

Theorem 3.1. *Suppose that* $\omega \in \sum$. *Then* w *satisfies*

$$\begin{aligned} w(x) \quad &- \quad \frac{\omega^2\mu_0}{2\pi}\int_{\mathbf{R}^2} F[(\varepsilon_2 - \varepsilon(x))w(x)\chi_\Omega]\phi(x,\xi)e^{ix_1\xi_1}d\xi \\ &= \frac{\omega^2\mu_0}{2\pi}\int_{\mathbf{R}^2} F[(\varepsilon_2 - \varepsilon(x))v(x)\chi_\Omega]\phi(x,\xi)e^{ix_1\xi_1}d\xi \ , \forall\, x \in \mathbf{R}^2, \qquad (3) \end{aligned}$$

where $F[\cdot]$ *is the Fourier transform.*

Remarks. The representation follows essentially Morgan and Babuška [31] in their study of homogenized solutions of elliptic PDE for periodic media. The integrals in (3) are defined as Fourier-Bochner integrals. The assumption on ω may be dropped, provided that either $\mathrm{Im}(\varepsilon_1) > 0$ or $\mathrm{Im}(\varepsilon_2) > 0$. The function ϕ is analytic with respect to $\xi = (\xi_1, \xi_2)$. Although the periodic radiation condition depends on ξ_2 the well-posedness of the periodic interface problem depends only on the variable ξ_1.

Making use of this representation we can prove that w is exponentially decaying with respect to the variable x_1. The well-posedness of the model problem is established in [4].

4. Chiral Gratings

Chiral gratings provide an exciting combination of the medium and structure, which gives rise to new features and applications. For instance, chiral gratings are capable of converting a linearly polarized incident field into two nearly circularly polarized diffracted modes in different directions. Other potential applications include antennas, microwave devices, waveguides, and many other fields. Mathematically, in chiral media, the electric and magnetic fields are no longer decoupled in the constitution equations unlike in the standard Maxwell's system. Therefore, the model system is always in vector form. Recently, a variational approach has been developed in [3], [5]. Results on the well-posedness of the model problem and finite element methods have been established.

The electromagnetic fields are governed by time harmonic Maxwell's equations:

$$\begin{aligned} \nabla \times E - i\omega B &= 0\,, \\ \nabla \times H + i\omega D &= 0\,, \end{aligned}$$

where E, H, D, and B denote the electric field, the magnetic field, the electric and magnetic displacement vectors in $\mathbf{R}^3$, respectively.

In addition, the following Drude-Born-Fedorov constitutive equations [30] hold:

$$\begin{aligned} D &= \varepsilon(x)\Big(E + \beta(x)\nabla \times E\Big)\,, \\ B &= \mu(x)\Big(H + \beta(x)\nabla \times H\Big)\,, \end{aligned}$$

where ε is the electric permittivity, μ is the magnetic permeability, and β is the chirality admittance. Similarly, the Maxwell equations may be rewritten as

$$\begin{aligned} \nabla \times E &= (\gamma(x))^2\,\beta(x)E + i\omega\mu(x)(\frac{\gamma(x)}{k(x)})^2\,H\,, \\ \nabla \times H &= (\gamma(x))^2\,\beta(x)\,H - i\omega\varepsilon(x)(\frac{\gamma(x)}{k(x)})^2\,E\,, \end{aligned}$$

where $k^2(x) = \omega^2\varepsilon\mu$. In these equations, the parameter $\gamma(x)$ is defined as :

$$(\gamma(x))^2 = \frac{(k(x))^2}{1-(\,k(x)\,\beta(x)\,)^2}\,.$$

Here it is always assumed that $(\,k(x)\,\beta(x)\,)^2 \neq 1, x \in \mathbf{R}^3$.

Moreover, the above system may be shown to be equivalent in a weak sense to

$$\nabla \times \Big(\frac{1-\omega^2\beta^2\varepsilon\mu}{\mu}\Big)\nabla \times E - \omega^2\nabla \times (\varepsilon\beta E) - \omega^2\varepsilon\beta\nabla \times E - \omega^2\varepsilon E = 0\,, \quad (4)$$

$$\nabla \times E = (\gamma(x))^2\,\beta(x)E + i\omega\mu(x)(\frac{\gamma(x)}{k(x)})^2\,H\,. \quad (5)$$

Hence in order to solve the Maxwell equations, it suffices to solve the equation (4) for E. Once the electric field E is determined, the magnetic field H follows from the equation (5).

As in the achiral case, the scattering problem may be reduced to a bounded domain by introducing a pair of transparent boundary conditions [6].

Theorem 4.1. *For all but possibly a discrete set of frequencies ω, the scattering problem admits a unique weak solution E in $H(curl, \Omega)$.*

The theorem may be proved by combining a variational approach and the Hodge decomposition of the electric field.

5. Surface Enhanced Nonlinear Optical Effects

A remarkable application of nonlinear diffractive optics is to generate coherent radiation at a frequency that is twice that of available lasers, so-called *second harmonic generation* (SHG). Recently, it has been found experimentally that diffraction gratings can greatly enhance nonlinear optical effects. The model for SHG is the system of nonlinear Maxwell's equations with quadratic nonlinear terms. Little is known mathematically on nonlinear optics in periodic structures. The model problem in the 2-D setting has been solved in [11], [12]. Currently, we are investigating the model problem in 3-D. The problem is difficult since it involves a complicated nonlinear P.D.E. in vector form.

The electromagnetic fields in a general medium are governed by the following Maxwell equations:

$$\begin{aligned} \nabla \times E &= i\omega\mu_0 H, \quad \nabla \cdot H = 0\,, \\ \nabla \times H &= -i\omega(E + 4\pi P), \ \nabla \cdot E = -4\pi P, \end{aligned}$$

where the new term P is the polarization vector. The medium is linear if $P = \chi^{(1)}(\omega) \cdot E$ where $\chi^{(1)}$ is the linear susceptibility tensor of the medium. The (linear) dielectric constant is defined by $\epsilon(w) = 1 + 4\pi\chi^{(1)}(\omega)$.

In general, P is a nonlinear function of E. The simplest nonlinear optical wave interaction deals with second harmonic generation — a special case of the second-order nonlinear optical effects. In this case, $P = \chi^{(1)}(\omega) \cdot E(x, \omega) + \chi^{(2)}(\omega)(\omega = \omega_i + \omega_j) : E(x, \omega_i)E(x, \omega_j)$, where $\chi^{(2)}$ is the second order nonlinear susceptibility tensor that measures the nonlinearity of the medium, and $\chi^{(2)} : E\ E$ is a vector whose i-th component is $\chi^{(2)}_{ijk} E_j E_k$. Note that new frequency components are present in the above expression, which is the most striking difference between nonlinear and linear optics. When a pumping wave with frequency $\omega_1 = \omega$ is incident on a nonlinear medium, second harmonic generation leads to two wave fields $E(x, \omega_1)$ and $E(x, \omega_2 = 2\omega_1)$.

In the nonlinear case, the group symmetry properties of $\chi^{(2)}$ have important effects on the polarization and thus on the model equations. By far, the most commonly used model is the two dimensional TM-TE model which assumes that the fundamental field is TM polarized; while the second harmonic field is TE polarized. The model is supported by nonlinear materials of cubic symmetry structures, *e.g.*, ZnS and ZnSe. In addition, it assumes that the depletion of energy from the pump waves at frequency ω_1 can be neglected – a process is often referred to as the undepleted pump approximation in nonlinear optics literature. Mathematically, it is a linearization procedure. The well-posedness of the above two-dimensional model has recently been established by Bao and Chen [11].

In the following, we consider the linearization procedure of nonlinear diffractive optics in biperiodic structures by ignoring the small nonlinear terms in the system at frequency ω_1. An attempt for solving the fully nonlinear model would certainly be of interest. At present, analysis and computation for the linearized (biperiodic) model are completely open. For the linearized model, at frequency

$\omega_1 = \omega$, we have

$$\nabla \times E(\omega_1) = -i\omega_1 H(\omega_1)\,, \quad \nabla \times H(\omega_1) = i\omega_1 \epsilon E(\omega_1)\,. \tag{6}$$

At frequency $\omega_2 = 2\omega$, the Maxwell equations become

$$\nabla \times E(\omega_2) = -i\omega_2 H(\omega_2)\,, \quad \nabla \times H(\omega_2) = i\omega_2[E(\omega_2) + 4\pi\chi^{(2)}(\omega_2) : E(\omega_1)E(\omega_1)]\,. \tag{7}$$

The system (6) may be shown to be equivalent in a weak sense to the following problem

$$\begin{aligned} \nabla \times (\frac{1}{\epsilon}\nabla \times H(\omega_1)) - \nabla \cdot (\frac{1}{\epsilon}\nabla \cdot H(\omega_1)) - \omega_1^2 H(\omega_1) &= 0\,, \\ \nabla \times H(\omega_1) + i\omega_1\epsilon E(\omega_1) &= 0\,. \end{aligned}$$

Similarly, the system (7) is equivalent to

$$\nabla \times \nabla \times E(\omega_2) - \omega_2^2 E(\omega_2) = 4\pi i\omega_2\chi^{(2)}(\omega_2) : E(\omega_1)E(\omega_1)\,, \tag{8}$$

$$\nabla \times E(\omega_2) + i\omega_2 H(\omega_2) = 0\,. \tag{9}$$

Evidently, in order to study the well-posedness of the model, we will need refined regularity results on $H(\omega_1)$ and hence on $E(\omega_1)$. Actually, since $H(\omega_1) \in W^{1,2}$, $E(\omega_1) \in L^2$, the right hand side of Equation (8) is at best in L^1, which presents a severe difficulty: No result on existence and uniqueness is available for such a L^1 right hand side. Recent progress has been made in resolving this difficulty. In [17], we obtain interior L^p estimates for Maxwell's equations with source terms under certain restrictions on the medium.

Another important problem is to analyze the dependence of wave fields on the nonlinear susceptibility tensors. In fact, the harmonics of order higher than three are almost always negligibly small. Perhaps, such an analysis could be done by performing a perturbation analysis of the solutions, which should also reveal important physical information on the models.

6. Inverse Diffraction

An inverse diffraction problem is to determine the periodic structure or the shape of the interface from the measured scattered fields.

Let the scattering profile be described by the periodic surface $S = \{x \in \mathbf{R}^3 : x_3 = f(x_1, x_2)\}$ of period $\Lambda = (\Lambda_1, \Lambda_2)$, that is, $f(x_1+n_1\Lambda_1, x_2+n_2\Lambda_2) = f(x_1, x_2)$ for integers n_1, n_2, and some positive constants Λ_1, Λ_2. The function f is supposed to be sufficiently smooth, for example of C^2. The space below S is filled with some perfectly reflecting material (a perfect conductor). Let $\Omega = \{(x \in \mathbf{R}^3 : x_3 > f(x_1, x_2)\}$ be filled with a nonmagnetic material ($\mu = \mu_0$) whose dielectric coefficient is a fixed constant $\varepsilon = \varepsilon_0 > 0$. Consider a plane wave in Ω of the same form as in Section 1, $E_I = se^{iq\cdot x}$, $H_I = pe^{iq\cdot x}$, incident on S.

From the Maxwell equations, it is easily seen that

$$(\Delta + \omega^2\varepsilon_0\mu_0)E = 0 \ \text{ in } \Omega. \tag{10}$$

Since the region below S is a perfect conductor, only reflected waves exist:

$$\nu \times E = 0 \text{ on } S , \tag{11}$$

where ν is the outward normal to the surface. The following boundary condition may be derived from the radiation condition:

$$e_3 \times (\nabla \times (E - E_I)) = B(P(E - E_I)) \text{ on } \Gamma, \tag{12}$$

where $e_3 = (0, 0, 1)$, B is a pseudo-differential operator, and P is the projection onto the plane orthogonal to e_3.

Therefore, the direct scattering problem can be formulated as follows: To find a quasiperiodic solution that solves the problem (10), (11), and (12). The inverse problem can be stated as follows: For a given incident plane wave E_I, determine $f(x_1, x_2)$ from the knowledge of $e_3 \times E|_\Gamma$.

Suppose that $E_{f_j}(x)$ $(j = 1, 2)$ are Λ-quasiperiodic and solve the scattering problem (10), (11), and (12) with respect to the profiles $f_j(x_1, x_2)$, where the functions f_j are Λ-periodic. Let $b > max\{f_1(x_1, x_2), f_2(x_1, x_2)\}$ be a fixed constant. Denote $D_j = \{f_j < x_3 < b\}$.

Two profiles Γ_1 and Γ_2 are said to satisfy Property (A) if there is a simply connected bounded domain U such that the following two conditions are satisfied: U is convex; $\partial U = \partial U_1 \cup \partial U_2$, $\partial U_1 \subset \Gamma_1$ and $\partial U_2 \subset \Gamma_2$; and ∂U is of C^2.

Theorem 6.1. *([20]) Assume that f_1, f_2 are Λ-periodic C^2 function and that the profiles $S_1 = \{x_3 = f_1(x_1, x_2)\}$ and $S_2 = \{x_3 = f_2(x_1, x_2)\}$ satisfy Property (A). Then there is a constant $\delta(k) > 0$ such that if the radius of $U \leq \delta(k)$ then $e_3 \times E_{f_1}|_{x_3=b} = e_3 \times E_{f_2}|_{x_3=b}$ implies $f_1(x_1, x_2) = f_2(x_1, x_2)$.*

The result extends earlier 2-D uniqueness results in [28] and [7]. In applications, it is impossible to make exact measurements. Thus stability results are crucial in the reconstruction of profiles. Regarding the stability of the inverse diffraction problem, only partial stability results are available in [15] (2-D) and [20] (3-D). Another direction is to study the inverse transmission problem or the non-conductor problem. In this case, one determines the structure from information on reflected and on transmitted waves. The problem becomes much more difficult. The only available results are some stability results proved in [15] for singly periodic structures.

7. Optimal Design

Given the incident field, the optimal design problem concerns the creation of grating profiles that give rise to some specified diffraction patterns. Numerical computation of the inverse and design problems in biperiodic structures is completely open. The problem can be posed as a nonlinear least-squares problem. Difficulties arise since the scattering pattern depends on the interface in a very implicit fashion and in general the set over which the function is minimized is neither convex nor closed. The formulation of the design problem is very close to similar problems in elasticity, for which fast and efficient algorithms have recently been developed. Initial progress on the design problem has been made via weak convergence analysis methods [2], [22], and the homogenization theory [9] along with "relaxation" technique [29]. The main idea is to allow the grating profiles to be highly oscillating and to use relaxed formulation of the optimization problem. The crucial step is to determine the relaxed formulation which involves materials and the effective dielectric properties [9]. While the relaxation method achieves certain success, it suffers from some severe difficulties. It enlarges the solution set and consequently could add new solutions to the relaxed problem. Unfortunately, there is often no guarantee that a feasible solution, *i.e.*, a structure that can be fabricated, may be obtained. One possible solution is to develop a local approach thet makes use of the a priori (engineering) knowledge of the grating structure. The advantages of this approach are that the number of unknowns could be significantly reduced and it is justifiable theoretically because of our local uniqueness and stability results. In addition, the approach allows one to restrict his attention to a family of curves often characterized by a finite number of parameters, for example the Fourier coefficients of the interface function or certain physical parameters. Thus, the structure shapes could be restricted to those which are feasible in practice. However, the method also has a drawback: the solutions (structures) of the design problem might be only suboptimal rather than optimal. We also call the reader's attention to a series of recent papers by Elschner and Schmidt [25], [26] for the optimal design problem in which a careful analysis and computation of the gradients with respect to the variations of grating parameters are presented.

Optimal design of nonlinear gratings is of critical importance in the study of second harmonic generation. It is well known that nonlinear optical effects are generally so weak that the observation of nonlinear phenomena in the optical region can only be made by using high intensity laser beams. Enhancement of nonlinear optical effects presents a great challenge to the optical science communities. Recently, it has been announced in [32] [33] [34] that SHG can be greatly enhanced by using diffraction gratings. Recently, a computational approach has been developed by Bao and Li [16] for solving the optimal design problem in a nonlinear layered medium (1-D). Very little is known on optimal design of nonlinear gratings. An effort to solve the problem by an optimization approach is currently in progress [18].

Acknowledgments

The research of the author was partially supported by the NSF Applied Mathematics Programs grant DMS 98-03604, the NSF University-Industry Cooperative Research Programs grant DMS 99-72292, the NSF Western Europe Programs grant INT 98-15798, and the ONR grant N000140010299.

References

[1] T. Abboud, *Formulation variationnelle des équations de Maxwell dans un réseau bipériodique de* $\mathbf{R}^3$, C. R. Acad. Sci. Paris, t. 317, Série I (1993), 245–248.

[2] Y. Achdou and O. Pironneau, *Optimization of a photocell,* Optimal Control Appl. Meth. **12** (1991), 221–246.

[3] H. Ammari and G. Bao, *Analysis of the diffraction by periodic chiral structures*, C.R. Acad. Sci., Paris, Série I, t. **326** (1998), 1371–1376.

[4] H. Ammari and G. Bao, *Scattering by a subwavelength object embedded in a periodic structure*, C.R. Acad. Sci., Paris, t. **330**, Série I (2000), 333–338.

[5] H. Ammari and G. Bao, *Coupling of finite element and boundary element methods for the electromagnetic diffraction by a periodic chiral structure*, College de France Seminar, Pitman Research Notes in Mathematics, Pitman Advanced Pub. Program, to appear.

[6] H. Ammari and G. Bao, *Maxwell's equations in periodic chiral structures*, preprint.

[7] G. Bao, *A uniqueness theorem for an inverse problem in periodic diffractive optics*, Inverse Problems **10** (1994), 335–340.

[8] G. Bao, *Variational approximation of Maxwell's equations in biperiodic structures*, SIAM J. Appl. Math. **57** (1997), 364–381.

[9] G. Bao and E. Bonnetier, *Optimal design of periodic diffractive structures in TM polarization*, Appl. Math. Opt., to appear.

[10] G. Bao, Y. Cao, and H. Yang, *Least-squares finite element computation of diffraction problems*, Math. Meth. in the Appl. Sci., Vol. 23 (2000), 1073–1092.

[11] G. Bao and Y. Chen, *A nonlinear grating problem in diffractive optics*, SIAM J. Math. Anal., Vol. 28, No. 2 (1997), 322–337.

[12] G. Bao and D. Dobson, *Diffractive optics in nonlinear media with periodic structure*, Euro. J. Appl. Math. **6** (1995), 573–590.

[13] G. Bao and Dobson, *On the scattering by biperiodic structures*, Proc. Am. Math. Soc., 128 (2000), 2715–2723.

[14] G. Bao, D. Dobson, and J. A. Cox, *Mathematical studies of rigorous grating theory*, J. Opt. Soc. Am. **A 12**(1995), 1029–1042.

[15] G. Bao and A. Friedman, *Inverse problems for scattering by periodic structures*, Arch. Rat. Mech. Anal. **132** (1995), 49–72.

[16] G. Bao and G. Li, *Optimal design in nonlinear optics*, in "Encyclopedia of Optimization", Ed. by P. M. Pardalos and C. A. Floudas, Kluwer Academic Pub., to appear.

[17] G. Bao, A. Minut, and Z. Zhou, L^p *estimates for Maxwell's equations in stratified media*, preprint.

[18] G. Bao and G. Schmidt, *Optimal design of nonlinear gratings*, in preparation.

[19] G. Bao and H. Yang, *A least-squares finite element analysis for diffraction problems*, SIAM J. Numer. Anal., Vol. 37, No. 2 (2000), 665–682.

[20] G. Bao and Z. Zhou, *An inverse problem for scattering by a doubly periodic structure*, Trans. Ameri. Math. Soc. **350** (1998), 4089–4103.

[21] O. Bruno and F. Reitich, *Numerical solution of diffraction problems: A method of variation of boundaries III. Doubly-periodic gratings*, J. Optical Soc. Amer., 10 (1993), 2551–2562.

[22] D. Dobson, *Optimal design of periodic antireflective structures for the Helmholtz equation*, Euro. J. Appl. Math. **4** (1993), 321–340.

[23] D. Dobson, *A variational method for electromagnetic diffraction in biperiodic structures*, Modél. Math. Anal. Numér. **28** (1994), 419–439.

[24] D. Dobson and A. Friedman, *The time-harmonic Maxwell equations in a doubly periodic structure*, J. Math. Anal. Appl. **166** (1992), 507–528.

[25] J. Elschner and G. Schmidt, *Numerical solution of optimal design problems for binary gratings*, J. Comput. Phys. **146** (1998), 603–626.

[26] J. Elschner and G. Schmidt, *Diffraction in periodic structures and optimal design of binary gratings. Part I: Direct problems and gradient formulas*, Math. Meth. Appl. Sci. **21** (1998), 1297–1342.

[27] C. Girard and A. Dereux, *Near-field optics theories*, Rep. Prog. Phys. **59** (1996), 657–699.

[28] A. Kirsch, *Uniqueness theorems in inverse scattering theory for periodic structures*, Inverse Problems **10** (1994), 145–152.

[29] R. Kohn and G. Strang, *Optimal design and relaxation of variational problems I, II, III*, Comm. Pure Appl. Math. **39**(1986), 113–137, 139–182, 353–377.

[30] A. Lakhtakia, V.K. Varadan, and V.V. Varadan, *Time-Harmonic Electromagnetic Fields in Chiral Media*, Lecture Notes in Physics, 355 (1989).

[31] R. C. Morgan and I. Babuška, *An approach for constructing families of homogenized equations for periodic media. I: An integral representation and its consequences; II: Properties of the kernel*, SIAM J. Math. Anal. **22** (1991), 1–15; 16–33.

[32] E. Popov and M. Nevière, *Surface-enhanced second-harmonic generation in nonlinear corrugated dielectrics: new theoretical approaches*, J. Opt. Soc. Am. B, Vol. 11, No. 9 (1994), 1555–1564.

[33] R. Reinisch and M. Nevière, *Electromagnetic theory of diffraction in nonlinear optics and surface-enhanced nonlinear optical effects*, Phys. Rev. B **28** (1983), 1870–1885.

[34] R. Reinisch, M. Nevière, H. Akhouayri, J. Coutaz, D. Maystre, and E. Pic, *Grating enhanced second harmonic generation through electromagnetic resonances*, Opt. Eng. **27** (1988), 961–971.

[35] *Electromagnetic Theory of Gratings,* Topics in Current Physics, Vol. 22, edited by R. Petit, Springer-Verlag, Heidelberg, 1980.

[36] *Mathematical Modeling in Optical Science*, the SIAM Frontiers in Applied Mathematics, edited by G. Bao, L. Cowsar, and W. Masters, SIAM, Philadelphia, to appear.

Department of Mathematics
Michigan State University
East Lansing, MI 48824-1027, USA
E-mail address: bao@math.msu.edu

International Series of Numerical Mathematics, Vol. 139, 31–42

A Linearized Model for Boundary Layer Equations

Jean-Marie Buchot and Jean-Pierre Raymond

Abstract. By using the so-called Crocco transformation, the two dimensional Prandtl equations, which are stated in an unbounded domain, are transformed into a nonlinear degenerate parabolic equation (the Crocco equation) stated in a domain $\Omega\times(0,T)=]0,L[\times]0,1[\times(0,T)$. In this paper, we study a degenerate parabolic equation in $\Omega\times(0,T)$ coming from the linearization of the Crocco equation. This is a crucial step to next construct feedback control laws to settle stabilization problems.

1. Introduction

We are interested in a stabilization problem for two dimensional boundary layer equations. We consider a flow with a longitudinal incident velocity on a flat plate. Due to the presence of the plate, the flow develops a boundary layer, which is laminar at the beginning of the plate, and which becomes turbulent before the end of the plate. In the laminar boundary layer the flow is described by the Prandtl equations [9], [3]. When the incident velocity is stationary, let us say (u_s, v_s), the laminar-turbulent transition is stationary. When the incident velocity varies the transition also varies. We are interested in the stabilization of the laminar-turbulent transition by suction or blowing through the plate. Using the so-called Crocco transformation, the system can be described by a degenerate parabolic equation (the Crocco equation, see [9]). Let us denote by Z_s the stationary solution of the Crocco equation corresponding to (u_s, v_s). We look for a control law in feedback form for the suction velocity. To achieve this goal, we linearize the Crocco equations around the stationary solution Z_s. We obtain a linear degenerate parabolic equation of the following form

$$\begin{aligned}
&z_t + az_x - bz_{yy} + cz = f \quad \text{in } \Omega\times(0,T) = (0,L)\times(0,1)\times(0,T),\\
&z_y(x,0,t) = v\chi_s \text{ for } (x,t)\in(0,L)\times(0,T),\\
&z(x,1,t) = 0 \text{ for } (x,t)\in(0,L)\times(0,T),\\
&\sqrt{a}z(0,y,t) = \sqrt{a}w_0 \text{ for } (y,t)\in(0,1)\times(0,T) \text{ and } z(x,y,0) = z_0 \text{ in } \Omega,
\end{aligned} \tag{1}$$

the domain $(0,L)$ represents a part of the plate where the flow is laminar, $(0,1)$ represents the thickness of the boundary layer in Crocco variables, the final time T can be finite or infinite, the terms f and w_0 depends on the incident velocity, the function v depends on the suction velocity through the plate, χ_s is the characteristic function of the interval $[x_0, x_1] \subset (0, L)$ where the suction is applied, the coefficients a, b, c depends on Z_s. We have denoted by z_t, z_x,..., the partial derivatives of z. We adopt this notation throughout the paper, and we shall continue to denote by z_y the derivative of z, even if z only depends on y. Due to the boundary conditions satisfied by Z_s, the coefficients a, b, c vanish on the boundary of Ω. More precisely, $a(y) = \alpha\, y$ with $\alpha > 0$ constant, $c(x, y) > 0$ in Ω and $c(x, 0) = 0$, $b(x, y) > 0$ in Ω, and there exist constants b_1 and b_2 such that $0 < b_1 \le \frac{b(x,y)}{(y-1)^2} \le b_2$. Numerical experiments based on this linearization technique, where a feedback law is designed by solving a Riccati equation, are presented in [5]. In particular, it is numerically shown that the feedback law applied to the initial Prandtl equations gives a very good stabilization performance. The main purpose of this paper is the mathematical analysis of the degenerate linear equation (1). This is a crucial step to next study the associated control problems. For simplicity, throughout the paper we set $a(y) = y$ and $b(y) = (y - 1)^2$, and we suppose that c only depends on y (recall that c is nonnegative and regular), but the results can be adapted to a more general situation.

The second author is personally grateful to the organizers of the Conference for inviting him.

2. A Degenerate Parabolic Equation

We can notice that a function z in $L^2(\Omega \times (0, T))$ satisfies the first equation in (1), in the sense of distributions in $\Omega \times (0, T)$, if and only if $\zeta = e^{-kx} z$ satisfies

$$\zeta_t + a\zeta_x - b\zeta_{yy} + c\zeta + ka\zeta = e^{-kx} f \quad \text{in } \Omega \times (0, T).$$

To study equation (1) we study the property of the mapping $f \mapsto z$ where z is the solution to the equation

$$\begin{aligned} & az_x - bz_{yy} + cz + kaz = f \quad \text{in } \Omega, \\ & z_y(x, 0) = 0 \quad \text{and} \quad z(x, 1) = 0 \text{ for } x \in (0, L), \\ & z(0, y) = 0 \text{ for } \; y \in (0, 1), \end{aligned} \tag{2}$$

where $k > 0$ will be chosen big enough (see Lemma 1). Since the coefficient b vanishes at $y = 1$, we must specify in which sense the boundary condition $z(x, 1) = 0$ is satisfied. For this we first study elliptic equations associated with (2).

2.1. Degenerate elliptic equations

We consider spaces of functions with values in $\mathbb{C}$. We denote by $C_c^\infty([0, 1))$ the space of C^∞ functions with compact support in $[0, 1)$. We introduce Sobolev's spaces with weight. We denote by $H^1(0, 1; d)$ (respectively $H^1_{\{1\}}(0, 1; d)$) the closure of $C^\infty([0, 1])$ (respectively $C_c^\infty([0, 1))$) in the norm

$$\|z\|_{H^1(0,1;d)} = \left(\int_0^1 (|z|^2 + |y - 1|^2 |z_y|^2) dy \right)^{1/2},$$

and by $H^2(0,1;d)$ (respectively $H^2_{\{1\}}(0,1;d)$) the closure of $C^\infty([0,1])$ (respectively $C_c^\infty([0,1)))$ in the norm

$$\|z\|_{H^2(0,1;d)} = \left(\int_0^1 (|z|^2 + |y-1|^2|z_y|^2 + |y-1|^4|z_{yy}|^2)dy\right)^{1/2}.$$

According to [11, Theorem 2.9.2/1], $H^1_{\{1\}}(0,1;d) \equiv H^1(0,1;d)$, and $H^2_{\{1\}}(0,1;d) \equiv H^2(0,1;d)$. We want to study the degenerate elliptic equations

$$-bz_{yy} + cz + kaz + \lambda az = f \quad \text{in } (0,1), \quad z_y(0) = 0, \quad z(1) = 0, \tag{3}$$

and

$$-(b\phi)_{yy} + c\phi + ka\phi + \lambda a\phi = \psi \quad \text{in } (0,1), \quad (b\phi)_y(0) = 0, \quad \phi(1) = 0, \tag{4}$$

where $\lambda \in \mathbb{C}$, $\mathrm{Re}\lambda > -r_0$, $r_0 > 0$ is given fixed, $k > 0$ depends on r_0 and is precisely defined in Lemma 1.

Consider the sesquilinear forms defined on $H^1(0,1;d)$ by

$$\beta_\lambda(z,\phi) = \int_0^1 (bz_y\bar{\phi}_y + b_yz_y\bar{\phi} + cz\bar{\phi} + kaz\bar{\phi} + \lambda az\bar{\phi})dy,$$

and

$$\beta^*_\lambda(\phi,z) = \int_0^1 (b\phi_y\bar{z}_y + b_y\phi\bar{z}_y + c\phi\bar{z} + ka\phi\bar{z} + \lambda a\phi\bar{z})dy.$$

A weak solution to equation (3) is a function $z \in H^1(0,1;d)$ satisfying

$$\beta_\lambda(z,\phi) = \int_0^1 f\bar{\phi}\,dy \quad \text{for every } \phi \in H^1(0,1;d). \tag{5}$$

A similar definition can be stated for equation (4).

Lemma 1. *For every $r_0 > 0$, there exists $k > 0$ such that*

$$\beta_0(z,z) \geq \frac{1}{2}\|z\|^2_{H^1(0,1;d)} + r_0\|z\|^2_{L^2} \quad \text{and} \quad \beta^*_0(z,z) \geq \frac{1}{2}\|z\|^2_{H^1(0,1;d)} + r_0\|z\|^2_{L^2},$$

for every $z \in H^1(0,1;d)$ (here $\|z\|_{L^2} = \|z\|_{L^2(0,1)}$).

Proof. We establish the inequality for β_0. Since $\int_0^1 b_yz_y\bar{z}dy = \int_0^1 2(y-1)z_y\bar{z}dy \leq \frac{1}{4}\int_0^1 (y-1)^2|z_y|^2dy + 4\int_0^1 |z|^2dy$, it is sufficient to prove that there exists $k > 0$ such that $\int_0^1 (\frac{1}{4}(y-1)^2|z_y|^2 + ka|z|^2)dy \geq (4 + \frac{1}{2} + r_0)\int_0^1 |z|^2dy$. This can be shown by arguing by contradiction. □

Throughout the sequel we suppose that $r_0 > 0$ is given fixed, and that $k > 0$ satisfies the conclusion of Lemma 1.

Theorem 2. *For all $f \in L^2(0,1)$, and all $\lambda \in \mathbb{C}$ such that $Re\,\lambda > -r_0$, equation (3) (respectively (4)) admits a unique solution in $H^1(0,1;d)$. This solution belongs to $H^2(0,1;d)$, it satisfies $z_y(0) = 0$, and the estimate*

$$|\lambda|\|\sqrt{a}z\|_{L^2(0,1)} + \|z\|_{H^2(0,1;d)} \leq C\|f\|_{L^2(0,1)}, \tag{6}$$

for all $\lambda \in \mathbb{C}$ such that $Re\,\lambda > -r_0$.

Proof. Due to Lemma 1 and to the inequality $\sqrt{a} \leq 1$, we have

$$\text{Re } \beta_\lambda(z,z) \geq \frac{1}{2}\|z\|^2_{H^1(0,1;d)} + (\text{Re } \lambda + r_0)\|\sqrt{a}z\|^2_{L^2(0,1)}. \tag{7}$$

Thus the existence of a weak solution in $H^1(0,1;d)$ to equation (3) can be proved with the Lax-Milgram theorem. Since $\text{Im}(\beta_\lambda(z,z)) = \text{Im}(\lambda)\int_0^1 az^2dy$, using (7) and equation (5), by classical calculations we can prove that

$$|\lambda|\|\sqrt{a}z\|_{L^2(0,1)} + \|z\|_{H^1(0,1;d)} \leq C\|f\|_{L^2(0,1)}.$$

Estimate (6) can finally be deduced by using equation (3). □

Following [1] we introduce other weighted Sobolev's spaces. We denote by $L^2_d(0,1)$ the space of functions z such that $z/|y-1|$ belongs to $L^2(0,1)$, and we endow $L^2_d(0,1)$ with the norm $\|z\|_{L^2_d(0,1)} = \|z/|y-1|\|_{L^2(0,1)}$. We denote by $H^1_d(0,1)$ the closure of $C^\infty_c([0,1))$ in the norm

$$\|z\|_{H^1_d(0,1)} = \left(\int_0^1 (|y-1|^{-2}|z|^2 + |z_y|^2)dy\right)^{1/2}.$$

We can prove the following regularity result for equations (3) and (4).

Theorem 3. *For all $f \in L^2_d(0,1)$, and all $\lambda \in \mathbb{C}$ such that Re $\lambda > -r_0$, the solution to equation (3) (respectively (4)) belongs to $H^1_d(0,1)$, in particular it satisfies $z(1) = 0$.*

Proof. We give the proof for equation (3). Due to the Lax-Milgram theorem, the variational equation

$$\int_0^1 \left(z_y\bar{\varphi}_y + \frac{c+ka+\lambda a}{(y-1)^2}\, z\,\bar{\varphi}\right)dy = \int_0^1 \frac{f}{(y-1)^2}\,\bar{\varphi}\,dy, \text{ for all } \varphi \in H^1_d(0,1), \tag{8}$$

admits a unique solution z in $H^1_d(0,1)$. Observe that if ϕ belongs to $H^1(0,1;d)$, then $\varphi = (y-1)^2\phi$ belongs to $H^1_d(0,1)$. By setting $\varphi = (y-1)^2\phi$ in equation (8), we can easily check that z is also the weak solution to equation (5). The proof is complete. □

Theorem 4. *For all $f \in (H^1(0,1;d))'$, and all $\lambda \in \mathbb{C}$ such that Re $\lambda > -r_0$, equation (3) (respectively (4)) admits a unique weak solution in $H^1(0,1;d)$. This solution satisfies the estimate*

$$|\lambda|\|\sqrt{a}z\|_{L^2(0,1)} + \|z\|_{H^1(0,1;d)} \leq C\|f\|_{(H^1(0,1;d))'}, \tag{9}$$

for all $\lambda \in \mathbb{C}$ such that Re $\lambda > -r_0$.

Proof. The proof is similar to that of Theorem 2. □

Let Λ and Λ^* be the unbounded operators in $L^2(0,1)$ defined by

$$\Lambda z = -bz_{yy} + cz + kaz, \quad D(\Lambda) = \{z \in H^2(0,1;d) \mid z_y(0) = 0\},$$

and

$$\Lambda^*\phi = -(b\phi)_{yy} + c\phi + ka\phi, \quad D(\Lambda^*) = \{\phi \in H^2(0,1;d) \mid (b\phi)_y(0) = 0\}.$$

We still denote by Λ (respectively Λ^*) the continuous extension of Λ (respectively Λ^*) to $(H^1(0,1;d))'$. Due to Theorem 4, Λ is an isomorphism from $H^1(0,1;d)$ to $(H^1(0,1;d))'$.

Remark 5. *The existence and uniqueness results stated in Theorems 2 and 4, together with estimates (6) and (9) are also true for all* $\lambda \in \Sigma(-r_0,\omega)$ *and some* $\omega \in (\pi, \frac{\pi}{2})$, *where* $\Sigma(-r_0,\omega) = \{\lambda \in \mathbb{C} \mid |arg(\lambda + r_0)| \leq \omega\}$.

We denote by M the bounded operator in $L^2(0,1)$ or in $(H^1(0,1;d))'$ defined by $Mz = az$ (M is the multiplication operator by the function a).

Theorem 6. *The operator* $(\Lambda, D(\Lambda))$ *(respectively* $(\Lambda^*, D(\Lambda^*))$ *) is a closed and densely defined operator in* $L^2(0,1)$. *For all* $\lambda \in \Sigma(-r_0,\omega)$, $\lambda M + \Lambda$ *and* $\lambda M + \Lambda^*$ *have bounded inverses from* $L^2(0,1)$ *to* $D(\Lambda)$ *and the following estimates hold*

$$|\lambda|^{\frac{1}{2}} \|M(\lambda M + \Lambda)^{-1} f\|_{L^2(0,1)} \leq C \|f\|_{L^2(0,1)}, \tag{10}$$

and

$$|\lambda|^{\frac{1}{2}} \|M(\lambda M + \Lambda^*)^{-1} f\|_{L^2(0,1)} \leq C \|f\|_{L^2(0,1)}, \tag{11}$$

for all $f \in L^2(0,1)$. *The operator* $(\Lambda, H^1(0,1;d))$ *(respectively* $(\Lambda^*, H^1(0,1;d)$ *) is a closed and densely defined operator in* $(H^1(0,1;d))'$. *For all* $\lambda \in \Sigma(-r_0,\omega)$, $\lambda M + \Lambda$ *and* $\lambda M + \Lambda^*$ *have bounded inverses from* $H^1(0,1;d)$ *to* $(H^1(0,1;d))'$, *and the following estimates hold*

$$|\lambda| \|M(\lambda M + \Lambda)^{-1} f\|_{(H^1(0,1;d))'} \leq C \|f\|_{(H^1(0,1;d))'}, \tag{12}$$

and

$$|\lambda| \|M(\lambda M + \Lambda^*)^{-1} f\|_{(H^1(0,1;d))'} \leq C \|f\|_{(H^1(0,1;d))'}, \tag{13}$$

for all $f \in (H^1(0,1;d))'$.

Proof. Due to Theorem 2, Theorem 4, and remark 5, we have only to prove estimates (10), (11), (12) and (13). The proof follows the lines of [7, Chapter 3]. Let us first prove (12) (the proof of (13) is similar). Let f be in $(H^1(0,1;d))'$, and let $z = (\lambda M + \Lambda)^{-1} f$. Due to estimate (9) and since Λ is an isomorphism from $H^1(0,1;d)$ to $(H^1(0,1;d))'$, we have

$$\|\Lambda z\|_{(H^1(0,1;d))'} \leq C \|z\|_{H^1(0,1;d)} \leq C \|f\|_{(H^1(0,1;d))'}.$$

Using the identity $\lambda M(\lambda M + \Lambda)^{-1} = I - \Lambda(\lambda M + \Lambda)^{-1}$, we finally obtain (12).

We prove (10). From (6), we deduce

$$|\lambda| \|az\|^2_{L^2(0,1)} \leq |\lambda| \|\sqrt{a} z\|^2_{L^2(0,1)} \leq C \|f\|^2_{L^2(0,1)}, \tag{14}$$

with $z = (\lambda M + \Lambda)^{-1} f$. Therefore we have

$$|\lambda|^{1/2} \|M(\lambda M + \Lambda)^{-1} f\|_{L^2(0,1)} \leq C \|f\|_{L^2(0,1)}, \tag{15}$$

for all $f \in L^2(0,1)$ and all $\lambda \in \Sigma(-r_0,\omega)$. □

We need an additional theorem which will be useful to establish regularity results for nondegenerate parabolic equations (see the proof of Theorem 15).

Theorem 7. *The operator $(-\Lambda, D(\Lambda))$ is the infinitesimal generator of analytic semigroup in $L^2(0,1)$, and the semigroup is exponentially stable.*

Proof. As in the proof of Theorem 2, we can establish that, for all $f \in L^2(0,1)$, and all $\lambda \in \mathbb{C}$ such that Re $\lambda > -r_0$, the equation

$$\beta_0(z,\phi) + \lambda \int_0^1 z\bar{\phi}\,dy = \int_0^1 f\bar{\phi}\,dy, \quad \text{for all } \phi \in H^1(0,1;d),$$

admits a unique solution $z \in H^1(0,1;d)$. This solution belongs to $H^2(0,1;d)$ and satisfies the estimates

$$|\lambda + r_0|\,\|z\|_{L^2(0,1)} + \|z\|_{H^2(0,1;d)} \le C\|f\|_{L^2(0,1)}, \tag{16}$$

for all $\lambda \in \mathbb{C}$ such that Re $\lambda > -r_0$, and

$$|\lambda + r_0|\|z\|_{L^2(0,1)} \le \|f\|_{L^2(0,1)}, \tag{17}$$

for all $\lambda \in \mathbb{R}$ such that $\lambda > -r_0$. Due to (16), $(-\Lambda, D(\Lambda))$ is the infinitesimal generator of an analytic semigroup in $L^2(0,1)$ (see also [12, Theorem 3.1], where a similar result is established for a slighly different operator). From (17), it follows that the semigroup is exponentially stable. □

Setting $D(\Lambda_d^*) = \{\phi \in H_d^1(0,1) \mid (y-1)\phi_y \in H^1(0,1),\ (b\phi)_y(0) = 0\}$, and using Theorem 3, we can prove the following

Theorem 8. *For all $\lambda \in \Sigma(-r_0,\omega)$, $\lambda M + \Lambda^*$ has a bounded inverse from $L_d^2(0,1)$ to $D(\Lambda_d^*)$ and the following estimate holds*

$$|\lambda|^{\frac{1}{2}}\|M(\lambda M + \Lambda^*)^{-1}f\|_{L_d^2(0,1)} \le C\|f\|_{L_d^2(0,1)}, \tag{18}$$

for all $f \in L_d^2(0,1)$.

2.2. Adjoint parabolic equation

With the results obtained in the previous subsection, we can study the following degenerate parabolic equation

$$\begin{aligned} &-a\phi_x - (b\phi)_{yy} + c\phi + ka\phi = \psi \quad \text{in } \Omega,\\ &(b\phi)_y(x,0) = 0 \quad \text{and} \quad \phi(x,1) = 0 \quad \text{for } x \in (0,L),\\ &\phi(L,y) = 0 \text{ for } \ y \in (0,1). \end{aligned} \tag{19}$$

Following [7], [8], equation (19) can be rewritten as an equation with a multivalued operator

$$-\frac{d\phi}{dx} + (\Lambda^* a^{-1})\phi \ni \psi, \quad \phi(L) = 0, \tag{20}$$

where $D(\Lambda^* a^{-1}) = \{av \mid v \in V\}$, $(\Lambda^* a^{-1})\phi = \{\Lambda^* v \mid v \in V,\ av = \phi\}$, with $V = H^1(0,1;d)$ if equation (20) is studied in $(H^1(0,1;d))'$, and $V = D(\Lambda^*)$ if equation (20) is studied in $L^2(0,1)$.

Theorem 9. *For any* $\psi \in C^1([0,L];(H^1(0,1;d))')$*, equation (19) admits a unique strict solution* ϕ *such that* $a\phi \in C^1([0,L);(H^1(0,1;d))')$ *and* $\Lambda^*\phi \in C([0,L);(H^1(0,1;d))')$.

Proof. The theorem is a direct consequence of estimate (13) and of [7, Theorem 3.8] (see also [7, Example 3.3], [8, Example 6.3]. The difference is that in [7] and [8] the elliptic operator Λ^* is not degenerate). □

Applying the maximal regularity results stated in [7, Chapter 3], we can prove the following theorem.

Theorem 10. *If* $\psi \in C^1([0,L];L^2(0,1))$ *and* $\psi(L,y) = 0$ *for all* $y \in [0,1]$*, then the strict solution* ϕ *to equation (19) satisfies* $a\phi \in C^1([0,L];L^2(0,1))$ *and* $\phi \in C([0,L];D(\Lambda^*))$.

Proof. The theorem is a direct consequence of estimate (11) and of [7, Theorem 3.26]. □

Similar results can be obtained for equation (2). In particular we have the following theorem.

Theorem 11. *If* $f \in C^1([0,L];L^2(0,1))$ *and* $f(0,y) = 0$ *for all* $y \in [0,1]$*, then equation (2) admits a unique strict solution* z *such that* $az \in C^1([0,L];L^2(0,1))$ *and* $z \in C([0,L];D(\Lambda))$*. Moreover* z *satisfies*

$$\|\sqrt{a}z\|_{C([0,L];L^2(0,1))} + \|z\|_{L^2(0,L;H^1(0,1;d))} \leq C\|f\|_{L^2(\Omega)}. \tag{21}$$

Proof. The existence of a unique strict solution z such that $az \in C^1([0,L];L^2(0,1))$ and $\Lambda z \in C([0,L];L^2(0,1))$ can be proved as in Theorem 10. To prove estimate (21), we multiply equation (2) by $\bar{z}$, we integrate over $(0,\ell) \times (0,1)$, and with integrations by parts we obtain

$$\int_0^1 a|z(\ell)|^2 dy + \int_0^\ell \beta_0(z,z)dx = \int_0^\ell \int_0^1 f\,\bar{z}\,dy\,dx \quad \text{for all } \ell \in [0,L].$$

We can conclude with Lemma 1. □

For every solution ϕ to equation (19), with $\psi \in C^1([0,L];L^2(0,1))$ and $\psi(L,\cdot) = 0$, we set $A^*\phi = \psi$. In the sequel we denote by H the space

$$H = \{\phi \mid \phi \text{ is the solution to (19) for some } \psi \in C_c^1([0,L) \times [0,1))\}.$$

Remark 12. *With Theorem 8, it can be shown that any function* $\phi \in H$ *belongs to* $C([0,L];D(\Lambda_d^*))$*. With* [7, Theorem 3.26]*, we can show that* $\phi \in C^{0,\alpha}([0,L];D(\Lambda_d^*))$ *for all* $0 < \alpha < 1$.

2.3. An existence and uniqueness result for equation (2)

To apply the theory in [7] to equation (2), the function f must be sufficiently regular with respect to the variable x. Now we are interested in equation (2) when f belongs to $L^2(\Omega)$.

Definition 13. *A weak solution to equation (2) is a function $z \in L^2(\Omega)$ satisfying*

$$\int_\Omega z\, \bar{\psi}\, dxdy = \int_\Omega f\, \bar{\phi}\, dxdy \tag{22}$$

for all $\psi \in C^1([0, L]; L^2(0,1))$ such that $\psi(L) = 0$, and where ϕ is the solution to equation (19) associated with ψ.

It is clear that $z \in L^2(\Omega)$ is a weak solution to equation (2) if and only if

$$\int_\Omega z\, A^* \bar{\phi}\, dxdy = \int_\Omega f\, \bar{\phi}\, dxdy \quad \text{for all } \phi \in H.$$

Proposition 14. *Equation (2) admits a unique weak solution in the sense of Definition 13. Moreover the following estimate holds*

$$\|\sqrt{a}z\|_{C([0,L];L^2(0,1))} + \|z\|_{L^2(0,L;H^1(0,1;d))} \le C\|f\|_{L^2(\Omega)}. \tag{23}$$

Proof. Let $(f_n)_n$ be a sequence of functions in $\mathcal{D}(\Omega)$ such that $(f_n)_n$ converges to f in $L^2(\Omega)$. Let z_n be the strict solution to equation (2) associated with f_n. From estimate (21), it follows that $(\sqrt{a}z_n)_n$ is a Cauchy sequence in $C([0, L]; L^2(0,1))$, and $(z_n)_n$ is a Cauchy sequence in $L^2(0, L; H^1(0,1;d))$. Thus the sequence $(z_n)_n$ converges to some z in $L^2(0, L; H^1(0,1;d))$, and $(\sqrt{a}z_n)_n$ converges to $\sqrt{a}z$ in $C([0, L]; L^2(0,1))$. Since z_n satisfies

$$\int_\Omega z_n\, A^* \bar{\phi}\, dxdy = \int_\Omega f_n\, \bar{\phi}\, dxdy \quad \text{for all } \phi \in H,$$

by passing to the limit in this variational equation, when n tends to ∞, we prove that z is a weak solution to equation (2). Estimate (23) is deduced from (21).

Let us prove the uniqueness. Suppose that $f \equiv 0$. Let $\psi \in \mathcal{D}(\Omega)$, and let ϕ be the solution to (19) corresponding to ψ. We have $\int_\Omega z\, \bar{\psi}\, dxdy = 0$ for every $\psi \in \mathcal{D}(\Omega)$. Thus $z \equiv 0$. □

If z is a weak solution to equation (2), then

$$\int_\Omega z(-a\bar{\phi}_x - (b\bar{\phi})_{yy} + c\bar{\phi} + ka\bar{\phi})dxdy = \int_\Omega f\bar{\phi}\, dxdy$$

for all $\phi \in \mathcal{D}(\Omega)$. Thus equation (2) is satisfied in the sense of distributions in Ω. Let us set

$$Az = az_x - bz_{yy} + cz + kaz,$$

where the derivatives are calculated in the sense of distributions. Therefore if $z \in L^2(0, L; H^1(0,1;d))$ is a weak solution to equation (2), then $(az, -bz_y)$ belongs to $(L^2(\Omega))^2$, and $\mathrm{div}(az, -bz_y) = Az - b_y z_y - cz - kaz$ belongs to $L^2(\Omega)$. Thus we can define the normal trace $N(az, -bz_y)$ of the vectorfield $(az, -bz_y)$ on the boundary Γ of Ω, in $H^{-1/2}(\Gamma)$ (see [10, Chapter 1]). Let us set $\Gamma_1 = [0, L[\times\{0\}$, and $\Gamma_2 = \{0\}\times]0,1[$. For any $z \in L^2(0, L; H^1(0,1;d))$ such that $(az, -bz_y) \in L^2(\Omega)$, and $\mathrm{div}(az, -bz_y) = Az - b_y z_y - cz - kaz \in L^2(\Omega)$, we denote by $N_{\Gamma_2}(az, -bz_y)$ the restriction of $N(az, -bz_y)$ to Γ_2. If z is a weak solution to equation (2), then $\sqrt{a}z \in C([0, L]; L^2(0,1))$, and we can prove that $N_{\Gamma_2}(az, -bz_y) = az(0, \cdot) = 0$. We would like to define the restriction of $N(az, -bz_y)$ to Γ_1. This is not easy since Γ_1 is not open in Γ. To overcome this difficulty, for $\varepsilon > 0$, we denote by z_ε the weak solution to the equation

$$\begin{aligned} &az_x - bz_{yy} + cz + kaz = f_\varepsilon \quad \text{in }]-\varepsilon, L[\times]0,1[, \\ &z_y(x,0) = 0 \quad \text{and} \quad z(x,1) = 0 \text{ for } x \in (-\varepsilon, L), \\ &z(-\varepsilon, y) = 0 \text{ for } \ y \in (0,1), \end{aligned} \tag{24}$$

where f_ε is the extension of f by zero on $]-\varepsilon, 0] \times (0,1)$. We can verify that $z_\varepsilon|_{(-\varepsilon,0)\times(0,1)} = 0$, and $z_\varepsilon|_{(0,L)\times(0,1)} = z$. We set $T(az, -bz_y) = N^\varepsilon(az_\varepsilon, -bz_{\varepsilon y})|_{]-\varepsilon,L[\times\{0\}}$, where N^ε is the operator of normal trace on the boundary of the domain $]-\varepsilon, L[\times(0,1)$. We set

$$\langle T(az,-bz_y), \phi\rangle_{\Gamma_1} = \langle N^\varepsilon(az,-bz_y)|_{]-\varepsilon,L[\times\{0\}}, \phi\rangle_{H^{-1/2}\times H^{1/2}(]-\varepsilon,L[\times\{0\})},$$

where $\phi \in C^{0,\alpha}([-\varepsilon, L]; D(\Lambda^*))$ for all $0 < \alpha < 1$ (see Remark 12). With this definition for T, we can establish the following integration by parts formula

$$\begin{aligned} &\int_\Omega \Big(\operatorname{div}(az, -bz_y) + b_y z_y\Big)\,\bar\phi\, dxdy \\ &= \langle T(az,-bz_y), \phi\rangle_{\Gamma_1} + \int_\Omega \Big(-az\,\bar\phi_x + bz_y\,\bar\phi_y + b_y z_y\,\bar\phi\Big) dxdy, \end{aligned} \tag{25}$$

for all $\phi \in H$, and all $z \in \{z \in L^2(0,L;H^1(0,1;d)) \mid \sqrt{a}z \in C([0,L];L^2(0,1)),\ az(0,\cdot) = 0,\ Az \in L^2(\Omega)\}$. Let us set

$$\begin{aligned} D_A = \{&z \in L^2(0,L;H^1(0,1;d)) \mid \sqrt{a}z \in C([0,L];L^2(0,1)),\ az(0,\cdot) = 0, \\ &T(az,-bz_y) = 0,\ Az \in L^2(\Omega),\ z \in L^2(0,L;H^2(\tfrac{1}{2},1;d))\}. \end{aligned}$$

Theorem 15. *If $z \in L^2(0,L;H^1(0,1;d))$ is a solution to equation (2) then z belongs to D_A. Conversely, if z belongs to D_A then z is the weak solution to equation (2) corresponding to $f = Az$.*

Proof. Let z be a solution to equation (2). To prove that $z \in D_A$, thanks to the previous remarks, it is sufficient to show that z belongs to $L^2(0,L;H^2(\frac{1}{2},1;d))$, $az(0,\cdot) = 0$, and $T(az,-bz_y) = 0$. Let us prove that z belongs to $L^2(0,L;$ $H^2(\frac{1}{2},1;d))$. To avoid the problem of degeneracy of the coefficient a for $y = 0$, we use a cut-off function. Let θ be a regular function on $[0,1]$ defined by

$$\theta(y) = \begin{cases} 0 & \text{if } 0 \le y \le \tau/2, \\ \theta(y) & \text{with } 0 \le \theta(y) \le 1 \text{ if } \ \tau/2 \le y \le \tau, \\ 1 & \text{if } \tau \le y \le 1, \end{cases}$$

with $0 < \tau < 1/2$. We suppose that z is a weak solution to equation (2) for $f \in L^2(\Omega)$. By a straightforward calculation, we verify that θz is a weak solution to the equation

$$\begin{aligned} &a(\theta z)_x - b(\theta z)_{yy} + c(\theta z) + ka(\theta z) = g \quad \text{in } \Omega, \\ &(\theta z)_y(x,0) = 0 \quad \text{and} \quad (\theta z)(x,1) = 0 \quad \text{for } x \in (0,L), \\ &(\theta z)(0,y) = 0 \text{ for } \ y \in (0,1), \end{aligned} \tag{26}$$

with $g = -2bz_y\theta_y - bz\theta_{yy} + \theta f$ and $g \in L^2(0,1)$. Thus θz is also the solution to

$$\begin{aligned} &a_\tau(y)(\theta z)_x - b(\theta z)_{yy} + c(\theta z) + ka(\theta z) = g \quad \text{in } \Omega, \\ &(\theta z)_y(x,0) = 0 \quad \text{and} \quad (\theta z)(x,1) = 0 \quad \text{for } x \in (0,L), \\ &(\theta z)(0,y) = 0 \text{ for } \ y \in (0,1), \end{aligned} \tag{27}$$

where $a_\tau(y) = \frac{y^2}{\tau} + \frac{\tau}{4}$ if $0 \le y \le \frac{\tau}{2}$, and $a_\tau(y) = a(y) = y$ if $y > \frac{\tau}{2}$. Since a_τ is regular and $a_\tau(y) \ge \frac{\tau}{4}$, with Theorem 7, we can apply maximal regularity results to equation (27) ([2, Chapter 1, Proposition 3.7]), and we deduce that $(\theta z) \in L^2(0,L;D(\Lambda))$. Hence $z|_{(0,L)\times(1/2,1)}$ belongs to $L^2(0,L;H^2(\frac{1}{2},1;d))$.

Let us prove that $az(0,\cdot) = 0$, and $T(az,-bz_y) = 0$. Using the divergence theorem, integrations by parts, and the identity $\operatorname{div}(az,-bz_y) = f - b_y z_y - cz - kaz$, we have

$$\int_\Omega f\,\bar\phi = \int_0^1 az(0,y)\,\bar\phi(0,y)\,dy + \langle T(az,-bz_y),\phi\rangle_{\Gamma_1} + \int_\Omega z\,A^*\bar\phi$$

for all $\phi \in H$. Thus $\int_0^1 az(0,y)\,\bar\phi(0,y)\,dy + \langle T(az,-bz_y),\phi\rangle_{\Gamma_1} = 0$. Let $\varphi \in \mathcal{D}(0,1)$. Choosing $\phi = 0$ in $[0,L] \times ([0,1] \setminus \operatorname{supp}\varphi)$ in the above equality, we prove that $az(0,\cdot) = 0$. Next for any $\varphi \in C_c^1([0,L))$, we choose $\phi \in H$ such that $\phi(0,y) = \phi_y(0,y) = \varphi(y)$. We obtain $\langle T(az,-bz_y),\varphi\rangle_{\Gamma_1} = 0$, for any $\varphi \in C_c^1([0,L))$. Hence $T(az,-bz_y) = 0$.

Now suppose that z belongs to D_A. Set $f = Az$. Due to (25), for every $\phi \in H$, we have

$$\begin{aligned} \int_\Omega f\,\bar\phi\,dxdy &= \int_\Omega \Big(\operatorname{div}(az,-bz_y) + b_y z_y + cz + kaz\Big)\,\bar\phi\,dxdy \\ &= \int_\Omega \Big(-az\,\bar\phi_x + bz_y\,\bar\phi_y + b_y z_y\,\bar\phi + cz\,\bar\phi + kaz\,\bar\phi\Big)dxdy. \end{aligned}$$

Integrating by parts, and taking remark 12 into account, we obtain $\int_\Omega (bz_y\bar\phi_y + b_y z_y\bar\phi)\,dxdy = \int_\Omega -z(b\bar\phi)_{yy}\,dxdy$ because $(b\phi)_y(0,\cdot) = 0$ and $(y-1)z|_{y=1} = 0$. Thus we have $\int_\Omega f\,\bar\phi\,dxdy = \int_\Omega z\,A^*\bar\phi\,dxdy$ for every $\phi \in H$, and the proof is complete. □

Theorem 16. *The domain D_A is dense in $L^2(\Omega)$. The operator (A, D_A) is a closed operator in $L^2(\Omega)$. The operator $-A$ with domain D_A is the generator of an exponentially stable semigroup in $L^2(\Omega)$.*

Proof. The first part of the theorem follows from Theorem 15. To prove the exponential stability, it is enough to observe that the weak solution z to the equation

$$\begin{aligned} &az_x - bz_{yy} + cz + kaz + \lambda z = f \quad \text{in } \Omega, \\ &z_y(x,0) = 0 \quad \text{and} \quad z(x,1) = 0 \quad \text{for } x \in (0,L), \\ &z(0,y) = 0 \text{ for } \ y \in (0,1), \end{aligned} \tag{28}$$

satisfies $(\lambda + r_0)\|z\|_{L^2(\Omega)} \le \|f\|_{L^2(\Omega)}$ for all $\lambda \in \mathbb{R}$ such that $\lambda > -r_0$. □

3. The Linearized Crocco Equation

In this section we consider functions with values in $\mathbb{R}$. Following the approach of the previous section, the domain of A^* is defined as follows

$$D(A^*) = \{\phi \in L^2(0,L;H^1(0,1;d)) \mid \sqrt{a}\phi \in C([0,L];L^2(0,1)),\ a\phi(L,\cdot) = 0,$$
$$\mathcal{T}(-a\phi, -(b\phi)_y) = 0, A^*\phi \in L^2(\Omega), \phi \in L^2(0,L;H^2(\frac{1}{2},1;d))\},$$

where $\mathcal{T}$ is defined by $\mathcal{T}(-a\phi, -(b\phi)_y) = N_\varepsilon(-a\phi_\varepsilon, -(b\phi_\varepsilon)_y)|_{(0,L+\varepsilon)\times\{0\}}$, N_ε is the operator of normal trace on the boundary of $(0,L+\varepsilon)\times(0,1)$, and ϕ_ε is the extension of ϕ by zero on $(L,L+\varepsilon)\times(0,1)$. Due to Theorem 16, if $f \in L^2(0,T;L^2(\Omega))$ and $z_0 \in L^2(\Omega)$, the equation

$$\frac{dz}{dt} + Az = f, \qquad z(0) = z_0, \tag{29}$$

admits a unique weak solution in $L^2(0,T;L^2(\Omega))$ in the sense of [2, Chapter 1]. Moreover we can prove that z belongs to $C([0,T];L^2(\Omega)) \cap L^2((0,T)\times(0,L);H^1(0,1;d))$, $\sqrt{a}z$ belongs to $C([0,L];L^2((0,T)\times(0,1)))$ and

$$\|z\|_{L^\infty(0,T;L^2(\Omega))} + \|\sqrt{a}z\|_{L^\infty(0,L;L^2((0,T)\times(0,1)))} + \|z\|_{L^2((0,T)\times(0,L);H^1(0,1;d))}$$
$$\le C\big(\|f\|_{L^2((0,T)\times\Omega)} + \|z_0\|_{L^2(\Omega)}\big).$$

A similar result also holds for the equation

$$-\frac{d\phi}{dt} + A\phi = \psi, \qquad \phi(T) = 0. \tag{30}$$

Now, we give an other definition of weak solution for equation (1).

Definition 17. *A function $z \in L^2(0,T;L^2(\Omega))$ is a weak solution to equation (1) if and only if,*

$$\int_0^T\int_\Omega z\psi = \int_0^T\int_\Omega f\phi + \int_0^T\int_{\chi_s} bv\phi + \int_0^T\int_0^1 aw_0\,\phi(0,\cdot) + \int_\Omega \phi(\cdot,0)z_0$$

for all $\psi \in L^2(0,T;L^2(\Omega))$, where ϕ is the solution to (30).

Theorem 18. *For all $f \in L^2(0,T;L^2(\Omega))$, all $v \in L^2(0,T;L^2(x_0,x_1))$, all $w_0 \in L^2(0,T;L^2(0,1))$, and all $z_0 \in L^2(\Omega)$, equation (1) admits a unique weak solution z in $L^2(0,T;L^2(\Omega))$ in the sense of Definition 17, moreover z belongs to $L^2((0,T)\times(0,L);H^1(0,1;d)) \cap L^\infty(0,L;L^2((0,T)\times(0,1)))$, $\sqrt{a}z$ belongs to $L^\infty(0,L;$*
$L^2((0,T)\times(0,1)))$, and we have

$$\|z\|_{L^\infty(0,T;L^2(\Omega))} + \|\sqrt{a}z\|_{L^\infty(0,L;L^2((0,T)\times(0,1)))} + \|z\|_{L^2((0,T)\times(0,L);H^1(0,1;d))}$$
$$\le C\Big(\|f\|_{L^2} + \|v\|_{L^2} + \|w_0\|_{L^2} + \|z_0\|_{L^2}\Big). \tag{31}$$

Proof. The uniqueness is obvious. To prove the existence, we proceed by approximation. We set $v_\varepsilon = \frac{1}{\varepsilon}v(x,t)K_\varepsilon(x,y)$, where K_ε is the characteristic function of $[x_0,x_1]\times[0,\varepsilon]$, and $w_{0\varepsilon} = \frac{1}{\varepsilon}w_0(y,t)\chi_\varepsilon(x,y)$, where χ_ε is the characteristic function of $[0,\varepsilon]\times[0,1]$. Let z_ε be the solution to (29) corresponding to $f_\varepsilon = f + v_\varepsilon + aw_{0\varepsilon}$. We can prove that (31) is satisfied by z_ε. Therefore the existence of z is obtained by passing to the limit when ε tends to zero. □

References

[1] V. Barbu, A. Favini and S. Romanelli, *Degenerate evolution equations and regularity of their associated semigroups,* Funk. Ekv., **39** (1996), 421–448.

[2] A. Bensoussan, G. Da Prato, M. C. Delfour, S. K. Mitter, *Representation and Control of Infinite Dimensional Systems, Volume 1,* 1992, Birkhäuser, Boston, Basel, Berlin.

[3] J.-M. Buchot, P. Villedieu, Construction de modèles pour le contrôle de la position de transition laminaire-turbulent sur une plaque plane, Technical report, ONERA, 1/3754.00 DTIM/T, 1999.

[4] J.-M. Buchot, *Stabilization of the laminar to turbulent transition location,* Proceedings MTNS 2000, El Jaï Ed., 2000.

[5] J.-M. Buchot, *PhD Thesis,* in preparation.

[6] A. Favini, J. A. Goldstein, and S. Romanelli, *Analytic semigroups on $L^p_w(0,1)$ and on $L^p(0,1)$ generated by some classes of second order differentail operators,* Taïwanese Journal of Math., **3** (1999), 181–210.

[7] A. Favini and A. Yagi, *Degenerate differential equations in Banach spaces,* Monographs and Textbooks in Pure and Applied Mathematics 215, 1999, Marcel Dekker, New York.

[8] A. Favini and A. Yagi, *Multivalued linear operators and degenerate evolution equations,* Ann. Math. Pura Appl. (IV), **163** (1993), 353–384.

[9] O. A. Oleinik, V. N. Samokhin, *Mathematical Models in Boundary Layer Try,* Applied Mathematics and Mathematical Computation 15, 1999, Chapman & Hall/CRC, Boca Raton, London, New York.

[10] R. Temam, *Navier Stokes Equations,* Noth-Holland, Amsterdam, 1979.

[11] H. Triebel, *Interpolation Theory, Function Spaces, Differential Operators,* North Holland, 1978.

[12] V. Vespri, *Analytic semigroups, Degenerate Elliptic Operators, and Applications to Nonlinear Cauchy Problems,* Ann. Math. Pura Appl. (IV), **155** (1989), 1073–1077.

Université Paul Sabatier,
Laboratoire MIP, UMR 5640
31062 Toulouse Cedex 4, France
E-mail address: `buchot@cert.fr`, `raymond@mip.ups-tlse.fr`

International Series of Numerical Mathematics, Vol. 139, 43–55

Numerical Solution of Optimal Control Problems Governed by the Compressible Navier–Stokes Equations

S. Scott Collis, Kaveh Ghayour, Matthias Heinkenschloss, Michael Ulbrich, and Stefan Ulbrich

Abstract. Theoretical and practical issues arising in optimal boundary control of the unsteady two–dimensional compressible Navier–Stokes equations are discussed. Assuming a sufficiently smooth state, formal adjoint and gradient equations are derived. For a vortex rebound model problem wall normal suction and blowing is used to minimize cost functionals of interest, here the kinetic energy at the final time.

1. Introduction

Recently, optimal control and optimal design problems governed by fluid flow models have received significant attention in the mathematical and in the engineering literature. See, e.g., the collections and reviews [7, 10, 11]. The coupling of accurate computational fluid dynamics analyses with optimal control theory holds the promise for modifying a wide-range of fluid flows to achieve enhancement of desirable flow characteristics. Reduction of skin-friction drag, separation suppression, and increased lift to drag ratios for airfoils are examples of the types of optimization that such an approach enables. Moreover, advances in smart materials and microelectromechanical systems (MEMS) have increased the possibilities to actually implement controllers in physical systems. Optimal control problems have been studied mathematically and numerically for steady and unsteady incompressible Navier–Stokes flow. The references [1, 2, 5, 11, 12] present a small sample of the work in this area. Compressible steady state Euler equations and Navier–Stokes equations have been used in the context of optimal design, see, e.g., [3, 14, 15, 16, 21].

In this paper we study the optimal control of two–dimensional unsteady compressible Navier–Stokes flows. To the best of the authors' knowledge, this is the first attempt to apply optimal control to problems governed by the *unsteady* compressible Navier–Stokes equations. Our research is motivated by the potential to develop novel and effective flow control strategies for inherently compressible phenomena including aeroacoustics and heat transfer by utilizing optimal control theory. Specifically, we plan to control the sound arising from Blade-Vortex Interaction (BVI) that can occur for rotorcraft in low speed, descending flight conditions, such as on approach to landing. When BVI occurs, tip vortices shed by a preceding blade interact with subsequent blades resulting in a high amplitude, impulsive noise that can dominate other rotorcraft noise sources. Reduction of the noise generated by this mechanism can alleviate restrictions on civil rotorcraft use near city centers and thereby enhance community acceptance. High frequency loading associated with this phenomenon also causes fatigue and hence reductions in BVI can have a direct impact on maintenance costs associated with blade failure in fatigue mode.

We use adjoint based gradient methods to solve the discretized optimal control problem. A critical issue in the numerical solution of optimal control problems is the accuracy of adjoint and gradient information. For successful optimization of the discretized problem, it is indispensable that the gradient approximation is sufficiently close to the derivative of the discretized objective function. To ensure that the solution of the discretized optimization problem approximates the infinite dimensional optimal control, it is also important that adjoint and gradient approximations used for the discretized problem converge towards their infinite dimensional counterparts as the discretization is refined. This requires a comprehensive view of the problem that integrates well posedness of the infinite dimensional problem, existence of adjoint equations and gradient equations, and properties of the discretization. Unfortunately, the mathematical foundation for optimal control problems governed by the unsteady compressible Navier–Stokes equations is not sufficiently developed to allow a rigorous and comprehensive study of gradient and adjoint accuracy in the previous sense. Even mathematical existence theories for the unsteady compressible Navier–Stokes equations are less developed than for the incompressible case.

In this paper, we discuss some of the theoretical issues arising in the formulation and solution of optimal boundary control problems governed by the compressible Navier–Stokes equations. Assuming a sufficiently smooth state, we derive formal adjoint and gradient equations. Finally, we present optimal control results for a vortex rebound test problem.

2. Problem Formulation

Let $\Omega = \{\mathbf{x} \in \mathbb{R}^2 : x_2 > 0\}$ denote the spatial domain occupied by the fluid and let Γ denote its spatial boundary. By

$$\mathbf{u} = (\rho, v_1, v_2, T)^T$$

we denote the primitive flow variables, where $\rho(t, \mathbf{x})$ is the density, $v_i(t, \mathbf{x})$ denotes the velocity in x_i-direction, $i = 1, 2$, $\mathbf{v} = (v_1, v_2)^T$, and $T(t, \mathbf{x})$ denotes the temperature. The pressure p and the total energy per unit mass E are given by

$$p = \frac{\rho T}{\gamma \mathsf{M}^2}, \quad E = \frac{T}{\gamma(\gamma-1)\mathsf{M}^2} + \frac{1}{2}\mathbf{v}^T\mathbf{v},$$

respectively, where γ is the ratio of specific heats and M is the reference Mach number. We write the conserved variables as functions of the primitive variables,

$$\mathbf{q}(\mathbf{u}) = (\rho, \rho v_1, \rho v_2, \rho E)^T$$

and we define the inviscid flux terms

$$\mathbf{F}^1(\mathbf{u}) = \begin{pmatrix} \rho v_1 \\ \rho v_1^2 + p \\ \rho v_2 v_1 \\ (\rho E + p)v_1 \end{pmatrix}, \quad \mathbf{F}^2(\mathbf{u}) = \begin{pmatrix} \rho v_2 \\ \rho v_1 v_2 \\ \rho v_2^2 + p \\ (\rho E + p)v_2 \end{pmatrix}, \tag{1}$$

and the viscous flux terms

$$\mathbf{G}^i(\mathbf{u}, \nabla\mathbf{u}) = \frac{1}{\mathsf{Re}} \begin{pmatrix} 0 \\ \tau_{1i} \\ \tau_{2i} \\ \tau_{1i}v_1 + \tau_{2i}v_2 + \frac{\kappa}{\mathsf{Pr}\mathsf{M}^2(\gamma-1)}T_{x_i} \end{pmatrix}, \tag{2}$$

$i = 1, 2$, where τ_{ij} are the elements of the stress tensor $\tau = \mu(\nabla\mathbf{v} + \nabla\mathbf{v}^T) + \lambda(\nabla\cdot\mathbf{v})I$. Here μ, λ are first and second coefficients of viscosity, κ is the thermal conductivity, Pr is the reference Prandtl number, and Re is the reference Reynolds number. For the demonstration problems presented here, constant Prandtl number and fluid properties (viscosities and thermal conductivity) are assumed along with Stokes hypothesis for the second coefficient of viscosity, $\lambda = -2\mu/3$. Variable fluid properties can be easily accommodated and these effects will be included in future studies.

The two–dimensional compressible Navier–Stokes equations for the time interval $[t_0, t_f]$ can now be written as

$$\mathbf{q}(\mathbf{u})_t + \sum_{i=1}^{2} \left(\mathbf{F}^i(\mathbf{u})_{x_i} - \mathbf{G}^i(\mathbf{u}, \nabla\mathbf{u})_{x_i}\right) = \mathbf{0} \quad \text{in } (t_0, t_f) \times \Omega, \tag{3}$$

$$\mathbf{B}(\mathbf{u}, \nabla\mathbf{u}, \mathbf{g}) = \mathbf{0} \quad \text{on } (t_0, t_f) \times \Gamma, \tag{4}$$

$$\mathbf{u}(t_0, \mathbf{x}) = \mathbf{u}_0(\mathbf{x}) \quad \text{in } \Omega. \tag{5}$$

The function $\mathbf{g}$ in the boundary conditions (4) acts as the control, which is taken to be suction and blowing in the wall normal direction on $\Gamma_c \subset \Gamma$. This is modeled by

$$\mathbf{v} = \mathbf{b} + \mathbf{g} \quad \text{on } \Gamma_c, \tag{6}$$

where $\mathbf{b}$ is a given boundary velocity that satisfies the compatibility condition $\mathbf{v}(t_0, \mathbf{x}) = \mathbf{b}(t_0, \mathbf{x})$ for $\mathbf{x} \in \Gamma$. Since $\Gamma_c \subset \{\mathbf{x} : x_2 = 0\}$, we have $\mathbf{g} = (0, g_2)^T$.

To the best of our knowledge, the question of existence and uniqueness of global solutions for the full compressible Navier–Stokes equations (3)–(5) in 2D and 3D is still open for large initial data. The existence and uniqueness of global solutions with $(\rho-\rho_0, \mathbf{v}, T-T_0) \in C(t_0, t_f; H^3) \cap C^1(t_0, t_f; H^1)$ for the initial value problem (3), (5) is shown in [18] if the initial data $\mathbf{u}_0$ are close in H^3 to a constant state $(\rho_0, 0, 0, T_0)^T$, $\rho_0, T_0 > 0$. Local existence in time can be shown also for large data. An analogous result for the initial-boundary value problem (3)–(5) on the half space or on the exterior of any bounded region with smooth boundary is shown in [19] for the boundary conditions $\mathbf{v}|_{(t_0,t_f)\times\Gamma} = 0$ and either $T|_{(t_0,t_f)\times\Gamma} = T_0$ or $\frac{\partial T}{\partial \mathbf{n}} = 0$, where $\mathbf{n}$ denotes the outward unit normal. Similar results can be found in the review article [22]. The global existence of weak solutions for the initial value problem is shown in [13], if $\rho(t_0,\cdot)-\rho_0$ is small in $L^2 \cap L^\infty$, $\mathbf{v}(t_0,\cdot)$ is small in $L^2 \cap L^4$, and $T(t_0,\cdot) - T_0$ is small in L^2 with constants $\rho_0, T_0 > 0$. It is shown that $\mathbf{v}, T$ are Hölder-continuous in space and time for $t > t_0$, $\mathbf{v}(t,\cdot), T(t,\cdot) - T_0 \in H^1$, but merely $\rho(t,\cdot) - \rho_0 \in L^2 \cap L^\infty$, $\rho - \rho_0 \in C(t_0, t_f; H^{-1})$. The question of uniqueness ist left open. Other results, mostly for the barotropic case in which pressure depends only on ρ and which decouples the energy equation from the remaining ones, can be found in [17, 22].

The optimal control problem treated in this paper is the minimization of kinetic energy in $\Omega_0 \subset \Omega$ at final time, more precisely,

$$\begin{aligned} \min_{\mathbf{g}\in\mathcal{G}} \widehat{J}(\mathbf{g}) &\stackrel{\text{def}}{=} \frac{1}{2}\int_{\Omega_0} \rho(t_f,\mathbf{x})\|\mathbf{v}(t_f,\mathbf{x})\|_2^2 d\mathbf{x} \\ &+ \int_{t_0}^{t_f}\int_{\Gamma_c} \left(\frac{\alpha_1}{2}\|\mathbf{g}_t\|_2^2 + \frac{\alpha_2}{2}\|\nabla \mathbf{g}\|_2^2 + \frac{\alpha_3}{2}\|\mathbf{g}\|_2^2\right) d\mathbf{x}dt. \end{aligned} \tag{7}$$

where $\alpha_1, \alpha_2, \alpha_3 > 0$ and where the control space is chosen to be

$$\mathcal{G} = \left\{\mathbf{g} : \mathbf{g} \in L^2(t_0, t_f; H_0^1(\Gamma_c)),\ \mathbf{g}_t \in L^2(t_0, t_f; L^2(\Gamma_c)),\ \mathbf{g}(t_0, \mathbf{x}) = 0 \text{ in } \Gamma_c\right\}.$$

Here $\nabla \mathbf{g}$ is the gradient of $\mathbf{g}$ on the boundary, in our case $\nabla \mathbf{g} = (0, (g_2)_{x_1})^T$. The second part of $\widehat{J}$ is a regularization term which, together with $\mathcal{G}$, must be chosen so that (7) is well–posed. In particular the regularity requirements on the control must be compatible with the regularity of the trace of $\mathbf{v}$ on Γ_c. For the incompressible Navier–Stokes equations such trace regularity estimates have been provided recently in [8, 9] and our choice of the control space and of the regularization term follows [12]. There is no theoretical justification yet that this choice is suitable for the compressible case. Significantly stronger regularity requirements on the controls seem necessary in connection with the theory in [18, 19]. Our regularity requirements are closer aligned with what one would expect from the theory in [13]. Application of general existence results such as those in [8] to (7) are not yet known. However, our numerical results indicate that (7) is well–posed for the flows we have considered. A relaxation of the regularity requirements, i.e., setting $\alpha_1 = 0$ or even $\alpha_1 = \alpha_2 = 0$ leads to highly oscillatory controls. More detailed grid convergence studies are under way. We remark that the choice of the control space and regularization term does not affect the adjoint equations computed in Sections 3.2, 3.3, it only effects how the gradient is computed given the adjoint (see (8)).

3. Adjoint Equation

3.1. Adjoint equation and gradient for an abstract problem

We first consider the gradient computation for an abstract functional whose evaluation involves implicitly defined functions. Let $\mathcal{G}$ be a Hilbert space with inner product $\langle\cdot,\cdot\rangle_{\mathcal{G}}$ and let $\mathcal{U}, \mathcal{C}$ be Banach spaces. We consider an equation $\mathbf{C}(\mathbf{u},\mathbf{g}) = \mathbf{0}$, where $\mathbf{C} : \mathcal{U} \times \mathcal{G} \to \mathcal{C}$. Suppose that for every $\mathbf{g} \in \mathcal{G}$ the equation $\mathbf{C}(\mathbf{u},\mathbf{g}) = 0$ has a unique solution $\mathbf{u}(\mathbf{g})$. We consider the abstract problem

$$\min_{\mathbf{g}\in\mathcal{G}} \widehat{J}(\mathbf{g}) = J(\mathbf{u}(\mathbf{g}),\mathbf{g}).$$

We assume the existence of neighborhoods $\bar{G}$, $\bar{U}$ of $\bar{\mathbf{g}}$ and $\bar{\mathbf{u}} = \mathbf{u}(\bar{\mathbf{g}})$, respectively, such that $\mathbf{C}$ is Fréchet–differentiable. Further, we assume that $\mathbf{u}(\mathbf{g})$ is differentiable on $\bar{G}$. This holds, e.g., if $\mathbf{C}$ is continuously Fréchet–differentiable on $\bar{U}\times\bar{G}$ and the partial derivative $\mathbf{C}_{\mathbf{u}}(\bar{\mathbf{u}},\bar{\mathbf{g}})$ is continuously invertible. However, in our context these latter requirements seem to be too restrictive, since the results in [18, Prop. 4.1] indicate that the solution of the linearized state equation is less regular than the state $\bar{\mathbf{u}}$ about which the linearization is done. Now suppose that J is Fréchet–differentiable on $\bar{U} \times \bar{G}$. Then the Fréchet–derivative $\widehat{J}_{\mathbf{g}}(\bar{\mathbf{g}}) \in \mathcal{G}^*$ and the gradient $\nabla\widehat{J}(\mathbf{g}) \in \mathcal{G}$ of $\widehat{J}$ can be computed from $\widehat{J}_{\mathbf{g}}(\bar{\mathbf{g}}) = J_{\mathbf{u}}(\bar{\mathbf{u}},\bar{\mathbf{g}}) \circ \mathbf{u}_{\mathbf{g}}(\bar{\mathbf{g}}) + J_{\mathbf{g}}(\bar{\mathbf{u}},\bar{\mathbf{g}})$ and $\langle\widehat{J}_{\mathbf{g}}(\bar{\mathbf{g}}),\mathbf{g}'\rangle_{\mathcal{G}^*\times\mathcal{G}} = \langle\nabla\widehat{J}(\bar{\mathbf{g}}),\mathbf{g}'\rangle_{\mathcal{G}}$ for all $\mathbf{g}' \in \mathcal{G}$, respectively, where $\mathcal{G}^*$ denotes the topological dual of $\mathcal{G}$ and $\langle\cdot,\cdot\rangle_{\mathcal{G}^*\times\mathcal{G}}$ denotes the duality pairing between $\mathcal{G}^*$ and $\mathcal{G}$.

It can be shown that the gradient $\nabla\widehat{J}(\bar{\mathbf{g}})$ can be computed from

$$\langle\nabla\widehat{J}(\bar{\mathbf{g}}),\mathbf{g}'\rangle_{\mathcal{G}} = \langle\mathbf{C}_{\mathbf{g}}(\bar{\mathbf{u}},\bar{\mathbf{g}})^*\bar{\boldsymbol{\lambda}},\mathbf{g}'\rangle_{\mathcal{G}^*\times\mathcal{G}} + \langle J_{\mathbf{g}}(\bar{\mathbf{u}},\bar{\mathbf{g}}),\mathbf{g}'\rangle_{\mathcal{G}^*\times\mathcal{G}} \tag{8}$$

for all $\mathbf{g}' \in \mathcal{G}$, if there exists an adjoint state $\bar{\boldsymbol{\lambda}} \in \mathcal{C}^*$ satisfying the adjoint equation $\mathbf{C}_{\mathbf{u}}(\bar{\mathbf{u}},\bar{\mathbf{g}})^*\bar{\boldsymbol{\lambda}} = -J_{\mathbf{u}}(\bar{\mathbf{u}},\bar{\mathbf{g}})$ in $\mathcal{U}^*$, i.e., if $\bar{\boldsymbol{\lambda}}$ satisfies

$$\langle\mathbf{C}_{\mathbf{u}}(\bar{\mathbf{u}},\bar{\mathbf{g}})^*\bar{\boldsymbol{\lambda}},\mathbf{u}'\rangle_{\mathcal{U}^*\times\mathcal{U}} = \langle -J_{\mathbf{u}}(\bar{\mathbf{u}},\bar{\mathbf{g}}),\mathbf{u}'\rangle_{\mathcal{U}^*\times\mathcal{U}} \tag{9}$$

for all $\mathbf{u}' \in \mathcal{U}$. The existence of $\bar{\boldsymbol{\lambda}}$ is ensured if $\mathbf{C}_{\mathbf{u}}(\bar{\mathbf{u}},\bar{\mathbf{g}})^*$ is onto. Note, however, that it is sufficient that the adjoint equation is solvable for the particular right-hand side $-J_{\mathbf{u}}(\bar{\mathbf{u}},\bar{\mathbf{g}})$.

3.2. Adjoint equation for the compressible Navier–Stokes equations

We carry out the formal derivation of the adjoint equation for the general situation that Ω is a domain with C^2-boundary Γ. According to (3)–(5) we define

$$\mathbf{C}(\mathbf{u},\mathbf{g}) = \begin{pmatrix} \mathbf{q}(\mathbf{u})_t + \sum_{i=1}^2 \left(\mathbf{F}^i(\mathbf{u}) - \mathbf{G}^i(\mathbf{u},\nabla\mathbf{u})\right)_{x_i} \\ \mathbf{B}(\mathbf{u},\nabla\mathbf{u},\mathbf{g}) \\ \mathbf{u}-\mathbf{u}_0 \end{pmatrix}. \tag{10}$$

To write the linearization of the Navier–Stokes equation it is useful to define $\mathbf{M} = \mathbf{q}_{\mathbf{u}}(\mathbf{u})$, $\mathbf{A}^i = \mathbf{F}^i_{\mathbf{u}}(\mathbf{u})$, $i = 1,2$, and to write

$$\mathbf{G}^i(\mathbf{u},\nabla\mathbf{u}) = \mathbf{K}^i_1(\mathbf{u})\mathbf{u}_{x_1} + \mathbf{K}^i_2(\mathbf{u})\mathbf{u}_{x_2}, \quad i = 1,2, \tag{11}$$

where

$$\mathbf{K}_1^1(\mathbf{u}) = \frac{1}{\mathsf{Re}}\begin{pmatrix} 0 & 0 & 0 & 0 \\ 0 & \widetilde{\mu} & 0 & 0 \\ 0 & 0 & \mu & 0 \\ 0 & \widetilde{\mu}v_1 & \mu v_2 & \frac{\kappa}{\mathsf{PrM}^2(\gamma-1)} \end{pmatrix}, \quad \mathbf{K}_2^1(\mathbf{u}) = \frac{1}{\mathsf{Re}}\begin{pmatrix} 0 & 0 & 0 & 0 \\ 0 & 0 & \lambda & 0 \\ 0 & \mu & 0 & 0 \\ 0 & \mu v_2 & \lambda v_1 & 0 \end{pmatrix},$$

$$\mathbf{K}_1^2(\mathbf{u}) = \frac{1}{\mathsf{Re}}\begin{pmatrix} 0 & 0 & 0 & 0 \\ 0 & 0 & \mu & 0 \\ 0 & \lambda & 0 & 0 \\ 0 & \lambda v_2 & \mu v_1 & 0 \end{pmatrix}, \quad \mathbf{K}_2^2(\mathbf{u}) = \frac{1}{\mathsf{Re}}\begin{pmatrix} 0 & 0 & 0 & 0 \\ 0 & \mu & 0 & 0 \\ 0 & 0 & \widetilde{\mu} & 0 \\ 0 & \mu v_1 & \widetilde{\mu}v_2 & \frac{\kappa}{\mathsf{PrM}^2(\gamma-1)} \end{pmatrix}$$

with $\widetilde{\mu} = 2\mu + \lambda$. From the representation (11) we obtain $\mathbf{G}^i_{\mathbf{u}_{x_j}}(\mathbf{u}, \nabla\mathbf{u}) = \mathbf{K}^i_j(\mathbf{u})$, $i, j = 1, 2$, and the definition (2) of the viscous terms implies

$$\mathbf{G}^i_{\mathbf{u}}(\mathbf{u}, \nabla\mathbf{u}) = \mathbf{D}^i(\nabla\mathbf{u}) = \frac{1}{\mathsf{Re}}\begin{pmatrix} 0 & 0 & 0 & 0 \\ 0 & 0 & 0 & 0 \\ 0 & 0 & 0 & 0 \\ 0 & \tau_{1i} & \tau_{2i} & 0 \end{pmatrix}, \qquad i = 1, 2.$$

In the following we simply write $\mathbf{K}^i_j$ instead of $\mathbf{K}^i_j(\mathbf{u})$, $i, j = 1, 2$, and $\mathbf{D}^i$ instead of $\mathbf{D}^i(\nabla\mathbf{u})$, $i = 1, 2$. With this notation, the linearized state equation can be written as

$$\mathbf{C}_{\mathbf{u}}(\mathbf{u}, \mathbf{g})\mathbf{u}' = \begin{pmatrix} (\mathbf{M}\mathbf{u}')_t + \sum_i \left(\mathbf{A}^i\mathbf{u}' - \mathbf{D}^i\mathbf{u}' - \sum_j \mathbf{K}^i_j \mathbf{u}'_{x_j}\right)_{x_i} \\ \mathbf{B}_{\mathbf{u}}(\mathbf{u}, \nabla\mathbf{u}, \mathbf{g})\mathbf{u}' + \sum_j \mathbf{B}_{\mathbf{u}_{x_j}}(\mathbf{u}, \nabla\mathbf{u}, \mathbf{g})\mathbf{u}'_{x_j} \\ \mathbf{u}' \end{pmatrix}. \tag{12}$$

The adjoint variables $\boldsymbol{\lambda}$ are partitioned according to the partition of $\mathbf{C}$ in (10) and are denoted by $\boldsymbol{\lambda} = (\boldsymbol{\lambda}^d, \boldsymbol{\lambda}^b, \boldsymbol{\lambda}^0)$.

We assume that

$$\begin{aligned} \langle D_{\mathbf{u}} J(\mathbf{u}, \mathbf{g}), \mathbf{u}' \rangle_{\mathcal{U}^* \times \mathcal{U}} = & \int_{t_0}^{t_f} \int_{\Omega} (\mathbf{u}')^T \mathbf{r} + \int_{\Omega} (\mathbf{u}'|_{t=t_f})^T \mathbf{r}_{t_f} \\ & + \int_{t_0}^{t_f} \int_{\Gamma} ((\mathbf{u}')^T \mathbf{r}_\Gamma + (\nabla\mathbf{u}'\mathbf{n})^T \mathbf{r}_{\Gamma,\mathbf{n}} + (\nabla\mathbf{u}'\mathbf{s})^T \mathbf{r}_{\Gamma,\mathbf{s}}), \end{aligned} \tag{13}$$

where $\mathbf{n} = (n_1, n_2)^T$ is the unit outward normal, $\mathbf{s} = (s_1, s_2)^T = (-n_2, n_1)^T$ is the unit tangential vector, and $\nabla\mathbf{u}'$ is the Jacobian of $\mathbf{u}'$. This is true for the objective function in (7), but also for many more general objective functions that involve distributed observations or observations of normal derivatives $\nabla\mathbf{u}\,\mathbf{n}$ or of tangential derivatives $\nabla\mathbf{u}\,\mathbf{s}$ of the state.

To derive the adjoint equations, we multiply $\mathbf{C}_{\mathbf{u}}(\mathbf{u}, \mathbf{g})\mathbf{u}'$ by $\boldsymbol{\lambda}$ and integrate the resulting terms over $(t_0, t_f) \times \Omega$, $(t_0, t_f) \times \Gamma$, and Ω, respectively. Integration by parts leads to (9), which in this case is given by

$$\begin{aligned}
&-\int_{t_0}^{t_f}\int_\Omega (\mathbf{u}')^T\mathbf{r} - \int_\Omega (\mathbf{u}'|_{t=t_f})^T\mathbf{r}_{t_f} \\
&\quad -\int_{t_0}^{t_f}\int_\Gamma ((\mathbf{u}')^T\mathbf{r}_\Gamma + (\nabla\mathbf{u}'\mathbf{n})^T\mathbf{r}_{\Gamma,\mathbf{n}} + (\nabla\mathbf{u}'\mathbf{s})^T\mathbf{r}_{\Gamma,\mathbf{s}}) \\
&= \int_{t_0}^{t_f}\int_\Omega (\mathbf{u}')^T\left(-\mathbf{M}^T\boldsymbol{\lambda}_t^d - \sum_i\left((\mathbf{A}^i-\mathbf{D}^i)^T\boldsymbol{\lambda}_{x_i}^d + \sum_j\left((\mathbf{K}_j^i)^T\boldsymbol{\lambda}_{x_i}^d\right)_{x_j}\right)\right) \\
&\quad + \int_{t_0}^{t_f}\int_\Gamma (\mathbf{u}')^T\left(\sum_i\left(n_i(\mathbf{A}^i-\mathbf{D}^i)^T\boldsymbol{\lambda}^d + \sum_j n_j(\mathbf{K}_j^i)^T\boldsymbol{\lambda}_{x_i}^d\right) + \mathbf{B}_\mathbf{u}^T\boldsymbol{\lambda}^b\right) \qquad (14) \\
&\quad + \int_{t_0}^{t_f}\int_\Gamma \sum_j (\mathbf{u}'_{x_j})^T\left(\mathbf{B}_{\mathbf{u}_{x_j}}^T\boldsymbol{\lambda}^b - \sum_i n_i(\mathbf{K}_j^i)^T\boldsymbol{\lambda}^d\right) \\
&\quad + \int_\Omega (\mathbf{u}')^T\mathbf{M}^T\boldsymbol{\lambda}^d|_{t=t_f} + \int_\Omega (\mathbf{u}')^T(\boldsymbol{\lambda}^0 - \mathbf{M}^T\boldsymbol{\lambda}^d)\Big|_{t=t_0} \qquad \forall \mathbf{u}'.
\end{aligned}$$

If we choose test functions $\mathbf{u}' \in C_0^\infty((t_0,t_f)\times\Omega)$, then (14) implies

$$\mathbf{M}^T\boldsymbol{\lambda}_t^d + \sum_i\left((\mathbf{A}^i-\mathbf{D}^i)^T\boldsymbol{\lambda}_{x_i}^d + \sum_j\left((\mathbf{K}_j^i)^T\boldsymbol{\lambda}_{x_i}^d\right)_{x_j}\right) = \mathbf{r} \qquad (15)$$

in $(t_0,t_f)\times\Omega$. If we choose test functions $\mathbf{u}'$ such that $\mathbf{u}'=0$ on $\{t_0\}\times\Omega$, $\mathbf{u}'=0$ and $\nabla\mathbf{u}'=0$ on $(t_0,t_f)\times\Gamma$, then (14) implies $(\mathbf{M}^T\boldsymbol{\lambda}^d)|_{t=t_f} = -\mathbf{r}_{t_f}$ in Ω or, equivalently,

$$\boldsymbol{\lambda}^d = -\mathbf{M}^{-T}\mathbf{r}_{t_f} \quad \text{in } \{t_f\}\times\Omega. \qquad (16)$$

Similarly, if we choose test functions $\mathbf{u}'$ such that $\mathbf{u}'=0$ on $\{t_f\}\times\Omega$, $\mathbf{u}'=0$ and $\nabla\mathbf{u}'=0$ on $(t_0,t_f)\times\Gamma$, then (14) implies $\boldsymbol{\lambda}^0 - (\mathbf{M}^T\boldsymbol{\lambda}^d)|_{t=t_0} = 0$ in Ω. This means that $\boldsymbol{\lambda}^0$ is determined by $\boldsymbol{\lambda}^d|_{t=t_0}$.

Next we choose test functions $\mathbf{u}'$ such that $\mathbf{u}'=0$ on $\{0,t_f\}\times\Omega$ and on $(t_0,t_f)\times\Gamma$. The tangential derivatives $(\nabla\mathbf{u}')\mathbf{s}$ of these test functions is zero. Using $\nabla\mathbf{u}' = \nabla\mathbf{u}'\mathbf{n}\mathbf{n}^T + \nabla\mathbf{u}'\mathbf{s}\mathbf{s}^T = \nabla\mathbf{u}'\mathbf{n}\mathbf{n}^T$, $\mathbf{u}'_{x_j} = \nabla\mathbf{u}'e_j = \nabla\mathbf{u}'\mathbf{n}n_j$ and the previous identities in (14), we obtain

$$\sum_j n_j\left(\mathbf{B}_{\mathbf{u}_{x_j}}^T\boldsymbol{\lambda}^b - \sum_i n_i(\mathbf{K}_j^i)^T\boldsymbol{\lambda}^d\right) = -\mathbf{r}_{\Gamma,n} \quad \text{on } (t_0,t_f)\times\Gamma. \qquad (17)$$

With (15)–(17) the identity (14) reduces to

$$
\begin{aligned}
&-\int_{t_0}^{t_f}\int_\Gamma \left((\mathbf{u}')^T \mathbf{r}_\Gamma + (\nabla \mathbf{u}'\mathbf{s})^T \mathbf{r}_{\Gamma,\mathbf{s}}\right) \\
&\quad = \int_{t_0}^{t_f}\int_\Gamma (\mathbf{u}')^T \left(\sum_i \left(n_i (\mathbf{A}^i - \mathbf{D}^i)^T \boldsymbol{\lambda}^d + \sum_j n_j (\mathbf{K}^i_j)^T \boldsymbol{\lambda}^d_{x_i} \right) + \mathbf{B}_\mathbf{u}^T \boldsymbol{\lambda}^b \right) \\
&\quad + \int_{t_0}^{t_f}\int_\Gamma (\nabla \mathbf{u}'\mathbf{s})^T \sum_j s_j \left(\mathbf{B}^T_{\mathbf{u}_{x_j}} \boldsymbol{\lambda}^b - \sum_i n_i (\mathbf{K}^i_j)^T \boldsymbol{\lambda}^d \right) \qquad \forall \mathbf{u}'.
\end{aligned}
$$

One can employ integration by parts over Γ on the integrals involving $(\nabla \mathbf{u}'\mathbf{s})$ to arrive at the more convenient identity

$$
\begin{aligned}
&\sum_i \left(n_i (\mathbf{A}^i - \mathbf{D}^i)^T \boldsymbol{\lambda}^d + \sum_j n_j (\mathbf{K}^i_j)^T \boldsymbol{\lambda}^d_{x_i} \right) + \mathbf{B}_\mathbf{u}^T \boldsymbol{\lambda}^b \\
&- \frac{\partial}{\partial \mathbf{s}} \left(\sum_j s_j \left(\mathbf{B}^T_{\mathbf{u}_{x_j}} \boldsymbol{\lambda}^b - \sum_i n_i (\mathbf{K}^i_j)^T \boldsymbol{\lambda}^d \right) + \mathbf{r}_{\Gamma,\mathbf{s}} \right) = -\mathbf{r}_\Gamma.
\end{aligned} \tag{18}
$$

3.3. Adjoint equation for the boundary control of final–time kinetic energy

In our model problem (7) we assume adiabatic boundary conditions for the temperature on the bottom wall $\Gamma = \{\mathbf{x} \in \mathbb{R}^2 : x_2 = 0\}$. The velocities on Γ are prescribed and are equal to $\mathbf{b}$ on $(t_0, t_f) \times (\Gamma \setminus \Gamma_c)$ and they are equal to $\mathbf{b} + \mathbf{g}$ on $(t_0, t_f) \times \Gamma_c$. The boundary condition operator $\mathbf{B}$ in (4) is

$$
\mathbf{B}(\mathbf{u}, \nabla \mathbf{u}, \mathbf{g}) = \begin{pmatrix} \mathbf{v} - \mathbf{g} - \mathbf{b} \\ -T_{x_2} \end{pmatrix}, \quad \mathbf{B}(\mathbf{u}, \nabla \mathbf{u}, \mathbf{g}) = \begin{pmatrix} \mathbf{v} - \mathbf{b} \\ -T_{x_2} \end{pmatrix}
$$

on $(t_0, t_f) \times \Gamma_c$ and on $(t_0, t_f) \times (\Gamma \setminus \Gamma_c)$, respectively. The partial Fréchet–derivative of the objective function J in (7) is given by (13) with $\mathbf{r} = \mathbf{r}_{\Gamma,\mathbf{n}} = \mathbf{r}_{\Gamma,\mathbf{s}} = \mathbf{0}$ and

$$
\mathbf{r}_{t_f}(\mathbf{x}) = \mathbf{1}_{\Omega_0}(\mathbf{x}) \left(\frac{1}{2} \|\mathbf{v}(\mathbf{x}, t_f)\|_2^2, \rho(\mathbf{x}, t_f)\mathbf{v}(\mathbf{x}, t_f), 0 \right)^T. \tag{19}
$$

Since $\Omega = \{\mathbf{x} \in \mathbb{R}^2 : x_2 > 0\}$, $\mathbf{n} = (0, -1)^T$ on Γ, the boundary condition (17) reads

$$
\mathbf{B}^T_{\mathbf{u}_{x_2}} \boldsymbol{\lambda}^b + (\mathbf{K}^2_2)^T \boldsymbol{\lambda}^d = 0, \quad \text{on } (t_0, t_f) \times \Gamma,
$$

which, using the definition of $\mathbf{B}$ and $\mathbf{K}^2_2$, is equivalent to

$$
\frac{1}{\mathsf{Re}} \begin{pmatrix} \mu & 0 & \mu v_1 \\ 0 & 2\mu + \lambda & (2\mu + \lambda) v_2 \\ 0 & 0 & \frac{\kappa}{\mathsf{Pr}\mathsf{M}^2(\gamma - 1)} \end{pmatrix} \begin{pmatrix} \lambda^d_2 \\ \lambda^d_3 \\ \lambda^d_4 \end{pmatrix} = \begin{pmatrix} 0 \\ 0 \\ \lambda^b_3 \end{pmatrix} \quad \text{on } (t_0, t_f) \times \Gamma.
$$

Hence, we obtain

$$
\lambda^b_3 = \frac{\kappa}{\mathsf{RePrM}^2(\gamma - 1)} \lambda^d_4 \text{ on } (t_0, t_f) \times \Gamma, \tag{20}
$$

and the boundary conditions

$$\lambda_2^d = -v_1 \lambda_4^d, \quad \lambda_3^d = -v_2 \lambda_4^d \quad \text{on } (t_0, t_f) \times \Gamma. \tag{21}$$

These boundary conditions imply $(\mathbf{K}_1^2)^T \boldsymbol{\lambda}^d = 0$. Hence, the boundary condition (18) on $(t_0, t_f) \times \Gamma$ reduces to

$$(\mathbf{A}^2 - \mathbf{D}^2)^T \boldsymbol{\lambda}^d + (\mathbf{K}_2^1)^T \boldsymbol{\lambda}_{x_1}^d + (\mathbf{K}_2^2)^T \boldsymbol{\lambda}_{x_2}^d = \mathbf{B}_{\mathbf{u}}^T \boldsymbol{\lambda}^b. \tag{22}$$

By definition, $\mathbf{A}^2 = \mathbf{F}_{\mathbf{u}}^2(\mathbf{u})$, where $\mathbf{F}^2(\mathbf{u})$ is one of the inviscid fluxes in (1). Hence,

$$\mathbf{A}^2 = \begin{pmatrix} v_2 & 0 & \rho & 0 \\ v_1 v_2 & \rho v_2 & \rho v_1 & 0 \\ v_2^2 + \frac{T}{\gamma \mathsf{M}^2} & 0 & 2\rho v_2 & \frac{\rho}{\gamma \mathsf{M}^2} \\ v_2 \left(\frac{T}{(\gamma-1)\mathsf{M}^2} + \frac{1}{2}\mathbf{v}^T\mathbf{v} \right) & \rho v_1 v_2 & \rho \left(\frac{T}{(\gamma-1)\mathsf{M}^2} + \frac{1}{2}\mathbf{v}^T\mathbf{v} \right) + \rho v_2^2 & \frac{\rho v_2}{(\gamma-1)\mathsf{M}^2} \end{pmatrix}.$$

Moreover, using the definition of $\mathbf{K}_2^1$ and (21) we obtain

$$(\mathbf{K}_2^1)^T \boldsymbol{\lambda}_{x_1}^d = \frac{1}{\mathsf{Re}} \begin{pmatrix} 0 \\ \mu (\lambda_3^d)_{x_1} + \mu v_2 (\lambda_4^d)_{x_1} \\ \lambda (\lambda_2^d)_{x_1} + \lambda v_1 (\lambda_4^d)_{x_1} \\ 0 \end{pmatrix} = \frac{1}{\mathsf{Re}} \begin{pmatrix} 0 \\ -\mu (v_2)_{x_1} \\ -\lambda (v_1)_{x_1} \\ 0 \end{pmatrix} \lambda_4^d.$$

If we insert the previous two equations and (21) into (22) we arrive at the condition

$$\begin{aligned} &\begin{pmatrix} v_2 & v_2 \left(\frac{T}{\gamma(\gamma-1)\mathsf{M}^2} - \frac{1}{2}\mathbf{v}^T\mathbf{v} \right) \\ 0 & -\frac{1}{\mathsf{Re}} (\mu (v_2)_{x_1} + \tau_{12}) \\ \rho & \rho \left(\frac{T}{(\gamma-1)\mathsf{M}^2} - \frac{1}{2}\mathbf{v}^T\mathbf{v} \right) - \frac{1}{\mathsf{Re}} (\lambda (v_1)_{x_1} + \tau_{22}) \\ 0 & \rho v_2 \end{pmatrix} \begin{pmatrix} \lambda_1^d \\ \lambda_4^d \end{pmatrix} \\ &\quad + \frac{1}{\mathsf{Re}} \begin{pmatrix} 0 & 0 & 0 \\ \mu & 0 & \mu v_1 \\ 0 & 2\mu + \lambda & (2\mu + \lambda) v_2 \\ 0 & 0 & \frac{\kappa}{(\gamma-1)\mathsf{M}^2 \mathsf{Pr}} \end{pmatrix} \begin{pmatrix} \lambda_2^d \\ \lambda_3^d \\ \lambda_4^d \end{pmatrix}_{x_2} = \begin{pmatrix} 0 \\ \lambda_1^b \\ \lambda_2^b \\ 0 \end{pmatrix} \end{aligned} \tag{23}$$

on $(t_0, t_f) \times \Gamma$. Equation (23) yields the boundary conditions

$$\begin{aligned} v_2 \lambda_1^d &= v_2 \left(-\frac{T}{\gamma(\gamma-1)\mathsf{M}^2} + \frac{1}{2}\mathbf{v}^T\mathbf{v} \right) \lambda_4^d, \\ \rho v_2 \lambda_4^d &= -\frac{\kappa}{(\gamma-1)\mathsf{M}^2 \mathsf{Pr} \mathsf{Re}} (\lambda_4^d)_{x_2} \end{aligned} \tag{24}$$

on $(t_0, t_f) \times \Gamma$. Thus, the adjoint boundary conditions for $\boldsymbol{\lambda}^d$ are given by (21), (24). If desired, the adjoint variables $\boldsymbol{\lambda}^b$ for the boundary data can be computed from $\boldsymbol{\lambda}^d$ using (20) and the second and third equation in (23).

We remark, that the general formulation of the adjoint equations (15)–(18) is also useful, when non-reflecting boundary conditions are introduced on the boundary $\partial\Omega_c \setminus \Gamma$ of the computational domain $\Omega_c \subset \Omega$.

3.4. Gradient computation

Given the adjoint $\boldsymbol{\lambda}$, the gradient can be computed from (8). As in [9, 12] this leads to an elliptic problem on the time–space boundary $(t_0, t_f) \times \Gamma$ for $\nabla \widehat{J}$. Due to space restrictions the details are omitted here. For the numerical solution of the optimal control problem the solution of this PDE can be avoided with a reformulation of the optimal control problem and working with a different, yet appropriate inner product. This is described in [4] for the semi–discrete case.

4. Results

We present results for a model problem consisting of two counter-rotating viscous vortices above an infinite wall which, due to the self-induced velocity field, propagate downward and interact with the wall. Our non–dimensionalization is based on initial vortex core radius and the maximum azimuthal velocities at the edge of the viscous cores. For the computations reported here, the Mach, Reynolds, and Prandtl numbers are $\mathsf{M} = 0.5, \mathsf{Re} = 25, \mathsf{Pr} = 1$, respectively. Our computational domain is $[-15, 15] \times [0, 15]$ in non–dimensional units. The compressible Navier–Stokes equations are discretized in space using fourth-order accurate central differences on a 128×128 uniform grid. Time integration is performed using the classical fourth–order Runge Kutta method with a fixed time step $\Delta t = 0.05$. To compute the initial conditions, two compressible Oseen vortices [6], located at $(\pm 2, 7.5)$, are superimposed at time $t = 0$. From this superposed field, which is not a solution to the equations, we advance 100 time steps until time $t_0 = 5$, and take the resulting flow field at time t_0 as the initial condition to our problem. The side and top boundaries are assumed to be located far enough from the main flow region to justify the imposition of a characteristics based inviscid far–field boundary condition. For details on the model problem and its discretization see [4].

We control the flow in the time window $t_0 = 5, t_f = 40$. Our control $\mathbf{g}$ is the wall normal velocity which for our geometry is given by $\mathbf{g} = (0, g_2)$. The following control plots show g_2. Positive g_2 represents injection (blowing) of fluid into the domain while negative g_2 corresponds to suction of fluid out of the domain.

Our numerical results are produced using a nonlinear conjugate gradient (NCG) algorithm [20] for the solution of the discretized problem. The inner products used in the NCG method are discretizations of the $\mathcal{G}$ inner product (see [4]) to minimize the mesh–dependent behavior of the cg method and to avoid artificial ill–conditioning due to discretization. All computations are performed in parallel on an SGI Origin 2000. Using four processors, one optimization run takes about 10hours.

We performed three runs, two include a regularization of the time derivative of the control, the third does not. The coefficients α_j in the regularization term, the value of the objective functional in (7) at the initial iterate, i.e., for zero control (J_0), at the final control iterate (J_{final}) and the terminal kinetic energy (the first integral in (7)) at the final control iterate ($\text{TKE}_{\text{final}}$) are shown in Table 1. We see

FIGURE 1. Optimal wall–normal velocity distributions

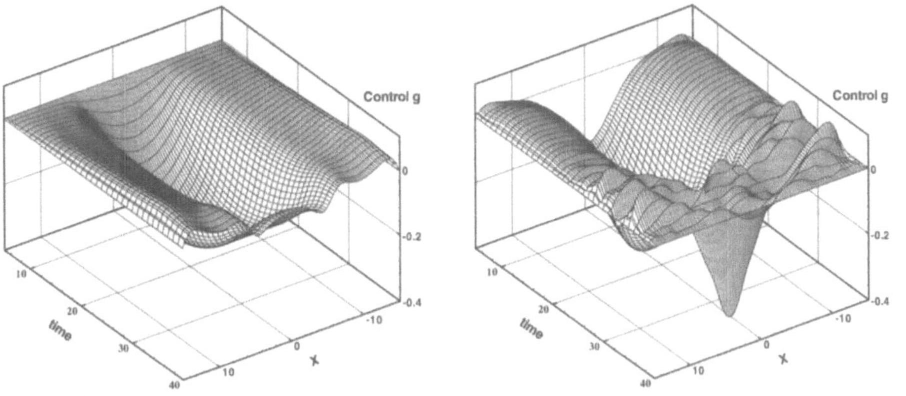

Optimal control run I — Optimal control run III

that because of the large regularization parameter α_1, the terminal kinetic energy reduction in run I is less than that for run III. However, a smaller $\alpha_1 > 0$ will give a smaller $\mathrm{TKE}_{\mathrm{final}}$, while maintaining temporal smoothness in the controls.

TABLE 1.

Run	α_1	α_2	α_3	J_0	J_{final}	$\mathrm{TKE}_{\mathrm{final}}$
I	0.5	0.005	0.005	12.43	0.48	0.42
II	0.05	0.005	0.005	12.43	0.37	0.32
III	0	0.005	0.005	12.43	0.24	0.20

In all cases the optimization is started with zero control. The optimal wall–normal velocity distributions g_2 for are plotted in Figure 1. The optimal controls for runs I and II are very similar, the amplitudes in the optimal control for runs II are slightly higher than those for run I, but no additional oscillations arise when α_1 is reduced to 0.05. The plots clearly show the effect of the regularization term $\int \frac{\alpha_1}{2}\|\mathbf{g}_t\|$. Without it, the control starts to oscillate in time in the second half of $[t_0, t_f]$ and it exhibits a large jump at t_0. If we even set $\alpha_1 = 0$ and $\alpha_2 = 0$, then controls are produced that exhibit strong spatial and temporal oscillations which frequently led to a failure in the compressible Navier–Stokes solver.

Figure 2 shows the contours of kinetic energy for the uncontrolled flow and the controlled flow, run I.

5. Acknowledgements

The work of S. Collis, K. Ghayour and M. Heinkenschloss was supported by Texas ATP grant 003604–0001, 1999. M. Ulbrich and S. Ulbrich were supported by DFG grants UL157/3-1 and UL158/2-1, respectively, and by CRPC grant CCR-9120008.

FIGURE 2. Kinetic energy contours for the uncontrolled flow (top row) and the controlled flow, run I (bottom row).

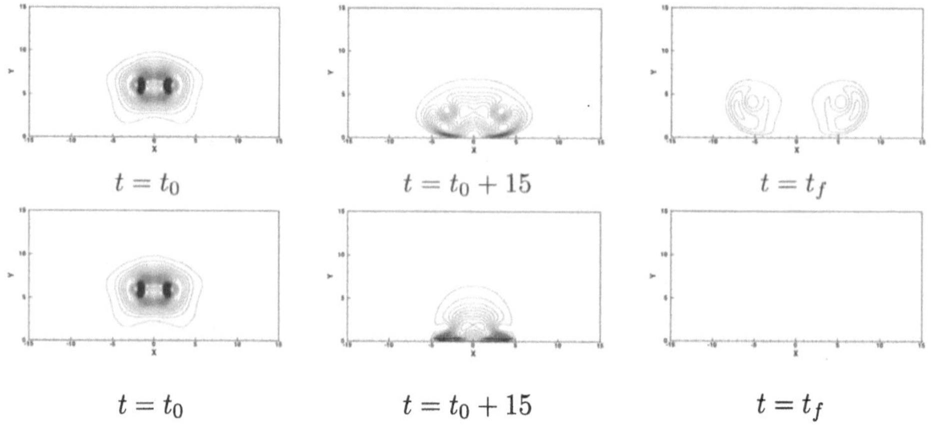

Computations were performed on an SGI Origin 2000 which was purchased with the aid of NSF SCREMS grant 98–72009.

References

[1] M. Berggren, *Numerical solution of a flow-control problem: Vorticity reduction by dynamic boundary action*, SIAM J. Scientific Computing, **19** (1998), 829–860.

[2] T. R. Bewley, P. Moin, and R. Teman, *DNS-based predictive control of turbulence: an optimal target for feedback algorithms*, Submitted to J. Fluid Mech., 2000.

[3] E. M. Cliff, M. Heinkenschloss, and A. Shenoy, *Airfoil design by an all–at–once method*, International Journal for Computational Fluid Mechanics, **11** (1998), 3–25.

[4] S. S. Collis, K. Ghayour, M. Heinkenschloss, M. Ulbrich, and S. Ulbrich, *Towards adjoint–based methods for aeroacoustic control*, in: 39th Aerospace Science Meeting & Exhibit, January 8–11, 2001, Reno, Nevada, AIAA Paper 2001–0821 (2001).

[5] S. S. Collis, Y. Chang, S. Kellogg, and R. D. Prabhu, *Large Eddy Simulation and Turbulence Control*, AIAA paper 2000 2564, (2000).

[6] T. Colonius, S. K. Lele, and P. Moin, *The Free Compressible Viscous Vortex,* J. Fluid Mech. **230** (1991), pp. 45–73.

[7] M. Gad el Hak, A. Pollard, and J.-P. Bonnet, editors, *Flow Control. Fundamental and Practices*, Springer Verlag, Berlin, Heidelberg, New York, 1998.

[8] A. V. Fursikov, *Optimal Control of Distributed Systems. Theory and Applications*, Translation of Mathematical Monographs **87**, American Mathematical Society, Providence, Rhode Island, 2000.

[9] A. V. Fursikov, M. D. Gunzburger, and L. S. Hou, *Boundary value problems and optimal boundary control for the Navier–Stokes systems: The two–dimensional case*, SIAM J. Control Optimization **36** (1998), 852–894.

[10] M. D. Gunzburger and L. S. Hou, editors, International Journal of Computational Fluid Dynamics, No. 1–2, **11** (1998).

[11] M. D. Gunzburger, L. S. Hou, and T. P. Svobotny, *Optimal control and optimization of viscous, incompressible flows*, in: M. D. Gunzburger and R. A. Nicolaides, editors, *Incompressible Computational Fluid Dynamics*, Cambridge University Press, New York, (1993) 109–150.

[12] M. D. Gunzburger and S. Manservisi, *The velocity tracking problem for Navier–Stokes flows with boundary control*, SIAM J. Control Optim., **39** (2000), 594–634.

[13] D. Hoff, *Discontinuous solutions of the Navier–Stokes equations for multidimensional flows of heat-conducting fluids*, Arch. Rational Mech. Anal., **139** (1997), 303–354.

[14] A. Jameson, L. Martinelli, and N. A. Pierce, *Optimum aerodynamic design using the Navier-Stokes equations*, Theor. and Comput. Fluid Dynamics, **10** (1998), 213–237.

[15] A. Jameson and J. Reuther, *Control theory based airfoil design using the Euler equations*, AIAA Paper 94-4272, (1994).

[16] W. H. Jou, W. P. Huffman, D. P. Young, R. G. Melvin, M. B. Bieterman, C. L. Hilems, and F. T. Johnson, *Practical considerations in aerodynamic design optimization*, in: Proceedings of the 12th AIAA Computational Fluid Dynamics Conference, San Diego, CA, June 19–22 1995. AIAA Paper 95-1730.

[17] P. L. Lions, *Mathematical Topics in Fluid Mechanics. Volume 2, Compressible Models*. Oxford Lecture Series in Mathematics and its Applications **10**, Claredon Press, Oxford, 1998.

[18] A. Matsumura and T. Nishida, *The initial value problem for the equations of motion of viscous and heat-conductive gases*, J. Math. Kyoto Univ., **20** (1980), 67–104.

[19] A. Matsumura and T. Nishida, *Initial-boundary value problems for the equations of motion of compressible viscous and heat-conductive fluids*, Comm. Math. Phys. **89** (1983), 445–464.

[20] J. Nocedal and S. J. Wright, *Numerical Optimization*, Springer Verlag, Berlin, Heidelberg, New York, 1999.

[21] B. Soemarwoto, *The variational method for aerodynamic optimization using the Navier–Stokes equations*, Technical Report 97–71, ICASE, NASA Langley Research Center, Hampton VA 23681–0001, 1997.

[22] A. Valli, *Mathematical results for compressible flows*, in: J. F. Rodriguez and A. Sequeira, editors, Mathematical Topics in Fluid Mechanics, Pitman Research Notes Mathematics **274**, Longman Scientific and Technical, Essex, (1992) 193–229.

Departments of Computational and Applied Mathematics and
of Mechanical Engineering and Materials Science
Rice University
Houston, TX 77005-1892, USA
E-mail address: `collis@rice.edu`, `kghayour@rice.edu`, `heinken@rice.edu`

Lehrstuhl für Angewandte Mathematik und Mathematische Statistik
Technische Universität München
D–80290 München, Germany
E-mail address: `mulbrich@ma.tum.de`, `sulbrich@ma.tum.de`

International Series of Numerical Mathematics, Vol. 139, 57–69

Intrinsic Asymptotic Model of Piezoelectric Shells

Michel C. Delfour and Michel Bernadou

Abstract. The object of this paper is to present new intrinsic linear models of thin and asymptotic *piezoelectric shells* starting from a three-dimensional model of piezoelectric material. It is based on the completely intrinsic methods developed and used in [2, 6] to obtain linear models of thin and asymptotic mechanical shells without local coordinates or Christoffel symbols. The *thin shell approximation* is achieved via an extended version of the $P(2,1)$ model studied in [2] and a scalar $P(2,1)$-approximation of the electrical potential. This type of approximation yields models of the Naghdi's type without the standard assumption on the stress tensor. We also present a new completely uncoupled system of two equations for the asymptotic model. The decoupling results from the choice of the scalar product used to define the projection. It is different from the one used in [2] where the second equation is coupled with the first one through a term which is zero for the plate and in the bending dominated case. Effective mechanical and electrical constitutive laws are obtained from arbitrary three-dimensional mechanical and electrical constitutive laws.

1. Introduction

The piezoelectricity can be viewed as an interaction between two phenomena [8]: (i) *the direct piezoelectric effect*: a mechanical deformation generates an electrical field in the material; (ii) *the inverse piezoelectric effect*: the application of some electric potentials generates a deformation of the material. The piezoelectric properties, discovered by Jacques and Pierre Curie, are more and more extensively used in industry, in particular to achieve active control of elastic structures. For instance, these materials are used to build sensors or actuators which are placed on or inside a structure in order to either damp or reduce its vibration or to improve its acoustical properties. It is worth noting that (i) these effects can be obtained without applying mechanical loading, and (ii) sensors and actuators are very light: they can be bonded to the structure without changing significantly its weight or its dynamical properties. Beyond vibration or acoustical control, such materials can be used to achieve shape control for airplane propellers, airplane wings, telescope mirrors as well as control of metal fatigue, artificial organs in biomechanics and much more. An extended presentation of such possibilities can be found for instance in [8, 11].

The potential of the intrinsic differential calculus on $C^{1,1}$ submanifolds developed in [6, 7] has been illustrated by studying some purely mechanical linear models of thin shells based on truncated series expansions with respect to the variable normal to the midsurface. In recent papers ([7, 2, 3, 4, 5]) it was established that for the mechanical material the polynomial $P(2,1)$ model is both pertinent and basic in the theory of *thin* and *asymptotic shells.* It yields by elimination of variables and approximation models of the *Naghdi's type* without using the standard a priori assumption on the stress tensor ([3]). Moreover it has been shown in [2] that, for an arbitrary constitutive law, the solution of the $P(2,1)$ model converges to the solution of an *asymptotic shell model* which consists of a coupled system of two variational equations. The first one yields the classical *membrane shell equation* and the Love-Kirchhoff terms, and the second one the bending equation for the projection of the asymptotic solution onto the space of inextensible deformations. It extends results obtained in special cases such as the *bending dominated case* where the extended membrane energy is assumed to be zero and in the case of *plates.*

The object of this paper is to use *intrinsic methods* in the study of linear models of thin and asymptotic *piezoelectric shells.* Starting from a three-dimensional linear model of piezoelectric material, a *global shell approximation* is done via the vector $P(2,1)$-approximation used in [2] for the mechanical part and a scalar $P(2,1)$-approximation for the electrical potential. We then construct the associated asymptotic model which consists of a coupled system of two equations. The first one is a generalized membrane piezoelectric shell equation and the second a generalized bending equation for the projection of the asymptotic solution onto the space of generalized inextensible deformations as in [2]. However there is an important difference. Here the bending equation for the projection of the asymptotic solution is not coupled with the solution of the membrane equation as in [2]. It is also independent of the electrical potential. The difference with [2] arises from the different and more natural choice of inner product to define the projection. Hence all the results and theorems of [2] for the mechanical shell remain true with the new projection.

2. Notation and Background Material

The inner product in $\mathbf{R}^3$ and the double inner product in $\mathcal{L}(\mathbf{R}^3;\mathbf{R}^3)$, the space of 3×3 matrices or tensors, are denoted as

$$x \cdot y = \sum_{i=1}^{3} x_i\, y_i, \quad A \cdot\cdot B = \sum_{i=1}^{3}\sum_{j=1}^{3} A_{ij}\, B_{ij}.$$

${}^*A$ denotes the transpose of A. Given Ω in $\mathbf{R}^3$, $\Omega \neq \varnothing$ (resp. $\Gamma \overset{\text{def}}{=} \partial\Omega \neq \varnothing$) the *distance function* (resp. *oriented distance function*) is defined as

$$d_\Omega(x) \overset{\text{def}}{=} \inf_{y\in\Omega} |y-x| \text{ (resp. } b_\Omega(x) = d_\Omega(x) - d_{\mathbf{R}^3-\Omega}(x)).$$

When Ω is of class $C^{1,1}$, $b = b_\Omega$ is $C^{1,1}$ in a neighborhood of every point of Γ and the converse is true. Its gradient ∇b coincides with the exterior unit normal n to the boundary on Γ and its Hessian matrix with the second fundamental form of Γ. The *projection* p onto Γ and the *orthogonal projection* P onto the *tangent plane* $T_x(\Gamma)$ are given by

$$p(x) \stackrel{\text{def}}{=} x - b(x)\,\nabla b(x), \quad P(x) \stackrel{\text{def}}{=} I - \nabla b(x)\;{}^*\nabla b(x),$$

where ${}^*V$ denotes the transpose of a column vector V in $\mathbf{R}^3$. Let ω be an open domain in Γ. When $\omega = \Gamma$, ω has no boundary; otherwise denote by γ the (relative) *boundary* of ω in Γ. Given $h > 0$, a *shell* and its *lateral boundary* are defined as

$$\begin{aligned} S_h(\omega) &\stackrel{\text{def}}{=} \{x \in \mathbf{R}^3 \colon |b_\Omega(x)| < h, p(x) \in \omega\} \\ \Sigma_h(\gamma) &\stackrel{\text{def}}{=} \{x \in \mathbf{R}^3 \colon |b_\Omega(x)| < h, p(x) \in \gamma\} \end{aligned} \tag{1}$$

The notation $S_h(\gamma)$ could also have been used, but $\Sigma_h(\gamma)$ emphasizes the fact that it is a part of the boundary of $S_h(\omega)$. Given any $\gamma' \subset \gamma$ also define $\Sigma_h(\gamma')$ as in (1) with γ' in place of γ.

Associate with $b \in C^{1,1}(\overline{S_h(\omega)})$, the bi-Lipschitzian map

$$(z, X) \mapsto T(z, X) \stackrel{\text{def}}{=} T_z(X) = X + z\,\nabla b(X) \,:\,]-h, h[\,\times\omega \to S_h(\omega) \tag{2}$$

and its inverse $T^{-1}(x) = (p(x), b(x))$. Define for all $|z| < h$ and $X \in \omega$,

$$j_z \stackrel{\text{def}}{=} \det DT_z(X) = \det\big[I + z\,D^2 b(X)\big] = 1 + z\,\kappa_1 + z^2\kappa_2,$$

where $\kappa_1 = \Delta b$ is twice the *mean curvature* and κ_2 is the *Gauss curvature.* It will be convenient to use the following notation for the decompositions of a vector or a matrix function into their tangential and normal parts with respect to ω. Given a vector function $V : \omega \to \mathbf{R}^3$

$$V_n(x) \stackrel{\text{def}}{=} V(x) \cdot n(x), \quad V_\Gamma(x) \stackrel{\text{def}}{=} P(x)V(x) = V(x) - V_n\, n(x).$$

By definition $V_\Gamma(x)$ belongs to the tangent plane $T_x(\Gamma)$ to Γ in x. For simplicity we shall drop the x wherever no confusion arises. Similarly for a matrix function $\tau : \omega \to \mathcal{L}(\mathbf{R}^3; \mathbf{R}^3)$ ($\mathcal{L}(\mathbf{R}^3; \mathbf{R}^3)$, the space of 3×3 matrices) define

$$\begin{gathered} \tau^P \stackrel{\text{def}}{=} P\tau P, \quad \tau_{nn} \stackrel{\text{def}}{=} \tau n \cdot n, \quad \tau^n \stackrel{\text{def}}{=} \tau - \tau_{nn}\, n\; {}^*n \\ \tau = \tau^P + (P\tau n)\; {}^*n + n\; {}^*(P\tau n) + \tau_{nn}\, n\; {}^*n = \tau^n + \tau_{nn}\, n\; {}^*n \end{gathered}$$

and the spaces of symmetric matrices

$$\begin{gathered} \mathrm{Sym}_3 \stackrel{\text{def}}{=} \{\tau \in \mathcal{L}(\mathbf{R}^3; \mathbf{R}^3) \colon {}^*\tau = \tau\}, \quad \mathrm{Sym}_3^n \stackrel{\text{def}}{=} \{\tau \in \mathrm{Sym}_3 \colon \tau_{nn} = 0\} \\ \mathrm{Sym}_3^P \stackrel{\text{def}}{=} \{\tau \in \mathrm{Sym}_3 \colon \tau\, n = 0\} \quad \Rightarrow \forall \tau \in \mathrm{Sym}_3,\ \tau^P \in \mathrm{Sym}_3^P \text{ and } \tau^n \in \mathrm{Sym}_3^n \end{gathered}$$

where it is understood that, as in the case of $T_x(\Gamma)$, Sym_3^n and Sym_3^P both depend on the point x of Γ. For the tangential calculus and the necessary supporting functional analysis, the reader is referred to [7, 6].

3. Basic Three-dimensional Equations

For simplicity consider the shell $S_h(\omega)$ of constant thickness $2h$ for a *bounded open connected* domain $\omega \neq \Gamma$. Its boundary is made up of three parts: the lateral boundary $\Sigma_h(\gamma)$, the upper surface $T_h(\omega)$ and the lower surface $T_{-h}(\omega)$.

Denote by V the *displacement vector* and by Φ the *electrical potential* defined on $S_h(\omega)$, and define the *linear deformation tensor* of V and the *electrical field* E

$$\varepsilon(V) \stackrel{\text{def}}{=} \frac{D(V) + {}^*D(V)}{2} \text{ and } E(\Phi) \stackrel{\text{def}}{=} -\nabla\Phi. \tag{3}$$

In the sequel it will be convenient to use the simpler notation $\varepsilon = \varepsilon(V)$ and $E = E(\Phi)$, $\varepsilon_h = \varepsilon(V_h)$ and $E_h = E(\Phi_h)$, $\overline{\varepsilon} = \varepsilon(\overline{V})$ and $\overline{E} = E(\overline{\Phi})$, and so forth.

Consider the following system of equations

i) *Fields equations*

$$-\overrightarrow{\operatorname{div}}\, \sigma = F \text{ and } \operatorname{div} D = 0 \text{ in } S_h(\omega)$$

where σ is the *stress tensor* and D the *electrical displacement vector.*

ii) *Mechanical boundary equations*

$$U = 0 \text{ on } \Sigma_h(\gamma_0),\ \sigma n_S = Q \text{ on } T_h(\omega) \cup T_{-h}(\omega) \cup \Sigma_h(\gamma\backslash\gamma_0),$$

where n_S denotes the unit exterior normal vector to $S_h(\omega)$ and some $\gamma_0 \subset \gamma$.

iii) *Electrical boundary conditions*

$$\Phi = \Phi_0 \text{ on } \partial_0 S_h(\omega) \text{ and } D \cdot n = 0 \text{ on } \partial S_h(\omega)\backslash\partial_0 S_h(\omega)$$

on some part $\partial_0 S_h(\omega)$ of the boundary of $S_h(\omega)$.

iv) *Mechanical-electrical constitutive laws*

$$\sigma = C\,\varepsilon - e\,E \text{ and } D = {}^*e\,\varepsilon + d\,E \text{ in } S_h(\omega) \tag{4}$$

where $C : \mathrm{Sym}_3 \to \mathrm{Sym}_3$, $e : \mathbf{R}^3 \to \mathrm{Sym}_3$ and $d : \mathbf{R}^3 \to \mathbf{R}^3$ are linear maps.

v) *Assumptions on the constitutive laws*

$$^*d = d \text{ and } \exists\alpha > 0 \text{ such that } \forall v \in \mathbf{R}^3, \quad (d\,v)\cdot v \geq \alpha\,|v|^2 \tag{5}$$

$$^*C = C \text{ and } \exists\beta > 0 \text{ such that } \forall\tau \in \mathrm{Sym}_3, \quad C\tau\cdot\cdot\,\tau \geq \beta\,\tau\cdot\cdot\,\tau. \tag{6}$$

For instance for the Lamé constants $\mu > 0$ and $\lambda \geq 0$, the special constitutive law $C\tau = 2\mu\,\tau + \lambda \operatorname{tr}\tau\, I$ verifies the assumptions with $\beta = 2\mu$.

Given a solution (V_h, Φ_h) of the above system of equations, inner product the first field equation by V and multiply the second one by $\Phi - \Phi_h$ for any (V, Φ) in

$$\mathcal{V}_h \stackrel{\text{def}}{=} \left\{(W,\Psi)\in H^1(S_h(\omega))^3 \times H^1(S_h(\omega)) : W|_{\Sigma_h(\gamma)} = 0 \text{ and } \Psi|_{\partial_0 S_h(\omega)} = \Phi_0\right\}. \tag{7}$$

Then assuming that $(V_h, \Phi_h) \in \mathcal{V}_h$ and using the fact that $\Phi - \Phi_h$ is zero on $\partial_0 S_h(\omega)$, we get upon integration by parts

$$\begin{aligned}\int_{S_h(\omega)} &\sigma_h\cdot\cdot\,\varepsilon(V) + D_h\cdot(E(\Phi) - E(\Phi_h))\,dx \\ &= \int_{S_h(\omega)} F\cdot V\,dx + \int_{T_{\pm h}(\omega)\cup\Sigma_h(\gamma\backslash\gamma_0)} Q\cdot V\,dS.\end{aligned} \tag{8}$$

Making use of the constitutive laws a solution of the above equations will verify the variational equation: Find (V_h, Φ_h) such that for all (V, Φ)

$$\int_{S_h(\omega)} [C\,\varepsilon_h - e\,E_h] \cdot\cdot\, \varepsilon(V) + [\,{}^*e\,\varepsilon_h + d\,E_h] \cdot (E(\Phi) - E(\Phi_h))\,dx$$
$$= \int_{S_h(\omega)} F \cdot V\,dx + \int_{T_h(\omega)} Q \cdot V\,d\Gamma_h + \int_{T_{-h}(\omega)} Q \cdot V\,d\Gamma_{-h} + \int_{\Sigma_h(\gamma\backslash\gamma_0)} Q \cdot V\,d\Sigma \tag{9}$$

where $\varepsilon_h = \varepsilon(V_h)$, $E_h = E(\Phi_h)$, $\varepsilon = \varepsilon(V)$, $E = E(\Phi)$ and $d\Gamma_h$ and $d\Gamma_{-h}$ are the respective surface (2-d Hausdorff) measures on $T_h(\omega)$ and $T_{-h}(\omega)$. Under the above assumptions equation (9) has a unique solution (V_h, Φ_h) in $\mathcal{V}_h$ for $\Phi_0 \in H^{1/2}(\partial_0 S_h(\omega))$, $F \in L^2(S_h(\omega))^3$ and $Q \in L^2(T_{\pm h}(\omega) \cup \Sigma_h(\gamma\backslash\gamma_0))^3$ by adapting the Lax-Milgram Theorem.

4. $P(2,1)$ Model

The *shell approximation* of the 3-dimensional variational equation (9) is done by the same technique as the one used in [2] for the purely mechanical part of the model: it is the so-called $P(2,1)$ model. First recall that the vector function V and the function Φ are transported by the map T defined in (2)

$$\Phi \mapsto \varphi \stackrel{\text{def}}{=} \Phi \circ T \colon H^1(S_h(\omega)) \to H^1(-h,h;L^2(\omega)) \cap L^2(-h,h;H^1(\omega))$$
$$V \mapsto v \stackrel{\text{def}}{=} V \circ T \colon H^1(S_h(\omega))^3 \to H^1(-h,h;L^2(\omega)^3) \cap L^2(-h,h;H^1(\omega)^3)$$

The approximation is made by polynomials of order 2 in the normal variable z for the functions and of order 1 for their derivatives

$$\Phi \circ T_z \simeq \varphi^0 + z\,\varphi^1 + z^2\,\varphi^2 \text{ and } E(\Phi) \circ T_z \simeq E^0(\varphi^0,\varphi^1) + z\,E^1(\varphi^0,\varphi^1,\varphi^2)$$
$$V \circ T_z \simeq v^0 + z\,v^1 + z^2\,v^2 \text{ and } \varepsilon(V) \circ T_z \simeq \varepsilon^0(v^0,v^1) + z\,\varepsilon^1(v^0,v^1,v^2)$$

where the $E^i(\varphi)$'s and the $\varepsilon^i(v)$'s are given by the expressions

$$-E^0(\varphi) \stackrel{\text{def}}{=} -E^0(\varphi^0,\varphi^1) = \varphi^1\,n + \nabla_\Gamma \varphi^0 \tag{10}$$
$$-E^1(\varphi) \stackrel{\text{def}}{=} -E^1(\varphi^0,\varphi^1,\varphi^2) = 2\,\varphi^2\,n + \nabla_\Gamma \varphi^1 - D^2 b\,\nabla_\Gamma \varphi^0 \tag{11}$$
$$\varepsilon^0(v) \stackrel{\text{def}}{=} \varepsilon^0(v^0,v^1) = \frac{1}{2}\,(v^1\ {}^*n + n\ {}^*v^1) + \varepsilon_\Gamma(v^0) \tag{12}$$
$$\varepsilon^1(v) \stackrel{\text{def}}{=} \varepsilon^1(v^0,v^1,v^2) = [v^2\ {}^*n + n\ {}^*v^2] + \varepsilon_\Gamma(v^1) - \frac{1}{2}[D_\Gamma(v^0)\,D^2 b + D^2 b\ {}^*D_\Gamma(v^0)]$$

and n is the normal to Γ as specified by b, $n = \nabla b$ on Γ. The vector v^0 is the *displacement*, v^1_Γ the *rotation* and v^1_n the *rotation around the normal* n. Substituting the above approximations in the three-dimensional variational equation (9) and dividing both sides by $2h$, we get the following variational equation: *Find* $v_h = (v^0_h, v^1_h, v^2_h)$ *and* $\varphi_h = (\varphi^0_h, \varphi^1_h, \varphi^2_h)$ *such that for all* $v = (v^0, v^1, v^2)$ *and*

$\varphi = (\varphi^0, \varphi^1, \varphi^2)$

P(2,1) piezoelectric shell model

$$\boxed{\begin{aligned}&\int_\omega \overline{\alpha}_0(h)\,\{[C\,\varepsilon_h^0 - e\,E_h^0]\cdot\cdot\,\varepsilon^0 + [{}^*e\,\varepsilon_h^0 + d\,E_h^0]\cdot(E^0 - E_h^0)\}\\ &\quad + \overline{\alpha}_1(h)\,\{[C\,\varepsilon_h^0 - e\,E_h^0]\cdot\cdot\,\varepsilon^1 + [{}^*e\,\varepsilon_h^0 + d\,E_h^0]\cdot(E^1 - E_h^1)\\ &\qquad + [C\,\varepsilon_h^1 - e\,E_h^1]\cdot\cdot\,\varepsilon^0 + [{}^*e\,\varepsilon_h^1 + d\,E_h^1]\cdot(E^0 - E_h^0)\}\\ &\quad + \overline{\alpha}_2(h)\,\{[C\,\varepsilon_h^1 - e\,E_h^1]\cdot\cdot\,\varepsilon^1 + [{}^*e\,\varepsilon_h^1 + d\,E_h^1]\cdot(E^1 - E_h^1)\}\,d\Gamma\\ &= \ell_h(v^0, v^1, v^2)\end{aligned}} \tag{13}$$

where $\varepsilon_h^i \stackrel{\text{def}}{=} \varepsilon^i(v_h)$, $\varepsilon^i \stackrel{\text{def}}{=} \varepsilon^i(v)$, $E_h^i \stackrel{\text{def}}{=} E^i(\varphi_h)$, $E^i \stackrel{\text{def}}{=} E^i(\varphi)$,

$$\frac{\overline{\alpha}_0(h) - 1}{h^2} = \frac{1}{3}\kappa_2, \quad \frac{\overline{\alpha}_1(h)}{h^2} = \frac{1}{3}\kappa_1, \quad \frac{\overline{\alpha}_2(h)}{h^2} = \frac{1}{3} + \frac{h^2}{5}\kappa_2 \tag{14}$$

$$\begin{aligned}\ell_h(v) \stackrel{\text{def}}{=} \ell_h(v^0, v^1, v^2) \stackrel{\text{def}}{=} &\frac{1}{2h}\int_{S_h(\omega)} F\cdot\left(v^0 \circ p + b\,v^1 \circ p + b^2 v^2 \circ p\right)\,dx\\ &+ \frac{1}{2h}\int_{T_h(\omega)} Q\cdot\left(v^0 \circ p + h\,v^1 \circ p + h^2 v^2 \circ p\right)\,d\Gamma_h\\ &+ \frac{1}{2h}\int_{T_{-h}(\omega)} Q\cdot\left(v^0 \circ p - h\,v^1 \circ p + h^2 v^2 \circ p\right)\,d\Gamma_{-h}\\ &+ \frac{1}{2h}\int_{\Sigma_h(\gamma\setminus\gamma_0)} Q\cdot\left(v^0 \circ p + b\,v^1 \circ p + b^2 v^2 \circ p\right)\,d\Sigma.\end{aligned} \tag{15}$$

Using Federer's formula the above expressions become integrals over ω:

$$\begin{aligned}\ell_h(v) = &\frac{1}{2h}\int_\omega\int_{-h}^{h} F\circ T_z\cdot\left(v^0 + z\,v^1 + z^2 v^2\right) j_z\,dz\,d\Gamma\\ &+ \frac{1}{2h}\int_\omega \Big(Q\circ T_h\cdot\left(v^0 + h\,v^1 + h^2 v^2\right) j_h\\ &\qquad + Q\circ T_{-h}\cdot\left(v^0 - h\,v^1 + h^2 v^2\right) j_{-h}\Big)\,d\Gamma\\ &+ \frac{1}{2h}\int_{\Sigma_h(\gamma\setminus\gamma_0)}\int_{-h}^{h} Q\circ T_z\cdot\left(v^0 + z\,v^1 + z^2 v^2\right) j_z\,dz\,d\gamma.\end{aligned} \tag{16}$$

It is readily seen that the natural norm associated with the $P(2,1)$-model,

$$\left\{\|\varepsilon^0(v)\|^2_{L^2(\omega)} + \|\varepsilon^1(v)\|^2_{L^2(\omega)} + \|E^0(\varphi)\|^2_{L^2(\omega)} + \|E^1(\varphi)\|^2_{L^2(\omega)}\right\}^{1/2},$$

splits into a norm on v and a norm on φ. For the part on v the associated space is $E^{01}_{\gamma_0}$ which is the completion of the space

$$H^1_{\gamma_0}(\omega)^3 \times H^1_{\gamma_0}(\omega)^3 \times L^2(\omega)^3$$

with respect to the norm

$$\left\{\left\|\varepsilon^0(v^0, v^1)\right\|^2_{L^2(\omega)} + \left\|\varepsilon^1(v^0, v^1, v^2)\right\|^2_{L^2(\omega)}\right\}^{1/2}. \tag{17}$$

This space has been characterized in [6] as follows. Assume that Γ is the boundary of a set of class $C^{1,1}$ in $\mathbf{R}^3$. Let ω be a bounded open domain with a Lipschitzian boundary γ in Γ, and let $\gamma_0 \subset \gamma$ have positive 2-dimensional Hausdorff measure

$$\boxed{E^{01}_{\gamma_0} = \left\{ (v^0, v^1, v^2) \; : \; \begin{array}{c} (v^0, v^1) \in Q^n_{\gamma_0},\; v^2_n \in L^2(\omega) \text{ and} \\ v^2_\Gamma \in H^{-1}(\omega)^3 \text{ such that} \\ P\varepsilon^1(0, v^1_n n, v^2_\Gamma) \in L^2(\omega)^3 \end{array} \right\}} \tag{18}$$

From the definitions (10)–(11) of E^0 and E^1, the associated norm on φ is

$$\|E^0(\varphi)\|^2_{L^2(\omega)} + \|E^1(\varphi)\|^2_{L^2(\omega)} = \|\varphi^1\|^2 + \|\nabla_\Gamma \varphi^0\|^2 + \|\varphi^2\|^2 + \|\nabla_\Gamma \varphi^1 - D^2 b \nabla_\Gamma \varphi^0\|^2$$

which means that

$$(\varphi^0, \varphi^1, \varphi^2) \in \frac{H^1(\omega)}{\mathbf{R}} \times H^1(\omega) \times L^2(\omega)$$

To get a unique solution we need at least one condition on φ^0 arising from the specification of the electric potential on $\partial_0 S_h(\omega) \subset \partial S_h(\omega)$, the boundary of $S_h(\omega)$.

Assume that $\partial_0 S_h(\omega)$ is made up of a finite number of components and that the potential is equal to a constant on each one:

$$\partial_0 S_h(\omega) = \Sigma_h(\gamma_1) \cup T_h(\omega^+) \cup T_{-h}(\omega^-)$$

for some $\gamma_1 \subset \gamma$ and

$$\omega^+ = \bigcup_{j=1}^{n^+} \omega_j^+, \quad \omega^- = \bigcup_{i=1}^{n^-} \omega_i^-$$

where $\omega_j^+ \subset \omega$ and $\omega_i^- \subset \omega$, and $\omega_j^+ \cap \omega_{j'}^+ = \varnothing$ and $\omega_j^- \cap \omega_{j'}^- = \varnothing$ for all $j \neq j'$. The boundary conditions for the electrical potential are

$$\Phi_h = \Phi_1 \text{ on } \Sigma_h(\gamma_1), \qquad \begin{array}{l} \Phi_h = \Phi_j^+ \text{ on } T_h(\omega_j^+),\ 1 \le j \le n^+ \\ \Phi_h = \Phi_i^- \text{ on } T_{-h}(\omega_i^-),\ 1 \le i \le n^- \end{array}$$

for some constants Φ_1, $\{\Phi_j^+\}$ and $\{\Phi_i^-\}$. Since, in the $P(2,1)$-approximation, $\varphi_h^2 \in L^2(\omega)$, these boundary conditions will yield boundary conditions on φ_h^0 and φ_h^1:

$$\varphi_h^0 + z\,\varphi_h^1 = \varphi_1 \stackrel{\text{def}}{=} \Phi_1 \circ T_z = \Phi_1 \text{ on }]-h, h[\times \gamma_1$$

$$\Rightarrow \boxed{\varphi_h^0 = \varphi_1 \text{ and } \varphi_h^1 = 0 \text{ on } \gamma_1} \tag{19}$$

$$\boxed{\varphi_h^0 + h\,\varphi_h^1 = \varphi_j^+ \stackrel{\text{def}}{=} \Phi_j^+ \circ T_h = \Phi_j^+ \text{ on } \omega_j^+,\ 1 \le j \le n^+} \tag{20}$$

$$\boxed{\varphi_h^0 - h\,\varphi_h^1 = \varphi_i^- \stackrel{\text{def}}{=} \Phi_i^- \circ T_{-h} = \Phi_i^- \text{ on } \omega_i^-,\ 1 \le i \le n^-} \tag{21}$$

When $\omega_j^+ \cap \omega_i^- \neq \varnothing$,

$$\varphi_h^0 = \frac{\varphi_j^+ + \varphi_i^-}{2} \text{ and } \varphi_h^1 = \frac{\varphi_j^+ - \varphi_i^-}{2h} \text{ on } \omega_j^+ \cap \omega_i^-$$

and φ_h^0 and φ_h^1 are completely specified on $\omega_j^+ \cap \omega_i^-$.

Finally, for h sufficiently small, (13) has a unique solution (v_h, φ_h) in

$$\mathcal{E}^{01}_{\gamma_0} \stackrel{\text{def}}{=} \left\{ (v,\varphi) : \begin{array}{l} v \in E^{01}_{\gamma_0} \text{ and } \varphi = (\varphi^0, \varphi^1, \varphi^2) \in H^1(\omega) \times H^1(\omega) \times L^2(\omega) \\ \text{such that } \varphi^0 = \varphi_1 \text{ and } \varphi^1 = 0 \text{ on } \gamma_1 \\ \varphi^0 + h\,\varphi^1 = \varphi_j^+ \text{ on } \omega_j^+,\ 1 \le j \le n^+ \\ \varphi^0 - h\,\varphi^1 = \varphi_i^- \text{ on } \omega_i^-,\ 1 \le i \le n^- \end{array} \right\} \tag{22}$$

5. Asymptotic Piezoelectric Shell Model

The asymptotic results of [2] can be extended to the piezoelectric case. Since the loading ℓ_h only depends on v, the assumptions under which the solution of the $P(2,1)$ model converges to the solution of the asymptotic model as h goes to zero will be the same as in [2]. In addition for all i and j such that $\overline{\omega_j^+} \cap \overline{\omega_i^-} \neq \varnothing$ the condition $\varphi_j^+ = \varphi_i^-$ must be verified. The associated solution spaces are characterized in [3, 4, 5]. As h goes to zero the $P(2,1)$ model (13) yields the following system of two asymptotic variational equations:

$$\boxed{\int_\omega \left[C\,\hat\varepsilon^0 - e\,\hat E^0\right] \cdot\cdot\, \varepsilon^0 + \left[{}^*e\,\hat\varepsilon^0 + d\,\hat E^0\right] \cdot (E^0 - \hat E^0)\, d\Gamma = I_0(v^0, v^1)} \tag{23}$$

and for all (v,φ) such that $v \in \ker \varepsilon^0$, and $\varphi - \hat\varphi \in \ker E^0$

$$\boxed{\begin{aligned} \frac{1}{3} \int_\omega \kappa_1 \Big\{ &\left[C\,\hat\varepsilon^0 - e\,\hat E^0\right] \cdot\cdot\, \varepsilon^1 + \left[{}^*e\,\hat\varepsilon^0 + d\,\hat E^0\right] \cdot (E^1 - \hat E^1)\Big\} \\ &+ \Big\{\left[C\,\hat\varepsilon^1 - e\,\hat E^1\right] \cdot\cdot\, \varepsilon^1 + \left[{}^*e\,\hat\varepsilon^1 + d\,\hat E^1\right] \cdot (E^1 - \hat E^1)\Big\}\, d\Gamma = J_0(v^0, v^1, v^2) \end{aligned}} \tag{24}$$

where $\hat\varepsilon^i \stackrel{\text{def}}{=} \varepsilon^i(\hat v)$, $\varepsilon^i \stackrel{\text{def}}{=} \varepsilon^i(v)$, $\hat E^i \stackrel{\text{def}}{=} E^i(\hat\varphi)$, $E^i \stackrel{\text{def}}{=} E^i(\varphi)$,

$$I_0(v^0, v^1) \stackrel{\text{def}}{=} \lim_{h \searrow 0} \ell_h(v^0, v^1, v^2),\ \forall (v^0, v^1) \in \ker \varepsilon^0 \tag{25}$$

$$J_0(v^0, v^1, v^2) \stackrel{\text{def}}{=} \lim_{h \searrow 0} \frac{\ell_h(v^0, v^1, v^2)}{h^2} \tag{26}$$

Define the spaces

$$F \stackrel{\text{def}}{=} \left\{ \varphi \in H^1(\omega) : \varphi = \varphi_1 \text{ on } \gamma_1, \begin{array}{l} \varphi = \varphi_j^+ \text{ on } \omega_j^+,\ 1 \le j \le n^+ \\ \varphi = \varphi_i^- \text{ on } \omega_i^-,\ 1 \le i \le n^- \end{array} \right\} \tag{27}$$

$$F_0 \stackrel{\text{def}}{=} \left\{ \varphi \in H^1(\omega) : \varphi = 0 \text{ on } \gamma_1, \begin{array}{l} \varphi = 0 \text{ on } \omega_j^+,\ 1 \le j \le n^+ \\ \varphi = 0 \text{ on } \omega_i^-,\ 1 \le i \le n^- \end{array} \right\} \tag{28}$$

and the projection π of $\mathcal{E}^{01}_{\gamma_0}$ onto the space

$$S \stackrel{\text{def}}{=} \left\{ (v,\varphi) : \begin{array}{l} (v^0, v^1, v^2) \in E^{01}_{\gamma_0} \text{ such that } \varepsilon^0(v^0, v^1) = 0 \\ (\varphi^0, \varphi^1, \varphi^2) \in F_0 \times L^2(\omega) \times L^2(\omega) \text{ such that } E^0(\varphi^0, \varphi^1) = 0 \end{array} \right\} \tag{29}$$

through the variational problem: *Find* $\pi(v,\varphi) \in S$ *such that for all* $(w,\psi) \in S$

equation for the projection π onto S

$$\begin{aligned}
&\int_\omega \{[C\,\varepsilon^0(v-\pi v) - e\,E^0(\varphi-\pi\varphi)] \cdot\cdot\, \varepsilon^0(w) \\
&\quad + [{}^*e\,\varepsilon^0(v-\pi v) + d\,E^0(\varphi-\pi\varphi)] \cdot E^0(\psi)\} \\
&+ \kappa_1 \frac{h^2}{3}\{ [C\,\varepsilon^0(v-\pi v) - e\,E^0(\varphi-\pi\varphi)] \cdot\cdot\, \varepsilon^1(w) \\
&\qquad + [{}^*e\,\varepsilon^0(v-\pi v) + d\,E^0(\varphi-\pi\varphi)] \cdot E^1(\psi) \\
&\qquad + [C\,\varepsilon^1(v-\pi v) - e\,E^1(\varphi-\pi\varphi)] \cdot\cdot\, \varepsilon^0(w) \\
&\qquad + [{}^*e\,\varepsilon^1(v-\pi v) + d\,E^1(\varphi-\pi\varphi)] \cdot E^0(\psi)\} \\
&+ \frac{h^2}{3}\{[C\,\varepsilon^1(v-\pi v) - e\,E^1(\varphi-\pi\varphi)] \cdot\cdot\, \varepsilon^1(w) \\
&\qquad + [{}^*e\,\varepsilon^1(v-\pi v) + d\,E^1(\varphi-\pi\varphi)] \cdot E^1(\psi)\}\, d\Gamma = 0
\end{aligned} \tag{30}$$

Notice that since $\varepsilon^0(w) = 0$, $E^0(\psi) = 0$, $\varepsilon^0(\pi v) = 0$ and $E^0(\pi\varphi) = 0$, the characterization of $\pi(v,\varphi) = (\pi(v), \pi(\varphi))$ is independent of h: for all $(w,\psi) \in S$

$$\begin{aligned}
&\int_\omega \kappa_1\{ [C\,\varepsilon^0(v) - e\,E^0(\varphi)] \cdot\cdot\, \varepsilon^1(w) + [{}^*e\,\varepsilon^0(v) + d\,E^0(\varphi)] \cdot E^1(\psi)\} \\
&\quad + \{[C\,\varepsilon^1(v-\pi v) - e\,E^1(\varphi-\pi\varphi)] \cdot\cdot\, \varepsilon^1(w) \\
&\qquad + [{}^*e\,\varepsilon^1(v-\pi v) + d\,E^1(\varphi-\pi\varphi)] \cdot E^1(\psi)\}\, d\Gamma = 0
\end{aligned} \tag{31}$$

With the projection π equation (24) can be rewritten as: for all $\psi^2 \in L^2\omega)$, $(\psi^0,\psi^1) \in \ker E^0$, $w^2 \in L^2\omega)^3$, and $(w^0,w^1) \in \ker \varepsilon^0$

$$\begin{aligned}
&\int_\omega \{[C\,\varepsilon^1(\pi\hat v) - e\,E^1(\pi\hat\varphi)] \cdot\cdot\, \varepsilon^1(w) + [{}^*e\,\varepsilon^1(\pi\hat v) + d\,E^1(\pi\hat\varphi)] \cdot E^1(\psi)\}\, d\Gamma \\
&= 3\, J_0(w^0, w^1, w^2)
\end{aligned} \tag{32}$$

It is important to note that this equation for the projection $\pi\hat v$ of the asymptotic solution $\hat v$ is not coupled with the first equation (23) as in [2]. This difference arises from the different and more natural choice of inner product to define the projection π. In fact all the results and theorems of [2] for the mechanical shell remain true with the new projection π.

The space of solution[1] is

$$\mathcal{E}^{01}_{\gamma_0\pi} \stackrel{\text{def}}{=} \{(v,\varphi) : (v^0,v^1,v^2) \in E^{01}_{\gamma_0,\pi} \text{ and } (\varphi^0,\varphi^1,\varphi^2) \in F \times L^2(\omega) \times L^2(\omega) \ \} \tag{33}$$

where $E^{01}_{\gamma_0,\pi}$ is defined as the completion of $E^{01}_{\gamma_0}$ with respect to the norm ([2])

$$\left\{ \int_\omega |\varepsilon^0(v)|^2 + |\varepsilon^1(\pi v)|^2\, d\Gamma \right\}^{1/2}.$$

With the above topology *there exists a unique solution in* $\mathcal{E}^{01}_{\gamma_0\pi}$ *to system* (23)–(32).

[1]There is an implicit assumption that the space F is not empty. This means that patches with a non-empty intersection or a common piece of boundary must be at the same potential and that patches with different potentials must be separated by a non-zero distance.

5.1. Decomposition of the first asymptotic equation

We need the following theorem from [6] for the purely mechanical part.

Theorem 5.1. *The transformation of* $\mathbf{R}^3$

$$\boxed{N(u) \stackrel{\text{def}}{=} [C(u \,{}^*n + n \,{}^*u)]n, \quad u \in \mathbf{R}^3} \tag{34}$$

and the effective constitutive law $C_{eP}\colon \mathrm{Sym}_3^P \to \mathrm{Sym}_3^P$ *defined as*

$$\boxed{C_{eP}\tau \stackrel{\text{def}}{=} C\{\tau - \{N^{-1}([C\tau]n) \,{}^*n + n \,{}^*N^{-1}([C\tau]n)\}\}, \ \tau \in \mathrm{Sym}_3^P} \tag{35}$$

are bijective, symmetrical and coercive. For all τ *and* σ *in* Sym^3

$$\boxed{\begin{aligned} C\tau &= C_{eP}\tau^P + C[N^{-1}([C\tau]n) \,{}^*n + n \,{}^*N^{-1}([C\tau]n)] \\ C\tau \cdot\cdot\, \sigma &= C_{eP}\tau^P \cdot\cdot\, \sigma^P + 2([C\tau]n)\cdot N^{-1}([C\sigma]n). \end{aligned}} \tag{36}$$

Here the three entities C, e and d are intertwined and we get mixed effective laws. Let

$$m \stackrel{\text{def}}{=} [en]\, n \in \mathbf{R}^3, \quad d_{nn} \stackrel{\text{def}}{=} dn\cdot n, \quad \overline{d} \stackrel{\text{def}}{=} d_{nn} + 2m\cdot N^{-1}m. \tag{37}$$

Then for all $\tau \in \mathrm{Sym}_3^P$ and $V_\Gamma \in T(\Gamma) \stackrel{\text{def}}{=} \{V \in \mathbf{R}^3 : V\cdot n = 0\}$

$$C_P\tau \stackrel{\text{def}}{=} C_{eP}\tau + \frac{{}^*e\tau\cdot n - 2N^{-1}m\cdot[C\tau]n}{\overline{d}}\left(en - C\{(N^{-1}m)\,{}^*n + n\,{}^*(N^{-1}m)\}\right) \tag{38}$$

$$\begin{aligned} e_P V_\Gamma \stackrel{\text{def}}{=}\ & eV_\Gamma - C\{N^{-1}([eV_\Gamma]n)\,{}^*n + n\,{}^*N^{-1}([eV_\Gamma]n)\} \\ & - \frac{2N^{-1}m\cdot[eV_\Gamma]n + dV_\Gamma\cdot n}{\overline{d}}\left(en - C\{(N^{-1}m)\,{}^*n + n\,{}^*(N^{-1}m)\}\right) \end{aligned} \tag{39}$$

$$\begin{aligned} d_P V_\Gamma \stackrel{\text{def}}{=}\ & dV_\Gamma + {}^*e\{N^{-1}([eV_\Gamma]n)\,{}^*n + n\,{}^*N^{-1}([eV_\Gamma]n)\} \\ & - \frac{2N^{-1}m\cdot[eV_\Gamma]n + dV_\Gamma\cdot n}{\overline{d}}\left({}^*e\,\{(N^{-1}m)\,{}^*n + n\,{}^*(N^{-1}m)\} + dn\right) \end{aligned} \tag{40}$$

Notice that C_P is equal to the old C_{eP} for the mechanical part plus a new term which is zero when $e = 0$. It is not difficult to check that

$$C_P : \mathrm{Sym}_3^P \to \mathrm{Sym}_3^P, \quad e_P : T(\Gamma) \to \mathrm{Sym}_3^P, \quad d_P : T(\Gamma) \to T(\Gamma) \tag{41}$$

and that C_P and d_P are symmetrical and coercive.

The first asymptotic equation (23) decomposes into three equations: the new *membrane piezoelectric shell equation* for $(\hat{v}^0, \hat{\varphi}^0)$ and two explicit expressions of $\hat{v}^1$ and $\hat{\varphi}^1$ in terms of $(\hat{v}^0, \hat{\varphi}^0)$. *Find* $(\hat{v}^0, \hat{\varphi}^0) \in \mathcal{E}_{\gamma_0}^P$ *such that for all* $(v^0, \varphi^0) \in \mathcal{E}_{\gamma_0}^P$

$$\boxed{\begin{aligned} \int_\omega & [C_P\, \varepsilon_\Gamma^P(\hat{v}^0) + e_P\, \nabla_\Gamma \hat{\varphi}^0] \cdot\cdot\, \varepsilon_\Gamma^P(v^0) \\ & + [-\,{}^*e_P\, \varepsilon_\Gamma^P(\hat{v}^0) + d_P\, \nabla_\Gamma\hat{\varphi}^0]\cdot\nabla_\Gamma(\varphi^0 - \hat{\varphi}^0)\, d\Gamma = \tilde{I}_0(v^0, \varphi^0 - \hat{\varphi}^0) \end{aligned}} \tag{42}$$

where the space $E^P_{\gamma_0}$ is defined in [2] and

$$\mathcal{E}^P_{\gamma_0} \stackrel{\text{def}}{=} \{(v^0,\varphi^0) \,:\, v^0 \in E^P_{\gamma_0} \text{ and } \varphi^0 \in F \quad\} \tag{43}$$

$$\tilde{I}_0(v^0,\varphi^0) \stackrel{\text{def}}{=} I_0(v^0, -2N^{-1}\Big\{[C\varepsilon_\Gamma(v^0) - e(\nabla_\Gamma\varphi^0)]n + m\,\frac{[\,{}^*e\,\varepsilon_\Gamma(v^0) + d\,(\nabla_\Gamma\varphi^0)]\cdot n}{\overline{d}}\Big\}) \tag{44}$$

Moreover $\hat{v}^1$ and $\hat{\varphi}^1$ are given by

$$\hat{\varphi}^1 = \frac{1}{\overline{d}}\{2N^{-1}m\cdot\big(q - [C\varepsilon^P_\Gamma(\hat{v}^0) + e(\nabla_\Gamma\hat{\varphi}^0)]n\big) + [\,{}^*e\,\varepsilon^P_\Gamma(\hat{v}^0) - d\,(\nabla_\Gamma\hat{\varphi}^0)]\cdot n\} \tag{45}$$

$$\begin{aligned}\frac{1}{2}\hat{v}^1 &+ \varepsilon_\Gamma(\hat{v}^0)n\\ =&N^{-1}\{q - [C\varepsilon^P_\Gamma(\hat{v}^0) + e(\nabla_\Gamma\hat{\varphi}^0)]n\}\\ &+N^{-1}m\,\frac{2N^{-1}m\cdot[C\varepsilon^P_\Gamma(\hat{v}^0) + e(\nabla_\Gamma\hat{\varphi}^0)]n - [\,{}^*e\,\varepsilon^P_\Gamma(\hat{v}^0) - d\,(\nabla_\Gamma\hat{\varphi}^0)]\cdot n}{\overline{d}}\end{aligned} \tag{46}$$

$$\begin{aligned}\hat{v}^1 =&2\,N^{-1}\{q - [C\varepsilon_\Gamma(\hat{v}^0) + e(\nabla_\Gamma\hat{\varphi}^0)]n\}\\ &-2N^{-1}m\,\frac{2N^{-1}m\cdot\{q - [C\varepsilon_\Gamma(\hat{v}^0) + e(\nabla_\Gamma\hat{\varphi}^0)]n\} + [\,{}^*e\,\varepsilon_\Gamma(\hat{v}^0) - d\,(\nabla_\Gamma\hat{\varphi}^0)]\cdot n}{\overline{d}}\end{aligned} \tag{47}$$

$$\exists q \in (L^2(\omega))^3,\ \forall v^1 \in (L^2(\omega))^3, \quad \int_\omega q\cdot v^1\,d\Gamma \stackrel{\text{def}}{=} I_0(0,v^1). \tag{48}$$

5.2. Decomposition of the second asymptotic equation

Equation (32) splits into a bending equation for $(\pi v)^0$ and explicit expressions for $(\pi v)^1$ and $(\pi v)^2$. This yields

$$\frac{1}{3}\int_\omega \big[C_P\,\varepsilon^{1P}(\pi\hat{v}) - e_P\,E^{1P}(\pi\hat{\varphi})\big]\cdot\cdot\,\varepsilon^{1P}(w)\,d\Gamma = \tilde{J}_0(w^0,w^1) \tag{49}$$

$$\tilde{J}_0(w^0,w^1) \stackrel{\text{def}}{=} J_0(w^0,w^1,-N^{-1}\Big\{[C\varepsilon^1(w^0,w^1,0) + m\,\frac{[\,{}^*e\,\varepsilon^1(w^0,w^1,0)]\cdot n}{\overline{d}}\Big\}) \tag{50}$$

From the definitions of E^0 and E^1 for all $(\psi^0,\psi^1) \in \ker E^0$

$$\begin{aligned}0 = -E^0(\psi^0,\psi^1) = \psi^1\,n + \nabla_\Gamma\psi^0 \quad &\Rightarrow \psi^1 = 0 \text{ and } \nabla_\Gamma\psi^0 = 0\\ \Rightarrow -E^1(\psi^0,\psi^1,\psi^2) = 2\psi^2\,n + \nabla_\Gamma\psi^1 - D^2b\nabla_\Gamma\psi^0 &= 2\psi^2\,n\end{aligned}$$

If, in addition, $\psi^0 \in F_0$, then

$$(\psi^0,\psi^1) \in \ker E^0 \quad \Rightarrow (\psi^0,\psi^1) = 0 \text{ and } -E^1(\psi) = 2\psi^2\,n \quad \Rightarrow E^{1P}(\psi) = 0.$$

By choosing test $w = 0$ and $\psi = (0,0,\psi^2)$ in (24) we get an equation for $(\pi\hat{\varphi})^2$

$$\big[\,{}^*e\,\varepsilon^1(\pi\hat{v}) + d\,E^1(\pi\hat{\varphi})\big]\cdot n = 0 \quad \Rightarrow 2\,dn\cdot n\,(\pi\hat{\varphi})^2 = {}^*e\,\varepsilon^1(\pi\hat{v})\cdot n$$

As in [2] we can further eliminate $\hat{v}^2$ by using test functions $w = (0, 0, w^2)$ for $w^2 \in L^2(\omega)^3$

$$\exists g \in L^2(\omega),\ \forall w^2 \in L^2(\omega),\ \int_\omega g \cdot w^2 d\Gamma \stackrel{\text{def}}{=} 3\, J_0(0,0,w^2)$$
$$2\left[C\varepsilon^1(\pi\hat{v}) - e\,E^1(\pi\hat{\varphi})\right] n = g$$
$$\Rightarrow 2\,(\pi\hat{v})^2 = N^{-1}\{g - 2\,C\,\varepsilon^1((\pi\hat{v})^0, (\pi\hat{v})^1, 0))n\}. \tag{51}$$

Moreover from [2] for $w \in E^{01}_{\gamma_0}$ such that $(w^0, w^1) \in \ker \varepsilon^0$

$$0 = \varepsilon^0(w^0, w^1) \quad \Rightarrow \varepsilon^P_\Gamma(w^0) = 0,\ w^1_n = 0,\ w^1_\Gamma + 2\varepsilon_\Gamma(w^0)n = 0$$
$$\Rightarrow \varepsilon^1(w^0, w^1, w^2) = w^2\,{}^*n + n\,{}^*w^2 + \varepsilon^1(w^0, w^1, 0)$$
$$= w^2\,{}^*n + n\,{}^*w^2 + \varepsilon^{1P}(w^0, -2\varepsilon_\Gamma(w^0)n, 0) = w^2\,{}^*n + n\,{}^*w^2 + \rho(w^0).$$

Introduce the modified *change of curvature tensor of the midsurface*

$$\boxed{\rho(v^0) \stackrel{\text{def}}{=} \varepsilon^{1P}(v^0, -2\varepsilon_\Gamma(v^0)n), 0).} \tag{52}$$

This tensor ρ was introduced in [9, 10] and its advantages were discussed in [1]. Equation (49) now reduces to the bending equation

$$\boxed{\forall w^0 \in \ker \varepsilon^P_\Gamma,\quad \frac{1}{3}\int_\omega \left[C_P\,\rho(\pi\hat{v}^0)\right] \cdot\cdot\, \rho(w^0)\, d\Gamma = \tilde{J}_0(w^0, -2\varepsilon_\Gamma(w^0))} \tag{53}$$

together with the two explicit expressions for $(\pi\hat{v})^2$ and $(\pi\hat{v})^1$

$$(\pi\hat{v})^1 = -2\varepsilon_\Gamma((\pi\hat{v})^0)n = 0, \quad \boxed{2\,(\pi\hat{v})^2 = N^{-1}\left\{g - 2\,C\,\rho((\pi\hat{v})^0)n\right\}} \tag{54}$$

and the two equations for $(\pi\hat{v})^2 - (\hat{v})^2$ and $(\pi\hat{\varphi})^2 - (\hat{\varphi})^2$ arising from equation (31) for the projection π by respectively using tests functions of the form $\psi = 0$ and $w = (0, 0, w^2)$ and $\psi = (0, 0, \psi^2)$ and $w = 0$. Eliminating the variables $(\pi\hat{v})^2$ and $(\pi\hat{\varphi})^2$ by using identities (51) and (54), we get the following two coupled equations for $\hat{v}^2$ and $\hat{\varphi}^2$

$$\left[C\,\varepsilon^1(\hat{v}) - e\,E^1(\hat{\varphi})\right] n = -\kappa_1\left[C\,\varepsilon^0(\hat{v}) - e\,E^0(\hat{\varphi})\right] n + g/2 \tag{55}$$
$$\left[{}^*e\,\varepsilon^1(\hat{v}) + d\,E^1(\hat{\varphi})\right] \cdot n = -\kappa_1\left[{}^*e\,\varepsilon^0(\hat{v}) + d\,E^0(\hat{\varphi})\right] \cdot n \tag{56}$$

The resulting *bending piezoelectric equation* (53) and the *membrane piezoelectric shell equation* (42) completely determine $(\hat{v}^0, \hat{\varphi}^0)$. The space of solution is $\mathcal{E}^0_{\gamma_0\pi}$ which is the completion of $H^1_{\gamma_0}(\omega)^3 \times F$ with respect to the norm

$$\int_\omega |\varepsilon^P_\Gamma(v^0)|^2 + |\rho(\pi v^0)|^2 + |E^{0P}(\varphi)|^2 + |E^{1P}(\pi\varphi)|^2\, d\Gamma$$
$$= \int_\omega |\varepsilon^P_\Gamma(v^0)|^2 + |\rho(\pi v^0)|^2 + |\nabla_\Gamma(\varphi^0)|^2\, d\Gamma$$

since $E^{1P}(\pi\varphi) = 0$ and $E^{0P}(\varphi) = -\nabla_\Gamma(\varphi^0)$. Hence $\mathcal{E}^0_{\gamma_0\pi} = E^0_{\gamma_0\pi} \times F$ where $E^0_{\gamma_0\pi}$ is the completion of $H^1_{\gamma_0}(\omega)^3$ with respect to the norm

$$\left\{\int_\omega |\varepsilon^P_\Gamma(v^0)|^2 + |\rho(\pi v^0)|^2\, d\Gamma\right\}^{1/2}.$$

References

[1] B. Budiansky and J.L. Sanders, *On the "best" first-order linear shell theory*, Progr. in Appl. Mech. (W. Prager Anniversary Volume), pp. 129–140, Macmillan, New York 1967.

[2] M.C. Delfour, *Intrinsic Differential Geometric Methods in the Asymptotic Analysis of Linear Thin Shells*, in "Boundaries, interfaces and transitions", M. Delfour, ed., pp. 19–90, CRM Proc. Lecture Notes, vol. 13, AMS Publ., Providence, R.I. 1998.

[3] M.C. Delfour, *Intrinsic $P(2,1)$ thin shell model and Naghdi's models without a priori assumption on the stress tensor*, in Proc International Conference on Optimal Control of Partial Differential Equations, K.H. Hoffmann, G. Leugering, F. Tröltzsch, eds., pp. 99–113, Int. Ser. of Numerical Mathematics, Vol. 133, Birkhäuser Verlag, Basel 1999.

[4] M.C. Delfour, *Membrane shell equation: characterization of the space of solutions*, in "Control of Distributed Parameter and Stochastic Systems", S. Chen, X. Li, J. Yong, X.Y. Zhou, eds., pp. 21–29, Chapman and Hall, 1999.

[5] M.C. Delfour, *Characterization of the space of solutions of the membrane shell equation for arbitrary $C^{1,1}$ midsurfaces*, Control and Cybernetics 4 (1999), pp. 481–501.

[6] M.C. Delfour, *Tangential differential calculus and functional analysis on a $C^{1,1}$ submanifold*, in "Differential-geometric methods in the control of partial differential equations", R. Gulliver, W. Littman and R. Triggiani, eds, pp. 83–115, Contemporary Mathematics, Vol. 268, AMS Publications, Providence, R.I., 2000.

[7] M.C. Delfour and J.-P. Zolésio, *Differential equations for linear shells: comparison between intrinsic and classical models*, Advances in Mathematical Sciences – CRM's 25 years (Luc Vinet, ed.), CRM Proc. Lecture Notes, Amer. Math. Soc., Providence, RI, 1997, pp. 42–124.

[8] T. Ikeda, *Fundamental of piezoelectricity*, Oxford Univ. Press, Oxford, 1990.

[9] W.T. Koiter, *On the nonlinear theory of thin elastic shells*, in Proc. Kon. Nederl. Akad. Wetensch. B59 (1966), 1–54; B73 (1970), 169–195.

[10] J.L. Sanders, *An improved first approximation theory of thin shells*, NASA Report 24, 1959.

[11] H.S. Tzou, *Piezoelectric shells: Distributed sensing and control of continua*, Kluwer Academic publishers, Dordrecht, The Netherlands, 1993.

MD: Centre de recherches mathématiques *and*
Département de Mathématiques et de statistique
Université de Montréal
C. P. 6128, succ. Centre-ville
Montréal QC, Canada H3C 3J7

MB: Pôle Universitaire Léonard dc Vinci
92916-Paris la Défense Cedex
France, *and*
INRIA-Rocquencourt
BP 105
78153 Le Chesnay Cedex, France

International Series of Numerical Mathematics, Vol. 139, 71–82

Modeling, Simulation and Control of Laser Heat Treatments

Jürgen Fuhrmann[1], Dietmar Hömberg[1], Jan Sokołowski[2]

Abstract. We investigate the optimal control of a laser surface hardening facility. The model consists of a nonlinear heat equation coupled with a system of ODEs to describe the occuring phase transitions. To avoid surface melting we include pointwise state constraints for the temperature.

We establish existence and stability results and derive necessary optimality conditions. Finally, some numerical results are presented to prove the validity of the model.

1. Introduction

The aim of surface hardening is to increase the hardness of the boundary layers of a workpiece by rapid heating and subsequent quenching. This heat treatment leads to a change in microstructure, which produces the desired hardening effect. When the workpiece is very big or the part of the surface to be hardened has a complicated shape, usually laser radiation is used as a heat source.

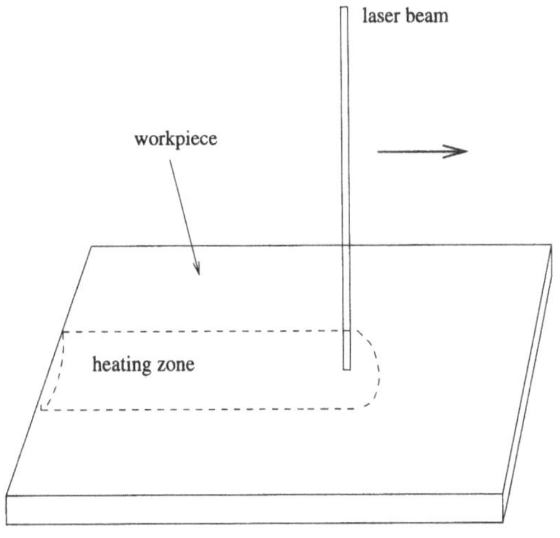

FIGURE 1. Sketch of a laser hardening process

The laser beam moves along the surface while its radiation is absorbed by the workpiece, leading to a rapid heating of its boundary layers (cf. Fig. 1). Then, the workpiece is quenched by "self–cooling" of the workpiece. To increase the scanning width, sometimes the laser beam performs an additional oscillating movement orthogonally to the principal moving direction.

Essential features of heat treatments of steel are solid–solid phase transitions and recalescence effects, caused by the latent heat of these transitions. Mathematical models for phase transitions in steel have been considered e.g.in [2], [3]–[6], [9], [12], [13]. For a survey on mathematical models for laser material treatments, we refer to [10]. Note that first results for the numerical optimal control of laser hardening have been obtained recently in [7], using the proper othogonal decomposition method.

In the next section we outline the mathematical model and formulate the control problem. Section 3 is devoted to the mathematical analysis of this problem and the derivation of optimality conditions. In last section we sketch a numerical algorithm and depict some numerical simulations of the state equations.

2. Problem Statement

2.1. Phase transitions

The reason why one can change the hardness of steel by thermal treatment lies in the occuring phase transitions. For reasons of space we can neither describe the crystallographic differences between these phases nor the mechanism of the transitions and refer to [3] for details. We just postulate that at room temperature, in general, steel is a mixture of four phases called *ferrite, pearlite, bainite* and *martensite*. Upon heating, these phases are transformed to the high temperature phase called *austenite*. Then, during cooling austenite is transformed back to a mixture of ferrite, pearlite, bainite and martensite.

The actual phase distribution at the end of the heat treatment depends on the cooling strategy. In the case of surface hardening, owing to high cooling rates most of the austenite is transformed to martensite by a diffusionless phase transition leading to the desired increase of hardness. For a detailed description of the following model for phase transitions during surface hardening, we refer to [2].

We introduce the following notations:

z_0:	volume fraction of austenite,
$z_1, \dots, z_4$:	relative volume fractions of ferrite, pearlite, bainite, martensite, which have been transformed from z_0,
A_s:	critical temperature, above which the formation of austenite starts,
M_s:	critical temperature, below which the formation of martensite starts ($M_s < A_s$),
$\mathcal{H}$:	smooth approximation of the Heaviside graph.

Then the evolution of volume fractions for given temperature evolution $\theta(.)$ can be described by the following initial–value problem:

$$z_0(0) = z_{00} \in (0,1), \tag{2.1a}$$
$$z_i(0) = 0, \qquad i = 1,\dots,4, \tag{2.1b}$$
$$z_{0,t}(t) = \frac{1}{\tau(\theta)}\Big(a_{eq}(\theta(t)) - z_0(t)\Big)\mathcal{H}(\theta(t) - A_s) - \sum_{j=1}^{4} z_{j,t}(t) \tag{2.1c}$$
$$z_{i,t}(t) = -z_0(t)\ln(z_0(t))\, g_i(t, z(t), \theta(t))\,\mathcal{H}(A_s - \theta(t)), \quad i = 1,\dots,3, \tag{2.1d}$$
$$z_{4,t}(t) = z_0(t)\mathcal{H}(-\theta_t) g_4(t, z(t), \theta(t))\,\mathcal{H}(M_s - \theta(t)), \tag{2.1e}$$

Remark 2.1.
(1) a_{eq} has been introduced by Leblond and Deveaux [9], to account for equilibrium fractions of austenite less than one.
(2) $\mathcal{H}(-\theta_t)$ prevents the formation of martensite, if the temperature is not decreasing.

2.2. Energy balance equation

We consider the following heat transfer equation:

$$\rho c\theta_t - k\Delta\theta = \tilde{F}_1[\theta] + \tilde{F}_2[\theta], \qquad \text{in } Q_T = \Omega \times (0,T),$$
$$\frac{\partial\theta}{\partial\nu} = 0 \qquad \text{in } \Sigma_T := \partial\Omega \times (0,T),$$
$$\theta(.,0) = \theta_0, \qquad \text{in } \Omega,$$

where $\Omega \subset \mathbb{R}^3$ with smooth boundary. The positive constants ρ, c and k denote density, specific heat at constant pressure and heat conductivity, respectively. The heat sources $\tilde{F}_1$, $\tilde{F}_2$ will take care of the latent heats of the phase transitions and the heating owing to laser radiation.

We assume that the laser radiation is volumetrically absorbed by the workpiece (see Sec. 4). Thus, we define

$$\tilde{F}_2[\theta] := \alpha(\theta)u,$$

where α measures the temperature dependent absorptivity of the workpiece's surface, and u is the radiation intensity inside the workpiece.

Moreover, we assume that the latent heat of all phase transitions has the same value L. Then, $\tilde{F}_1$ can be written as follows:

$$\tilde{F}_1[\theta] = -\rho L z_{0,t} = \rho L\Big(-F_1[\theta]A(\theta_t) + F_2[\theta]\Big),$$

with

$$
\begin{aligned}
F_1[\theta] &:= z_0 g_4(t,z,\theta)\mathcal{H}(M_s-\theta),\\
F_2[\theta] &:= -\frac{1}{\tau(\theta)}\Big(a_{eq}(\theta)-z_0\Big)\mathcal{H}(\theta-A_s)-z_0\ln(z_0)\mathcal{H}(A_s-\theta)\sum_{i=1}^{3}g_i(t,z,\theta),\\
A(\theta_t) &:= -\mathcal{H}(-\theta_t).
\end{aligned}
$$

Norming all physical constants to one, we finally obtain the following nonlinear parabolic problem:

$$
\begin{aligned}
\theta_t + F_1[\theta]A(\theta_t) - \Delta\theta &= F_2[\theta] + \alpha(\theta)u, \qquad \text{in } Q_T, &(2.2a)\\
\frac{\partial\theta}{\partial\nu} &= 0, \qquad \text{in } \Sigma_T, &(2.2b)\\
\theta(.,0) &= \theta_0, \qquad \text{in } \Omega. &(2.2c)
\end{aligned}
$$

2.3. Control problem

The aim of laser heat treatments is to increase the surface hardness of the workpiece. Therefore we have to control the volume fraction of martensite, i.e. we consider the following cost functional

$$
J(u) = \frac{\beta_1}{2}\int_\Omega \Big(z_4(x,T)-\tilde{m}(x)\Big)^2 dx + \frac{\beta_2}{2}\int_0^T\int_\Omega u^2 dxdt.
$$

In order to maintain the quality of the workpiece, it is important to avoid surface melting. To this end, we have to introduce the state constraint

$$
\theta(x,t) \le \theta_m, \qquad \text{a.e. in } Q_T, \tag{2.3}
$$

where θ_m is the melting temperature of the workpiece.

Then, the control problem for laser surface hardening takes the following form:

$$
\textbf{(CP)} \qquad \left\{
\begin{array}{l}
\text{Minimize } J(u)\\
\text{subject to (2.1a–e), (2.2a–c)}\\
\text{the constraint (2.3), and}\\
u \in U_{ad},
\end{array}\right.
$$

with the closed, convex set of admissible controls $U_{ad} \subset W^{1,4}(0,T;L^4(\Omega)) \cap L^9(Q_T)$, satisfying $u(.,0)=0$ a.e. in Ω for all $u \in U_{ad}$.

3. Optimality Conditions

3.1. Existence and stability results

In the sequel, we will extensively use Sobolev spaces $W_q^{2,1}(Q_T)$, $q \ge 1$ (cf. [8]), defined by

$$
W_q^{2,1}(Q_T) := W^{1,q}(0,T;L^q(\Omega)) \cap L^q(0,T;W^{2,q}(\Omega)).
$$

Note that in three space dimensions, for $q > 5/2$ we have the continuous embedding

$$W_q^{2,1}(Q_T) \subset C^\beta(\bar{Q_T}) \qquad \text{with } 0 \le \beta < 2 - 5/q. \tag{3.1}$$

We assume

(A1) $\mathcal{H} \in C^\infty(\mathbb{R})$, monotone regularization of the heaviside graph, satisfying $\mathcal{H}(0) = 0$ (cf. [11]),
(A2) $a_{eq} \in C^{1,1}(\mathbb{R})$, $a_{eq}(x) \in [0,1]$ for all $x \in \mathbb{R}$,
(A3) $\tau \in C^{1,1}(\mathbb{R})$, $m \le \tau(x) \le M$ for all $x \in \mathbb{R}$, and constants $0 < m < M$,
(A4) $g_i \in C^{1,1}(D)$, $i = 1, \ldots, 4$, $D = [0,T] \times [0,1]^5 \times \mathbb{R}$, moreover $0 \le g_i \le M$, for all $(t, z, \theta) \in D$ and a constant $M > 0$,
(A5) $\alpha \in C^{1,1}(\mathbb{R})$,
(A6) $u \in W^{1,4}(0,T; L^4(\Omega)) \cap L^9(Q_T)$, $u(.,0) = 0$, a.e. in Ω,
(A7) θ_0 constant,

then we have the following result:

Theorem 3.1. *Assume (A1)–(A7), then the following are valid:*

(1) (2.2a–c) has a unique solution satisfying $\theta \in W_9^{2,1}(Q)$, $\theta_t \in W_4^{2,1}(Q)$.

(2) Let θ_i, $i = 1, 2$, *be the solution to (2.2a–c) with respect to* $u_i \in W^{1,4}(0,T; L^4(\Omega)) \cap L^9(Q)$.
Then, there exists a constant $C > 0$ *such that*

$$\|\theta_1 - \theta_2\|_{W_2^{2,1}(Q)} + \|\theta_{1,t} - \theta_{2,t}\|_{W_2^{1,1}(Q)} \le C\|u_1 - u_2\|_{W^{1,4}(0,T;L^4(\Omega))}.$$

To prove the result, we need the following

Lemma 3.2. *Assume (A1)–(A4), and let* $\theta \in W^{1,p}(0,T; L^p(\Omega))$, $p \in [1, \infty]$, *then the following are valid:*
(1) (2.1a–e) has a unique solution $z \in [W^{1,\infty}(0,T; L^\infty(\Omega))]^5$.
(2) There are constants c_*, c^*, *independent of* θ, *such that*

$$0 < c_* < z_0(x,t) < c^* < 1, \qquad \text{a.e. in } Q_T.$$

(3) Let $\theta_i \in W^{1,p}(0,T; L^p(\Omega))$, *i=1,2, and* z^i *the corresponding solutions to (2.1a–e), then there exist constants* $L_i, \hat{L}_i > 0$, *such that*

$$\sup_{t \in (0,T)} \|F_i[\theta_1](.,t) - F_i[\theta_2](.,t)\|_{L^p(\Omega)}^q \le L_i\|\theta_1 - \theta_2\|_{W^{1,p}(0,T;L^p(\Omega))}^q,$$

for any $q \in [1, \infty)$, *and*

$$\int_0^t \left\| \frac{\partial F_i[\theta_1]}{\partial s}(.,s) - \frac{\partial F_i[\theta_2]}{\partial s}(.,s) \right\|_{L^2(\Omega)}^2 ds \le \hat{L}_i \|\theta_1 - \theta_2\|_{H^1(0,T;L^2(\Omega))}^2 ds,$$

The existence result in Theorem 3.1 can be proved using a straightforward contraction mapping argument in the space $H^1(0,T;L^2(\Omega))$. The regularity and stability results are obtained by formally differentiating (2.2) with respect to t and using Theorem IV.9.1 in [8].

3.2. Differentiability of the solution operator

In view of Lemma 3.2, the solution to (2.1a–e) defines an operator

$$\tilde{z} : \theta \mapsto \Big(\tilde{z}[\theta]\Big)(x,t) = z(x,t),$$

where z is the unique solution to (2.1a–e) for given temperature evolution θ. In the sequel, we will identify $\tilde{z}$ with z.

Using the implicit function theorem (cf. [14]) and the product rule, we obtain

Lemma 3.3. *Assume (A1)–(A4), then* $F_i : W^p(0,T;L^p(\Omega)) \longrightarrow W^{\frac{p}{2}}(0,T;L^{\frac{p}{2}}(\Omega))$, $i = 1,2$, *and* $p \in [2,\infty)$, *are Fréchet-differentiable, satisfying*

$$F_{i,\theta}[h] = g_{i1} \cdot h + g_{i2} \cdot z_\theta[h],$$

where $g_{i1} \in L^\infty(Q)$ *and* $g_{i2} \in [L^\infty(Q)]^5$ *for* $i = 1,2$.

Now, we can prove the differentiability of the solution operator.

Theorem 3.4. *Assume (A1)–(A7) and let* $S : U_{ad} \longrightarrow W_3^{2,1}(Q_T)$, $u \mapsto S(u) = \theta$ *be the solution operator to (2.2a–c).*

Then, S is differentiable, and for any h satisfying $u+h \in U_{ad}$, its directional derivative $\psi = S_u(u)[h]$ is the solution to the following linear problem:

$$\begin{aligned} (1 + F_1[\theta]A'(\theta_t))\psi_t - \Delta\psi \qquad & & & \\ +\Big(A(\theta_t)F_{1,\theta}[\theta] - F_{2,\theta}[\theta] - u\alpha'(\theta)\Big)[\psi] &=& \alpha(\theta)h, & \quad (3.2a) \\ \frac{\partial\psi}{\partial\nu} &=& 0, & \quad (3.2b) \\ \psi(.,0) &=& 0. & \quad (3.2c) \end{aligned}$$

For the proof one can use again Theorem IV.9.1 in [8], Lemma 3.3 and the stability estimates of Theorem 3.1.

3.3. Necessary conditions of optimality

We begin with some notations. Let

$$K = \{\eta \in C(\bar{Q_T})) \,\Big|\, \eta \le \theta_m\}.$$

For a control $u + h$, with $u, u+h \in U_{ad}$, we denote by $[\theta^h, z^h]$ the solution to (2.1a–e) and (2.2a–c), and by $[\psi, w]$ the solution to the linearized system (3.2a–c) and

$$\begin{aligned} w_t &= f_z w + f_\theta \psi, \qquad \text{in } Q_T, & \quad (3.3a) \\ w(.,0) &= 0, \qquad \text{in } \Omega, & \quad (3.3b) \end{aligned}$$

where f is the right–hand side of (2.1c–e). According to Theorem 3.4 and (3.1), the solution operator $S : U_{ad} \to C(\bar{Q_T})$ and the cost functional J are differentiable with

$$S'(u)[h] = \psi,$$

and

$$J'(u)[h] = \beta_1 \int_\Omega \Big(z_4(x,T) - \tilde{m}(x)\Big) w_4(x,T)\, dx + \beta_2 \int_0^T \int_\Omega uh\, dxdt.$$

We will now derive optimality conditions for the non–convex optimization problem under consideration using an abstract result for the existence of Lagrange multipliers by Casas (cf. [1], Theorem 5.2).

From the abstract result it follows that there exist

$$\lambda \geq 0 \text{ and a Borel measure } \mu \geq 0 \text{ such that}$$

$$\lambda + \|\mu\|_{\mathcal{M}(\bar{Q_T})} > 0,$$

$$\int (\eta - \theta) d\mu \leq 0 \text{ for all } \eta \in K,$$

$$\lambda J'(u)[v-u] + \int S'(u)[v-u] d\mu \geq 0 \text{ for all } v \in U_{ad}.$$

Here $u \in U_{ad}$ denotes an optimal control, $\theta = \theta(u), z_4 = z_4(u)$ and the second conditon means that the measure μ is supported on the set

$$\Xi = \{(x,t) \in \bar{Q} \,\Big|\, \theta(x,t) = \theta_m\} \tag{3.4}$$

which is closed since $\theta \in C(\bar{Q})$.

The multiplier $\lambda = 1$ provided there exists an admissible control $v \in U_{ad}$ such that the following Slater condition is satisfied,

$$\theta(u;x,t) + \psi(u;x,t)(v-u) < \theta_m$$

for all $(x,t) \in \bar{Q} = \bar{Q_T}$, where $\psi(u)(v-u) = S'(u)[v-u]$. In the sequel, we assume for simplicity that the Slater condition is satisfied.

To simplify the third optimality condition we introduce the adjoint state equations. To this end, for any given $\eta \in V_1, \xi = (\xi_0, \ldots, \xi_4) \in V_2$, we denote by

$$\Psi[(\eta,\xi)] = \beta_1 \int_\Omega \Big(z_4(x,T) - \tilde{m}(x)\Big) \xi_4(x,T)\, dx + \int \eta d\mu$$

the linear form which is defined on the space $V = V_1 \times V_2$, which will be specified below. We assume that the spaces V_1 and V_2 are selected in such a way that the linear form $\Psi[(\eta,\xi)]$ is continuous on the space V, i.e.

$$\left| \int \eta d\mu \right| \leq C_1 \|\eta\|_{V_1},$$

$$\left| \int_\Omega \Big(z_4(x,T) - \tilde{m}(x) \Big) \xi_4(x,T)\, dx \right| \le C_2 \|\xi\|_{V_2}.$$

The linearized state equations are rewritten in the following form

$$\psi \in V_1 \;:\; \mathcal{L}_{11}(\psi) = \alpha(\theta) h \text{ in } Q_T,$$
$$w \in V_2 \;:\; \mathcal{L}_{21}(\psi) + \mathcal{L}_{22}(w) = 0 \text{ in } Q_T,$$

where

$$\mathcal{L}_{11}(\psi) = (1 + F_1[\theta] A'(\theta_t))\psi_t - \Delta\psi + \Big(A(\theta_t) F_{1,\theta}[\theta] - F_{2,\theta}[\theta] - u\alpha'(\theta) \Big)[\psi],$$
$$\mathcal{L}_{21}(\psi) = -f_\theta \psi,$$
$$\mathcal{L}_{22}(w) = w_t - f_z w,$$

and we set

$$V_1 = \{\eta \in W_3^{2,1}(Q_T) | \eta(0) = 0 \text{ in } \Omega, \frac{\partial \eta}{\partial \nu} = 0 \text{ in } \Sigma_T\}.$$

For the choice made for the space V_1, the space V_2 can be defined e.g. in the following way. We have to satisfy two conditions by the definition. First, that the linearized state $w(h) \in V_2$, the second that the linear form

$$\xi \;\to\; \int_\Omega \Big(z_4(x,T) - \tilde{m}(x) \Big) \xi_4(x,T)\, dx$$

is continuous on the space V_2. Since the linearized state is regular, i.e. satisfies the equation

$$\mathcal{L}_{22}(w) = \mathcal{L}_{21}(\psi) \text{ in } Q \text{ with the initial condition } w(0) = 0 \text{ in } \Omega,$$

we can select

$$V_2 = \{\xi \in C(0,T;[L^2(\Omega)]^5) | \xi(0) = 0, \mathcal{L}_{22}(\xi) \in [L^2(Q)]^5\},$$

with the norm

$$\|\xi\|_{V_2} = \|\mathcal{L}_{22}(\xi)\|_{[L^2(Q)]^5},$$

therefore

$$\mathcal{L}_{21}(\eta) + \mathcal{L}_{22}(\xi) \in [L^2(Q)]^5 \text{ for all } \eta \in V_1, \xi \in V_2.$$

We introduce the linear mapping

$$\mathcal{L} \;:\; V \to W$$

of the following form

$$\mathcal{L}(\eta,\xi) = \begin{pmatrix} \mathcal{L}_{11}(\eta) \\ \mathcal{L}_{21}(\eta) + \mathcal{L}_{22}(\xi) \end{pmatrix} \tag{3.5}$$

where $V = V_1 \times V_2$, $W = W_1 \times W_2$ and $W_1 = L^3(Q)$, $W_2 = [L^2(Q)]^5$. Then an adjoint state $(p,r) \in W' = L^{\frac{3}{2}}(Q) \times [L^2(Q)]^5$ satisfies the following equation

$$\langle (p,r), \mathcal{L}(\eta,\xi) \rangle_{W' \times W} = \Psi[(\eta,\xi)] \text{ for all } (\eta,\xi) \in V_1 \times V_2.$$

The existence and uniqueness of the pair $(p,r) \in W'$ follows by an application of the representation theorem for linear and continuous functionals on the space V.

Using the adjoint state, it follows that

$$J'(u)[h] + \int S'(u)[h]d\mu = \Psi[(\psi[h], w[h])] + \beta_2 \int_0^T \int_\Omega uh\, dxdt$$

$$= \langle (p,r), \mathcal{L}(\psi[h], w[h]) \rangle_{W' \times W} + \beta_2 \int_0^T \int_\Omega uh\, dxdt$$

in view of (3.2a)

$$= \int_0^T \int_\Omega \alpha(\theta) hp dxdt + \beta_2 \int_0^T \int_\Omega uh\, dxdt.$$

The adjoint state $(p,r) \in L^{\frac{3}{2}}(Q_T) \times [L^2(Q_T)]^5$ is given by a solution to the following system

$$\int_0^T \int_\Omega \Big[(1 + F_1[\theta] A'(\theta_t))\eta_t - \Delta\eta + \Big(A(\theta_t) F_{1,\theta}[\theta] - F_{2,\theta}[\theta] - u\alpha'(\theta)\Big)\eta\Big] pdxdt$$
$$- \int_0^T \int_\Omega f_\theta \eta r dxdt = \int \eta d\mu \qquad (3.6a)$$

$$\int_0^T \int_\Omega [\xi_t - f_z \xi] r dxdt + \beta_1 \int_\Omega \Big(z_4(x,T) - \tilde{m}(x)\Big) \xi_4(x,T)\, dx = 0 \qquad (3.6b)$$

for all $\eta \in V_1$, $\xi \in V_2$.

Since the existence of an optimal control is a consequence of standard compactness arguments, which we omit here, the following necessary optimality conditions hold for **(CP)**:

Theorem 3.5. *There exists an optimal control $u \in U_{ad}$ which minimizes the cost functional $J(u)$ over the set of admissible controls and subject to the state constraint $\theta \in K$.*

If the Slater condition is satisfied, then there exists a Borel measure $\mu \geq 0$ supported on the set Ξ and the adjoint state $(p,r) \in L^{\frac{3}{2}}(Q_T) \times [L^2(Q_T)]^5$ such that the state equations (2.2a–c), the adjoint state equations (3.6a-b) and the following optimality condition is satisfied:

$$\int_0^T \int_\Omega \alpha(\theta) p(v-u) dxdt + \beta_2 \int_0^T \int_\Omega u(v-u)\, dxdt \geq 0$$

for all $v \in U_{ad}$.

4. Numerical Simulations

To demonstrate the validity of the model, we present some numerical simulations of the state equations (2.1a–e), (2.2a–c). For details concerning the algorithm and the physical data as well as for further results we refer to [2].
Let the part of the workpiece surface to be hardened lie in the plane $z = 0$. Then the laser radiation penetrates into the workpiece according to the radiation transfer equation

$$G = \alpha G_f e^{\kappa z}, \qquad z \leq 0.$$

Here, G is the radiation intensity of the laser beam, G_f the radiation intensity in the focal plane, κ the absorption coefficient and α the absorptivity of the surface, depending on the angle of incidence, the surface constitution (smoothness, cleanliness) and on the temperature. In applications, the laser beam moves along the workpiece surface according to a curve $t \longrightarrow r(t) \in \mathbb{R}^2$, $t \in [0, T]$, hence we have

$$G_f(x, y, t) = G_0 e^{-\frac{(x - r_1(t))^2 + (y - r_2(t))^2}{2R^2}},$$

where R is the radius of the focusing spot and G_0 its intensity in the spot center. The heat source then takes the form

$$\tilde{F}_2 = \kappa G.$$

We simulate the hardening along a strip around the y–axis on the upper face ($z = 0$) of the cube $\bar{\Omega} = [-2.5, 2.5] \times [0, 10.0] \times [-1.0, 0]$. Figure 2 shows the time evolution at the point $x = (0.0, 5.0, -0.01) \in \Omega$. Owing to the oscillations of the laser beam, the point is heated by steps. Austenite is formed, and during cooling this austenite is transformed to martensite and a fairly small amount of bainite.

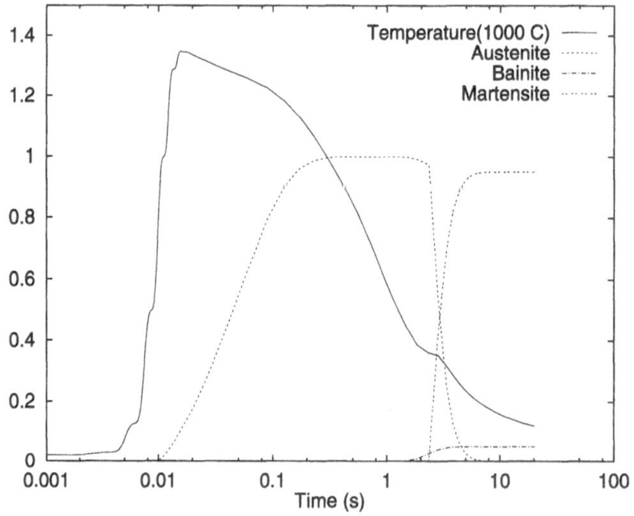

FIGURE 2. Time evolution of temperature, austenite, bainite and martensite fraction for $x = (0.0, 5.0, -0.01) \in \Omega$.

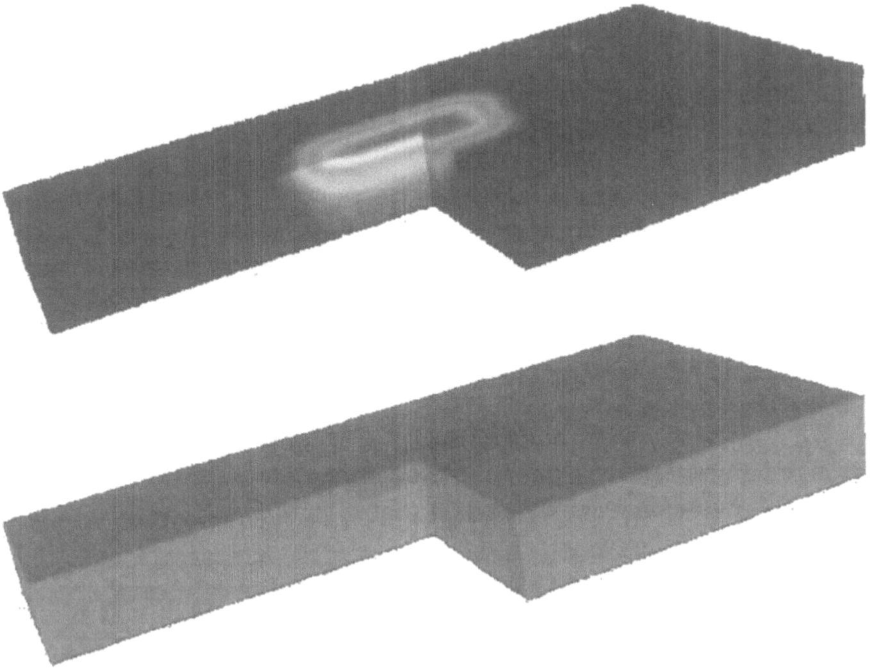

FIGURE 3. Temperature distribution inside Ω (above) and the resulting hardening profile (below).

In the course of martensite growth, the cooling process is slowed down by the release of latent heat. Finally, Fig. 3 depicts the temperature distribution inside the workpiece during the heating process and the resulting hardening profile.

References

[1] Casas, E, *Boundary control of semilinear elliptic equations with pointwise state constraints*, SIAM J. Control and Optimization **31** (1993) 993–1006.

[2] Fuhrmann, J., Hömberg, D., *Numerical simulation of surface heat treatments*, Num. Meth. Heat & Fluid Flow **9** (1999), 705–724.

[3] Hömberg, D., *A mathematical model for the phase transitions in eutectoid carbon steel*, IMA J. Appl. Math., **54** (1995), 31–57.

[4] Hömberg, D., *Irreversible phase transitions in steel*, Math. Methods Appl. Sci., **20**, (1997), 59–77.

[5] Hömberg, D., *A numerical simulation of the Jominy end–quench test*, Acta Mater., **44**, (1996), 4375–4385.

[6] Hömberg, D., Sokolowski, J., *Optimal control of laser hardening*, Adv. Math. Sci. Appl., **8** (1998), 911–928.

[7] Hömberg, D., Volkwein, S., *Suboptimal control of laser surface hardening using proper orthogonal decomposition*, WIAS Preprint 639(2001).

[8] Ladyenskaja, O.A., Solonnikov, V.A., Ural'ceva, N.N., *Linear and quasilinear equations of parabolic type*, Amer.Math.Soc.Transl. , Vol. 23, Providence, 1968.

[9] Leblond, J.–B., Devaux, J.,*A new kinetic model for anisothermal metallurgical transformations in steels including effect of austenite grain size*, Acta Metall. **32** (1984) 137–146.

[10] Mazhukin, V.I., Samarskii, A.A., *Mathematical modeling in the technology of laser treatments of materials*, Surv. Math. Ind. **4** (1994) 85–149.

[11] Tiba, D., Neittaanmaki, P., *Optimal control of nonlinear parabolic systems*, Monographs Textbooks Pure Appl. Math. 179, M. Dekker, New York, 1994.

[12] Verdi, C., Visintin, A., *A mathematical model of the austenite-pearlite transformation in plain steel based on the Scheil's additivity rule*, Acta Metall., **35** (1987), 2711–2717.

[13] Visintin, A., *Mathematical models of solid–solid phase transitions in steel*, IMA J. Appl. Math., **39** (1987), 143–157.

[14] Zeidler, E., *Nonlinear Functional Analysis and its Applications, Vol. I*, Springer–Verlag, New York, 1987.

[1] Weierstrass Institute for Applied Analysis and Stochastics
Mohrenstraße 39
D–10117 Berlin, Germany

[2] Institut Elie Cartan, Laboratoire de Mathématiques
Université Henri Poincaré Nancy I, BP 239
54506 Vandoeuvre lès Nancy Cedex, France and
Systems Research Institute of the Polish Academy of Sciences
ul. Newelska 6
01-447 Warszawa, Poland

E-mail address: hoemberg@wias-berlin.de

International Series of Numerical Mathematics, Vol. 139, 83–93

Numerical Analysis in Optimal Control

William W. Hager

Abstract. In this paper we explain and exemplify how one goes about analyzing the convergence of algorithms and discrete approximations in optimal control.

1. Introduction

The techniques used to analyse the convergence of penalty methods, multiplier methods, sequential quadratic programming methods, and discrete approximations in optimal control are closely related. In each case, we wish to approximate a local minimizer $\mathbf{w}^*$ of an optimization problem, where the approximation is the solution to a problem of the form:

$$\text{Find } \mathbf{w} \in \mathcal{X} \text{ such that } \mathcal{T}(\mathbf{w}) \in \mathcal{F}(\mathbf{w}). \tag{1}$$

Here $\mathcal{X}$ is a Banach space, $\mathcal{T} : \mathcal{X} \to \mathcal{Y}$, $\mathcal{Y}$ is a linear normed space, and $\mathcal{F} : \mathcal{X} \to 2^{\mathcal{Y}}$. Think of (1) as the first-order optimality system associated with the numerical approximating problem.

Typically, the solution $\mathbf{w}^*$ of the original optimization problem is not a solution of (1), rather it is an approximate solution. Let $\boldsymbol{\delta}$ be chosen as small as possible so that

$$\mathcal{T}(\mathbf{w}^*) + \boldsymbol{\delta} \in \mathcal{F}(\mathbf{w}^*). \tag{2}$$

Given that $\mathbf{w}^*$ is almost a solution of (1), we try to show that (1) has a solution $\mathbf{w}$ close to $\mathbf{w}^*$ which satisfies an estimate of the form $\|\mathbf{w} - \mathbf{w}^*\| \le c\|\boldsymbol{\delta}\|$, where c is a constant independent of $\boldsymbol{\delta}$ for $\boldsymbol{\delta}$ sufficiently small, and $\|\cdot\|$ denotes the norm in the appropriate space.

An existence result for (1), together with an estimate for the distance to $\mathbf{w}^*$, is gotten from a generalization of the implicit function theorem. In this generalization, the usual surjectivity property for the derivative of $\mathcal{T}$ at $\mathbf{w}^*$ is replaced by a Lipschitz property for an associated linearized problem of the form:

$$\text{Find } \mathbf{w} \in \mathcal{X} \text{ such that } \mathcal{L}(\mathbf{w}) + \boldsymbol{\pi} \in \mathcal{F}(\mathbf{w}). \tag{3}$$

Here $\mathcal{L}$ is a linear operator and $\boldsymbol{\pi} \in \mathcal{Y}$ stands for a "parameter."

This work, supported by the National Science Foundation, was presented at the Conference on Optimal Control of Complex Structures, June 4–10, 2000, Oberwolfach, Germany, organized by K.-H. Hoffmann, I. Lasiecka, G. Leugering, J. Sprekels and F. Troeltzsch.

In the typical first-order Taylor expansion, we would approximate $\mathcal{T}$ by $\mathcal{T}(\mathbf{w}^*) + \mathcal{T}'(\mathbf{w}^*)(\mathbf{w} - \mathbf{w}^*)$, in which case $\mathcal{L}$ would be $\mathcal{T}'(\mathbf{w}^*)$. If (3) has a unique solution for $\boldsymbol{\pi}$ near $\boldsymbol{\pi}^* = \mathcal{T}(\mathbf{w}^*) - \mathcal{T}'(\mathbf{w}^*)(\mathbf{w}^*)$ satisfying

$$\|\mathbf{w}_1 - \mathbf{w}_2\| \leq \lambda \|\boldsymbol{\pi}_1 - \boldsymbol{\pi}_2\|,$$

where $\mathbf{w} = \mathbf{w}_i$ is the solution of (3) corresponding to $\boldsymbol{\pi} = \boldsymbol{\pi}_i$, then under suitable assumptions, (1) has a solution $\mathbf{w}$, and the distance from $\mathbf{w}$ to $\mathbf{w}^*$ is very nearly bounded by $\lambda\|\boldsymbol{\delta}\|$. The precise estimate, given shortly, involves an additional factor $1/(1 - \lambda\epsilon)$ where ϵ is typically small. The bound $\lambda\|\boldsymbol{\delta}\|$ for the distance from $\mathbf{w}$ to $\mathbf{w}^*$ yields an error estimate for the numerical algorithm that (1) represents.

2. Abstract Estimate

The following result, given in a slightly more general form in [2, Thm. 3.1], is a version of the implicit function theorem for inclusions alluded to in the previous section. See [3, 4, 5, 10] for other related results.

Theorem 2.1. *Let $\mathcal{X}$ be a Banach space and let $\mathcal{Y}$ be a linear normed space with the norms in both spaces denoted $\|\cdot\|$. Let $\mathcal{F} : \mathcal{X} \mapsto 2^{\mathcal{Y}}$, let $\mathcal{L} : \mathcal{X} \mapsto \mathcal{Y}$ be a bounded linear operator, and let $\mathcal{T} : \mathcal{X} \mapsto \mathcal{Y}$ with $\mathcal{T}$ continuously Frechét differentiable in $B_r(\mathbf{w}^*)$ for some $\mathbf{w}^* \in \mathcal{X}$ and $r > 0$, where $B_r(\mathbf{w}^*)$ is the ball with center $\mathbf{w}^*$ and radius r. Suppose that the following conditions hold for some $\boldsymbol{\delta} \in \mathcal{Y}$ and scalars ϵ, λ, and $\sigma > 0$:*

(P1) $\mathcal{T}(\mathbf{w}^*) + \boldsymbol{\delta} \in \mathcal{F}(\mathbf{w}^*)$.
(P2) $\|\nabla\mathcal{T}(\mathbf{w}) - \mathcal{L}\| \leq \epsilon$ *for all* $\mathbf{w} \in B_r(\mathbf{w}^*)$.
(P3) *The map* $(\mathcal{F} - \mathcal{L})^{-1}$ *is single-valued and Lipschitz continuous in* $B_\sigma(\boldsymbol{\pi}^*)$, $\boldsymbol{\pi}^* = (\mathcal{T} - \mathcal{L})(\mathbf{w}^*)$, *with Lipschitz constant* λ.

If $\epsilon\lambda < 1$, $\epsilon r \leq \sigma$, $\|\boldsymbol{\delta}\| \leq \sigma$, and

$$\|\boldsymbol{\delta}\| \leq (1 - \lambda\epsilon)r/\lambda,$$

then there exists a unique $\mathbf{w} \in B_r(\mathbf{w}^*)$ such that $\mathcal{T}(\mathbf{w}) \in \mathcal{F}(\mathbf{w})$. Moreover, we have the estimate

$$\|\mathbf{w} - \mathbf{w}^*\| \leq \frac{\lambda}{1 - \lambda\epsilon}\|\boldsymbol{\delta}\|. \tag{4}$$

Proof. Let us define $\Phi(\mathbf{w}) = (\mathcal{F} - \mathcal{L})^{-1}(\mathcal{T}(\mathbf{w}) - \mathcal{L}(\mathbf{w}))$. By a Taylor expansion around $\mathbf{w}^*$, with integral remainder term, we have

$$\mathcal{T}(\mathbf{w}) - \mathcal{L}(\mathbf{w}) = \mathcal{T}(\mathbf{w}^*) - \mathcal{L}(\mathbf{w}^*) + \int_0^1 (\nabla\mathcal{T}(s\mathbf{w} + (1 - s)\mathbf{w}^*) - \mathcal{L})\, ds\, (\mathbf{w} - \mathbf{w}^*).$$

Hence, (P2) implies that $\|\mathcal{T}(\mathbf{w}) - \mathcal{L}(\mathbf{w}) - \boldsymbol{\pi}^*\| \leq \epsilon r$ for all $\mathbf{w} \in B_r(\mathbf{w}^*)$. By (P3), it follows that for all $\mathbf{w}_1, \mathbf{w}_2 \in B_r(\mathbf{w}^*)$,

$$\begin{aligned}
\|\Phi(\mathbf{w}_1) - \Phi(\mathbf{w}_2)\| &= \|(\mathcal{F} - \mathcal{L})^{-1}(\mathcal{T} - \mathcal{L})(\mathbf{w}_1) - (\mathcal{F} - \mathcal{L})^{-1}(\mathcal{T} - \mathcal{L})(\mathbf{w}_2)\| \\
&\leq \lambda\|(\mathcal{T} - \mathcal{L})(\mathbf{w}_1) - (\mathcal{T} - \mathcal{L})(\mathbf{w}_2)\| \\
&\leq \lambda\epsilon\|\mathbf{w}_1 - \mathbf{w}_2\|.
\end{aligned}$$

Since $\lambda\epsilon < 1$, Φ is a contraction on $B_r(\mathbf{w}^*)$. Since $\|\boldsymbol{\delta}\| \le \sigma$, we conclude that $(\mathcal{T} - \mathcal{L})(\mathbf{w}^*) + \boldsymbol{\delta} \in B_\sigma(\boldsymbol{\pi}^*)$. By (P3), $(\mathcal{F} - \mathcal{L})^{-1}$ is single-valued on $B_\sigma(\boldsymbol{\pi}^*)$, and by (P1) we have

$$\mathbf{w}^* = (\mathcal{F} - \mathcal{L})^{-1}[(\mathcal{T} - \mathcal{L})(\mathbf{w}^*) + \boldsymbol{\delta}].$$

It follows from (P2) and (P3) that

$$\begin{aligned} \|\Phi(\mathbf{w}) - \mathbf{w}^*\| &= \|(\mathcal{F} - \mathcal{L})^{-1}[(\mathcal{T} - \mathcal{L})(\mathbf{w})] - (\mathcal{F} - \mathcal{L})^{-1}[(\mathcal{T} - \mathcal{L})(\mathbf{w}^*) + \boldsymbol{\delta}]\| \\ &\le \lambda\|(\mathcal{T} - \mathcal{L})(\mathbf{w}) - (\mathcal{T} - \mathcal{L})(\mathbf{w}^*) - \boldsymbol{\delta}\| \\ &\le \lambda(\epsilon\|\mathbf{w} - \mathbf{w}^*\| + \|\boldsymbol{\delta}\|) \qquad (5) \\ &\le \lambda(\epsilon r + \|\boldsymbol{\delta}\|) \end{aligned}$$

for all $\mathbf{w} \in B_r(\mathbf{w}^*)$. The condition $\lambda\|\boldsymbol{\delta}\|/(1-\epsilon\lambda) \le r$ implies that $\lambda(\epsilon r + \|\boldsymbol{\delta}\|) \le r$, and hence, $\|\Phi(\mathbf{w}) - \mathbf{w}^*\| \le r$. Since Φ maps $B_r(\mathbf{w}^*)$ into itself and Φ is a contraction on $B_r(\mathbf{w}^*)$, the contraction mapping principle yields the existence of a unique fixed point $\mathbf{w} \in B_r(\mathbf{w}^*)$. Since $\|\Phi(\mathbf{w}) - \mathbf{w}^*\| = \|\mathbf{w} - \mathbf{w}^*\|$ for this fixed point, (5) gives (4). □

Theorem 2.1 says roughly

Consistency + Stability ⇒ Convergence,

where consistency is assumption (P1) and the bounds on the norm of $\boldsymbol{\delta}$, stability is assumption (P3) and the bound on the Lipschitz constant λ for the linearization, and convergence is (4).

3. Penalty Methods

We first illustrate the analysis using the penalty approximation to the following control problem:

$$\begin{aligned} &\text{minimize } C(\mathbf{x}, \mathbf{u}) = \int_0^1 \varphi(\mathbf{x}(t), \mathbf{u}(t))dt \qquad (6) \\ &\text{subject to } \quad \dot{\mathbf{x}}(t) = \mathbf{f}(\mathbf{x}(t), \mathbf{u}(t)), \quad \mathbf{u}(t) \in U \quad \text{a. e. } t \in [0,1], \\ &\qquad \mathbf{x}(0) = \mathbf{a}, \quad \mathbf{x} \in W^{1,\infty}, \quad \mathbf{u} \in L^\infty, \end{aligned}$$

where the state $\mathbf{x}(t) \in \mathbf{R}^n$, $\dot{\mathbf{x}}$ stands for $\frac{d}{dt}\mathbf{x}$, the control $\mathbf{u}(t) \in \mathbf{R}^m$, $\mathbf{f} : \mathbf{R}^n \times \mathbf{R}^m \mapsto \mathbf{R}^n$, $\varphi : \mathbf{R}^n \times \mathbf{R}^m \mapsto \mathbf{R}$, and $U \subset \mathbf{R}^m$ is closed and convex. Of course, L^p denotes the usual Lebesgue space of measurable functions with p-th power integrable, and $W^{m,p}$ is the Sobolev space consisting of vector-valued functions whose j-th derivative lies in L^p for all $0 \le j \le m$. Assume that (6) has a local minimizer $(\mathbf{x}^*, \mathbf{u}^*)$ and that φ and $\mathbf{f}$ are twice continuously differentiable.

Enforcing the differential equation constraint with a quadratic penalty term involving a "large" penalty parameter τ, we are led to the following approximating problem:

$$\text{minimize } C(\mathbf{x}, \mathbf{u}) + \frac{\tau}{2}\langle \mathbf{f}(\mathbf{x}, \mathbf{u}) - \dot{\mathbf{x}}, \mathbf{f}(\mathbf{x}, \mathbf{u}) - \dot{\mathbf{x}} \rangle \tag{7}$$

$$\text{subject to} \quad \mathbf{u}(t) \in U \quad \text{a .e. } t \in [0, 1],$$

$$\mathbf{x}(0) = \mathbf{a}, \quad \mathbf{x} \in W^{1,\infty}, \quad \mathbf{u} \in L^\infty.$$

Instead of studying (7) directly, we examine the first-order optimality system associated with (7):

$$\dot{\boldsymbol{\psi}} + \nabla_x H(\mathbf{x}, \mathbf{u}, \boldsymbol{\psi}) = \mathbf{0}, \quad \boldsymbol{\psi}(1) = \mathbf{0}, \tag{8}$$

$$\dot{\mathbf{x}} - \mathbf{f}(\mathbf{x}, \mathbf{u}) + \boldsymbol{\psi}/\tau = \mathbf{0}, \quad \mathbf{x}(0) = \mathbf{a}, \tag{9}$$

$$\nabla_u (H(\mathbf{x}(t), \mathbf{u}(t), \boldsymbol{\psi}(t))(\mathbf{v} - \mathbf{u}(t)) \geq 0 \quad \text{for all } \mathbf{v} \in U. \tag{10}$$

Here H is the Hamiltonian defined by $H(\mathbf{x}, \mathbf{u}, \boldsymbol{\psi}) = \varphi(\mathbf{x}, \mathbf{u}) + \boldsymbol{\psi}^\mathsf{T}\mathbf{f}(\mathbf{x}, \mathbf{u})$. The first two equations combine to give the usual Euler equation describing a minimizer of (7) over $\mathbf{x}$, assuming $\mathbf{u}$ is fixed. The last inequality describes a minimizer over $\mathbf{u}$, assuming $\mathbf{x}$ is fixed. Letting $\mathbf{w}$ denote the triple $(\mathbf{x}, \mathbf{u}, \boldsymbol{\psi})$, the system (8)–(10) of equalities and inequalities corresponds to the abstract inclusion (1). Note that the inequality (10) ie equivalent to the inclusion $\nabla_u H(\mathbf{x}, \mathbf{u}, \boldsymbol{\psi}) \in \mathcal{N}(\mathbf{u})$ where

$$\mathcal{N}(\mathbf{u}) = \{\chi \in L^\infty : \langle \chi, \mathbf{v} - \mathbf{u} \rangle \geq 0 \text{ for all } \mathbf{v} \in L^\infty, \mathbf{v}(t) \in U \text{ a. e. } t \in [0, 1]\}.$$

If $\boldsymbol{\psi}^*$ is the costate variable given by the Pontryagin minimum principle, then $\mathbf{w}^* = (\mathbf{x}^*, \mathbf{u}^*, \boldsymbol{\psi}^*)$ does not satisfy (9) due to the $\boldsymbol{\psi}/\tau$ term. That is, $\mathbf{x}^*$ and $\mathbf{u}^*$ satisfy the state equation $\dot{\mathbf{x}} = \mathbf{f}(\mathbf{x}, \mathbf{u})$, not (9). Hence, the term $\boldsymbol{\psi}^*/\tau$ would be put in the $\boldsymbol{\delta}$ of (2): $\boldsymbol{\delta} = (\mathbf{0}, -\boldsymbol{\psi}^*, \mathbf{0})/\tau$. With this choice for $\boldsymbol{\delta}$, we have $\mathcal{T}(\mathbf{w}^*) + \boldsymbol{\delta} \in \mathcal{F}(\mathbf{w}^*)$.

To apply Theorem 2.1, we need to analyse a linearization of (8)–(10). The linearization is gotten by neglecting the penalty term (this term is small when the penalty is large), and differentiating the other terms. More precisely, the linearized problem is the following:

$$\dot{\boldsymbol{\psi}} + \mathbf{A}^\mathsf{T}\boldsymbol{\psi} + \mathbf{Q}\mathbf{x} + \mathbf{S}\mathbf{u} + \boldsymbol{\alpha} = \mathbf{0}, \quad \boldsymbol{\psi}(1) = \mathbf{0},$$

$$L(\mathbf{x}, \mathbf{u}) + \boldsymbol{\beta} = \mathbf{0}, \quad \mathbf{x}(0) = \mathbf{a},$$

$$\mathbf{B}^\mathsf{T}\boldsymbol{\psi} + \mathbf{S}^\mathsf{T}\mathbf{x} + \mathbf{R}\mathbf{u} + \boldsymbol{\gamma} \in \mathcal{N}(\mathbf{u}),$$

where $\boldsymbol{\pi} = (\boldsymbol{\alpha}, \boldsymbol{\beta}, \boldsymbol{\gamma})$ is the parameter, $L(\mathbf{x}, \mathbf{u}) = \dot{\mathbf{x}} - \mathbf{A}\mathbf{x} - \mathbf{B}\mathbf{u}$ is the linearized system dynamics, and

$$\mathbf{A}(t) = \nabla_x \mathbf{f}(\mathbf{x}^*(t), \mathbf{u}^*(t)), \quad \mathbf{B}(t) = \nabla_u \mathbf{f}(\mathbf{x}^*(t), \mathbf{u}^*(t)),$$

$$\mathbf{Q}(t) = \nabla_{xx} H(\mathbf{w}^*(t)), \quad \mathbf{R}(t) = \nabla_{uu} H(\mathbf{w}^*(t)), \quad \mathbf{S}(t) = \nabla_{xu} H(\mathbf{w}^*(t)).$$

To apply Theorem 2.1, we need to verify (P3), which amounts to proving that the linearized problem has a solution depending Lipschitz continuously on the parameter. A natural space for $\mathbf{w} = (\mathbf{x}, \mathbf{u}, \boldsymbol{\psi})$ is $\mathcal{X} = W_0^{1,\infty} \times L^\infty \times W_1^{1,\infty}$, where

$$W_0^{1,\infty} = \{\mathbf{x} \in W^{1,\infty} : \mathbf{x}(0) = \mathbf{a}\} \quad \text{and} \quad W_1^{1,\infty} = \{\boldsymbol{\psi} \in W^{1,\infty} : \boldsymbol{\psi}(1) = \mathbf{0}\}.$$

Hence, a natural space for the image of $\mathcal{T}$ is $\mathcal{Y} = L^\infty$, and (P3) amounts to a regularity property for the linearized problem: For each $\boldsymbol{\pi}_1$ and $\boldsymbol{\pi}_2 \in L^\infty$, there exist associated solutions, $(\mathbf{x}_1, \mathbf{u}_1, \boldsymbol{\psi}_1)$ and $(\mathbf{x}_2, \mathbf{u}_2, \boldsymbol{\psi}_2)$ respectively, of the linearized problem such that

$$\|\mathbf{x}_1 - \mathbf{x}_2\|_{W^{1,\infty}} + \|\mathbf{u}_1 - \mathbf{u}_2\|_{L^\infty} + \|\boldsymbol{\psi}_1 - \boldsymbol{\psi}_2\|_{W^{1,\infty}} \le \lambda \|\boldsymbol{\pi}_1 - \boldsymbol{\pi}_2\|_{L^\infty}.$$

It turns out that this Lipschitz property holds when the matrices in the linearized problem possess a coercivity property (that also arises in second-order sufficient optimality conditions): There exists a constant $\alpha > 0$ such that

$$\mathcal{B}(\mathbf{x}, \mathbf{u}) = \langle \mathbf{Qx}, \mathbf{x}\rangle + 2\langle \mathbf{Su}, \mathbf{x}\rangle + \langle \mathbf{Ru}, \mathbf{u}\rangle \ \ge \ \alpha \|\mathbf{u}\|_{L^2}^2 \quad \text{for all } (\mathbf{x}, \mathbf{u}) \in \mathcal{M},$$

where

$$\mathcal{M} = \{(\mathbf{x}, \mathbf{u}) : \mathbf{x} \in W^{1,2}, \mathbf{u} \in L^2,\ \dot{\mathbf{x}} = \mathbf{Ax} + \mathbf{Bu},$$
$$\mathbf{x}(0) = \mathbf{0},\ \mathbf{u}(t) \in U - U \text{ a. e. } t \in [0, 1]\}.$$

Notice that the coercivity condition is formulated in L^2 spaces while the original control problem is formulated in L^∞ spaces. In the literature, this difference in spaces is called the 2-norm discrepancy. We need to formulate the original problem in L^∞, to ensure continuity of the functions defining the problem, but the coercivity condition should be formulated in L^2, ensuring Lipschitz stability for the linearized problem. For a proof of Lipschitz stability for the linearized control problem, see [10].

Next, we verify the conditions of Theorem 2.1. First, choose ϵ small enough that $\epsilon\lambda < 1$; then choose r small enough and τ large enough that (P2) holds; finally, choose τ large enough that

$$\|\boldsymbol{\delta}\| = \|\boldsymbol{\psi}^*\|_{L^\infty}/\tau \le (1 - \epsilon\lambda)r/\lambda.$$

Since σ is $+\infty$, all the assumption of Theorem 2.1 hold. Hence, (8)–(10) has a solution $(\mathbf{x}_\tau, \mathbf{u}_\tau, \boldsymbol{\psi}_\tau)$ and

$$\|\mathbf{x}_\tau - \mathbf{x}^*\|_{W^{1,\infty}} + \|\mathbf{u}_\tau - \mathbf{u}^*\|_{L^\infty} + \|\boldsymbol{\psi}_\tau - \boldsymbol{\psi}^*\|_{W^{1,\infty}} \le \frac{\lambda}{\tau(1 - \epsilon\lambda)} \|\boldsymbol{\psi}^*\|_{L^\infty}.$$

As τ tends to infinity, the solution of the penalized problem approaches the original local minimizer.

In the final phase of the analysis, it should be shown that this solution of (8)–(10) is a local minimizer of (7). This is done by expanding the cost function in a Taylor series. The first-order terms either vanish by (8) or are nonnegative by (10), and the second-order term is positive when the coercivity condition holds (see [10, Thm. 3] for the details). Penalty methods applied to terminal constraints are studied in [11].

4. SQP Methods

If $(\mathbf{x}_k, \mathbf{u}_k, \boldsymbol{\psi}_k)$ is an approximation to a solution of the control problem (6), then the next SQP iterate $(\mathbf{x}_{k+1}, \mathbf{u}_{k+1}, \boldsymbol{\psi}_{k+1})$ is a solution, and the associated costate variable, for the linear-quadratic problem

$$\text{minimize } \langle \nabla_x \varphi_k, \mathbf{x} - \mathbf{x}_k \rangle + \langle \nabla_u \varphi_k, \mathbf{u} - \mathbf{u}_k \rangle + \frac{1}{2} \mathcal{B}_k(\mathbf{x} - \mathbf{x}_k, \mathbf{u} - \mathbf{u}_k) \quad (11)$$

$$\text{subject to} \quad L_k(\mathbf{x} - \mathbf{x}_k, \mathbf{u} - \mathbf{u}_k) = \mathbf{f}_k - \dot{\mathbf{x}}_k, \quad \mathbf{u}(t) \in U \quad \text{a. e. } t \in [0,1],$$
$$\mathbf{x}(0) = \mathbf{a}, \quad \mathbf{x} \in W^{1,\infty}, \quad \mathbf{u} \in L^\infty,$$

where the k subscript means that the associated expression is evaluated at $\mathbf{x}_k$, $\mathbf{u}_k$, and $\boldsymbol{\psi}_k$. A bit more smoothness is needed in this section; for example, φ and $\mathbf{f} \in C^3$. As with the penalty method, we apply Theorem 2.1 to the first-order optimality conditions for (11). These conditions are the following:

$$\begin{aligned} \dot{\boldsymbol{\psi}} + \mathbf{A}_k^\mathsf{T} \boldsymbol{\psi} + \nabla_x \varphi_k + \mathbf{Q}_k(\mathbf{x} - \mathbf{x}_k) + \mathbf{S}_k(\mathbf{u} - \mathbf{u}_k) &= \mathbf{0}, \qquad \boldsymbol{\psi}(1) = \mathbf{0}, \\ L_k(\mathbf{x} - \mathbf{x}_k, \mathbf{u} - \mathbf{u}_k) &= \mathbf{f}_k - \dot{\mathbf{x}}_k, \quad \mathbf{x}(0) = \mathbf{a}, \\ \mathbf{B}_k^\mathsf{T} \boldsymbol{\psi} + \nabla_u \varphi_k + \mathbf{S}_k^\mathsf{T}(\mathbf{x} - \mathbf{x}_k) + \mathbf{R}_k(\mathbf{u} - \mathbf{u}_k) &\in \mathcal{N}(\mathbf{u}). \end{aligned}$$

The matrices here are the same as those of Section 3 except that they are evaluated at $\mathbf{w}_k = (\mathbf{x}_k, \mathbf{u}_k, \boldsymbol{\psi}_k)$ instead of at $\mathbf{w}^* = (\mathbf{x}^*, \mathbf{u}^*, \boldsymbol{\psi}^*)$.

The linearized problem is exactly the same as that used in the previous section; hence, when the coercivity property is satisfied, (P3) of Theorem 2.1 holds with $\sigma = +\infty$. By taking $\mathbf{w}_k$ close to $\mathbf{w}^*$, we can make $\|\nabla \mathcal{T}(\mathbf{w}) - \mathcal{L}\|$ in (P2) as small as we like. To evaluate the $\boldsymbol{\delta}$ of (2), we insert $\mathbf{w} = \mathbf{w}^* = (\mathbf{x}^*, \mathbf{u}^*, \boldsymbol{\psi}^*)$ in the left side of the first-order optimality system; by inspection, we see that (P1) holds for the following choice:

$$\boldsymbol{\delta} = -\begin{pmatrix} \dot{\boldsymbol{\psi}}^* + \nabla_x H(\mathbf{x}_k, \mathbf{u}_k, \boldsymbol{\psi}^*) + \mathbf{Q}_k(\mathbf{x}^* - \mathbf{x}_k) + \mathbf{S}_k(\mathbf{u}^* - \mathbf{u}_k) \\ \dot{\mathbf{x}}^* - \mathbf{f}_k - \mathbf{A}_k(\mathbf{x}^* - \mathbf{x}_k) - \mathbf{B}_k(\mathbf{u}^* - \mathbf{u}_k) \\ \nabla_u H(\mathbf{x}_k, \mathbf{u}_k, \boldsymbol{\psi}^*) + \mathbf{S}_k^\mathsf{T}(\mathbf{x}^* - \mathbf{x}_k) + \mathbf{R}_k(\mathbf{u}^* - \mathbf{u}_k) - \nabla_u H(\mathbf{x}^*, \mathbf{u}^*, \boldsymbol{\psi}^*) \end{pmatrix}$$

By Theorem 2.1, the first-order optimality system has a solution

$$\mathbf{w}_{k+1} = (\mathbf{x}_{k+1}, \mathbf{u}_{k+1}, \boldsymbol{\psi}_{k+1}),$$

and the distance from $\mathbf{w}_{k+1}$ to $\mathbf{w}^*$ is bounded by a constant times $\|\boldsymbol{\delta}\|$. Expanding the terms of $\boldsymbol{\delta}$ in a Taylor series around $\mathbf{x}^*$, $\mathbf{u}^*$, and $\boldsymbol{\psi}^*$, everything cancels but the quadratic terms to give us the following estimate:

$$\|\mathbf{x}_{k+1} - \mathbf{x}^*\|_{W^{1,\infty}} + \|\mathbf{u}_{k+1} - \mathbf{u}^*\|_{L^\infty} + \|\boldsymbol{\psi}_{k+1} - \boldsymbol{\psi}^*\|_{W^{1,\infty}} \le c \|\mathbf{w}_k - \mathbf{w}^*\|_{L^\infty}^2,$$

where c is independent of $\mathbf{w}_k = (\mathbf{x}_k, \mathbf{u}_k, \boldsymbol{\psi}_k)$ in a neighborhood of $\mathbf{w}^*$. In [7] we analyse problems that also include inequality control constraints and endpoint constraints on the state.

5. Discrete Approximations

For simplicity, we consider the discretization of the following unconstrained control problem:

$$\begin{aligned} &\text{minimize} \quad C(\mathbf{x}(1)) \\ &\text{subject to} \quad \dot{\mathbf{x}}(t) = \mathbf{f}(\mathbf{x}(t), \mathbf{u}(t)) \quad \text{a. e. } t \in [0,1], \\ &\qquad \mathbf{x}(0) = \mathbf{a}, \quad \mathbf{x} \in W^{1,\infty}, \quad \mathbf{u} \in L^\infty, \end{aligned} \tag{12}$$

where $C : \mathbf{R}^n \mapsto \mathbf{R}$. We study control constrained problems in [5, 8], state constrained problems in [2], and mixed control/state constraints in [6]. Suppose the differential equation in (12) is solved using a Runge-Kutta integration scheme. For convenience, we consider a uniform mesh of width $h = 1/N$ where N is a natural number, and we let $\mathbf{x}_k$ denote the approximation to $\mathbf{x}(t_k)$ where $t_k = kh$. An s-stage Runge-Kutta scheme [1] with coefficients a_{ij} and b_i, $1 \le i, j \le s$, is given by

$$\mathbf{x}'_k = \sum_{i=1}^{s} b_i \mathbf{f}(\mathbf{y}_i, \mathbf{u}_{ki}), \tag{13}$$

where

$$\mathbf{y}_i = \mathbf{x}_k + h \sum_{j=1}^{s} a_{ij} \mathbf{f}(\mathbf{y}_j, \mathbf{u}_{kj}), \quad 1 \le i \le s, \tag{14}$$

and prime denotes, in this discrete context, the forward divided difference:

$$\mathbf{x}'_k = \frac{\mathbf{x}_{k+1} - \mathbf{x}_k}{h}.$$

In (13) and (14), $\mathbf{y}_j$ and $\mathbf{u}_{kj}$ are the intermediate state and control variables on the interval $[t_k, t_{k+1}]$. The dependence of the intermediate state variables on k is not explicit in our notation even though these variables have different values on different intervals. With this notation, the discrete control problem is the following:

$$\begin{aligned} &\text{minimize} \quad C(\mathbf{x}_N) \\ &\text{subject to} \quad \mathbf{x}'_k = \textstyle\sum_{i=1}^{s} b_i \mathbf{f}(\mathbf{y}_i, \mathbf{u}_{ki}), \quad \mathbf{x}_0 = \mathbf{a}, \\ &\qquad \mathbf{y}_i = \mathbf{x}_k + h \textstyle\sum_{j=1}^{s} a_{ij} \mathbf{f}(\mathbf{y}_j, \mathbf{u}_{kj}), \quad 1 \le i \le s, \quad 0 \le k \le N-1. \end{aligned} \tag{15}$$

We apply Theorem 2.1 to the first-order optimality system (Kuhn-Tucker conditions) associated with (15). Suppose that a multiplier $\boldsymbol{\lambda}_i$ is introduced for the i-th intermediate equation (14) in addition to the multiplier $\boldsymbol{\psi}_{k+1}$ for the equation (13). Taking into account these additional multipliers, the Kuhn-Tucker conditions are the following:

$$\boldsymbol{\psi}_k - \boldsymbol{\psi}_{k+1} = \sum_{i=1}^{s} \boldsymbol{\lambda}_i, \quad \boldsymbol{\psi}_N = \nabla C(\mathbf{x}_N), \tag{16}$$

$$h\nabla_x \mathbf{f}(\mathbf{y}_j, \mathbf{u}_{kj})^{\mathsf{T}} (b_j \boldsymbol{\psi}_{k+1} + \sum_{i=1}^{s} a_{ij} \boldsymbol{\lambda}_i) = \boldsymbol{\lambda}_j, \tag{17}$$

$$\nabla_u \mathbf{f}(\mathbf{y}_j, \mathbf{u}_{kj})^{\mathsf{T}} (b_j \boldsymbol{\psi}_{k+1} + \sum_{i=1}^{s} a_{ij} \boldsymbol{\lambda}_i) = \mathbf{0}, \tag{18}$$

$1 \le j \le s$ and $0 \le k \le N-1$.

To apply Theorem 2.1, we should insert the continuous solution in these discrete first-order conditions and estimate a residual. Note though that the discrete first-order conditions seem to have no connection to the continuous first-order condition, the Pontryagin minimum principle. However, we first showed in [12] and more recently in [9], that when $b_j \neq 0$ for each j, the first-order conditions make more sense (and are more useful) when reformulated in terms of the variables $\boldsymbol{\chi}_j$ defined by

$$\boldsymbol{\chi}_j = \boldsymbol{\psi}_{k+1} + \sum_{i=1}^{s} \frac{a_{ij}}{b_j} \boldsymbol{\lambda}_i, \quad 1 \le j \le s. \tag{19}$$

With this definition, (16) and (17) are equivalent to the following scheme:

$$\boldsymbol{\psi}_{k+1} = \boldsymbol{\psi}_k - h \sum_{i=1}^{s} b_i \nabla_x \mathbf{f}(\mathbf{y}_i, \mathbf{u}_{ki})^{\mathsf{T}} \boldsymbol{\chi}_i, \quad \boldsymbol{\psi}_N = \nabla C(\mathbf{x}_N), \tag{20}$$

$$\boldsymbol{\chi}_i = \boldsymbol{\psi}_k - h \sum_{j=1}^{s} \bar{a}_{ij} \nabla_x \mathbf{f}(\mathbf{y}_j, \mathbf{u}_{kj})^{\mathsf{T}} \boldsymbol{\chi}_j, \quad \bar{a}_{ij} = \frac{b_i b_j - b_j a_{ji}}{b_i}. \tag{21}$$

This is a Runge-Kutta scheme applied to the adjoint equation, but the coefficients of this scheme are typically different from those of the original scheme.

This reformulation of the first-order optimality system is important not only for the analysis of the discretization, but also for numerical computations since it provides an efficient way to compute the gradient of the discrete cost function with respect to the control. Let $\mathbf{u} \in \mathbf{R}^{smN}$ denote the vector of intermediate control values for the entire interval $[0, 1]$, and let $C(\mathbf{u})$ denote the value $C(\mathbf{x}_N)$ of the discrete cost function associated with these controls. From the results of [13], we have

$$\nabla_{u_{kj}} C(\mathbf{u}) = h b_j \nabla_u \mathbf{f}(\mathbf{y}_j, \mathbf{u}_{kj})^{\mathsf{T}} \boldsymbol{\chi}_j, \tag{22}$$

where the intermediate values for the discrete state and costate variables are gotten by first solving the discrete state equations (13) and (14), for $k = 0, 1, \dots, N-1$, using the given values for the controls, and then using these computed values for both the state and intermediate variables in (20) and (21) when computing the values of the discrete costate for $k = N-1, N-2, \dots, 0$. Thus the discrete state

Order	Conditions ($c_i = \sum_{j=1}^{s} a_{ij}$, $d_j = \sum_{i=1}^{s} b_i a_{ij}$)
1	$\sum b_i = 1$
2	$\sum d_i = \frac{1}{2}$
3	$\sum c_i d_i = \frac{1}{6}$, $\sum b_i c_i^2 = \frac{1}{3}$, $\sum d_i^2/b_i = \frac{1}{3}$
4	$\sum b_i c_i^3 = \frac{1}{4}$, $\sum d_i^3/b_i^2 = \frac{1}{4}$, $\sum b_i c_i a_{ij} d_j/b_j = \frac{5}{24}$, $\sum c_i d_i^2/b_i = \frac{1}{12}$,
	$\sum d_i a_{ij} c_j = \frac{1}{24}$, $\sum b_i c_i a_{ij} c_j = \frac{1}{8}$, $\sum d_i c_i^2 = \frac{1}{12}$, $\sum d_i a_{ij} d_j/b_j = \frac{1}{8}$

Table 1. Order of a Runge-Kutta discretization for optimal control.

equation is solved by marching forward from $k = 0$, while the discrete costate equation is solved by marching backward from $k = N-1$.

In applying Theorem 2.1 to the first-order order conditions, we need to estimate the residual $\boldsymbol{\delta}$, and we need to analyse a linearized problem. Our linearization corresponds to the choice

$$\mathcal{L}(\mathbf{w}) = \begin{pmatrix} \mathbf{x}_k' - \mathbf{A}_k \mathbf{x}_k - \mathbf{B}_k \mathbf{u}_k \mathbf{b}, \quad 0 \le k \le N-1 \\ \boldsymbol{\psi}_k' + \mathbf{A}_k^{\mathsf{T}} \boldsymbol{\psi}_{k+1} + \mathbf{Q}_k \mathbf{x}_k + \mathbf{S}_k \mathbf{u}_k \mathbf{b}, \quad 0 \le k \le N-1 \\ b_j(\mathbf{R}_k \mathbf{u}_{kj} + \mathbf{S}_k \mathbf{x}_k + \mathbf{B}_k^{\mathsf{T}} \boldsymbol{\psi}_{k+1}), \quad 1 \le j \le s,\ 0 \le k \le N-1 \\ \boldsymbol{\psi}_N + \mathbf{V}\mathbf{x}_N \end{pmatrix}.$$

Here $\mathbf{V} = \nabla^2 C(\mathbf{x}^*(1))$, and the various matrices are the same as those introduced in Section 3 except that they are evaluated at $\mathbf{x}^*(t_k)$, $\mathbf{u}^*(t_k)$, and $\boldsymbol{\psi}^*(t_k)$. In [8, Lem. 6.1], we show that when the coercivity assumption holds, $b_j > 0$ for each j, and $\sum_{j=1}^{s} b_j = 1$, then the linearized problem is invertible, with norm of the inverse bounded by a constant independent of h for h sufficiently small.

To analyse the residual $\boldsymbol{\delta}$, we need to determine the order of the Runge-Kutta schemes (13), (14), (20), and (21), where $\mathbf{u}$ is chosen so that $\nabla_u \mathbf{f}(\mathbf{y}_j, \mathbf{u}_{kj})^{\mathsf{T}} \boldsymbol{\chi}_j = \mathbf{0}$. In Table 1 we give the order of these schemes. The conditions for any given order are those listed in Table 1 for that specific order along with those for all lower orders. We employ the following:

Summation Convention. *If an index range does not appear on a summation sign, then the summation is over each index, taking values from* 1 *to* s.

Notice that the order conditions of Table 1 are not the usual order conditions [1, p. 170] associated with a Runge-Kutta discretization of a differential equation. The conditions of Table 1 were gotten in [9] by checking the tree-based order conditions in [1]. However, it was pointed out by Peter Rentrop at the June 4–10, 2000, conference in Oberwolfach, Germany, that these conditions should also follow from the general theory developed for partitioned Runge-Kutta methods (see [14, II.15], [15]). In [14, Thm. 15.9] it is shown that a partitioned Runge-Kutta method is of order p if and only if certain equations hold for all P-trees of order up to p.

These order conditions for partitioned Runge-Kutta schemes when applied to (13), (14), (20), and (21), should also lead to the conditions of Table 1.

Finally, applying Theorem 2.1 as in [9], it follows that if $(\mathbf{x}^*, \mathbf{u}^*)$ is a local minimizer for (12), the Runge-Kutta scheme is order order p (see Table 1), $b_j > 0$ for each j, and the coercivity condition holds, then when $\mathbf{f}$ is sufficiently smooth, the discrete problem (15) has a local minimizer $(\mathbf{x}^h, \mathbf{u}^\mathbf{h})$, and we have

$$\max_{0 \le k \le N} |\mathbf{x}_k^h - \mathbf{x}^*(t_k)| + |\psi_k^h - \psi^*(t_k)| + |\mathbf{u}(\mathbf{x}_k^h, \psi_k^h) - \mathbf{u}^*(t_k)| \le ch^p,$$

where ψ^h is the solution of the discrete costate equations (20)–(21) and $\mathbf{u}(\mathbf{x}, \psi)$ denotes a minimizer of the Hamiltonian $H(\mathbf{x}, \mathbf{u}, \psi)$ over $\mathbf{u}$ (not one of the discrete controls). The order of approximation of the discrete controls in (15) is typically less than p. To obtain an approximation to an optimal control with the same order as that of the Runge-Kutta scheme, the Hamiltonian should be minimized over the control, using the computed discrete state and costate at each time level.

In [9] we show that the following scheme is 3-rd order accurate for differential equations, but only second order accurate for optimal control:

$$\mathbf{A} = \begin{bmatrix} 0 & 0 & 0 \\ \frac{1}{2} & 0 & 0 \\ 0 & \frac{3}{4} & 0 \end{bmatrix}, \quad \mathbf{b} = \begin{bmatrix} \frac{2}{9} \\ \frac{1}{3} \\ \frac{4}{9} \end{bmatrix}.$$

The following scheme, with $b_1 = 0$, is 2-nd order accurate for differential equations, but divergent for optimal control:

$$\mathbf{A} = \begin{bmatrix} 0 & 0 \\ \frac{1}{2} & 0 \end{bmatrix}, \quad \mathbf{b} = \begin{bmatrix} 0 \\ 1 \end{bmatrix}.$$

Although it appears difficult to construct a 4-th order Runge-Kutta scheme (13 conditions in Table 1 must be satisfied), it is shown in [9, Prop. 6.1] that every 4-stage explicit 4-th order Runge-Kutta scheme for differential equations, with $b_j > 0$ for each j, is 4-th order accurate for optimal control. This surprising result is due to the following identity, established by Butcher [1, p. 178], for 4-stage explicit 4-th order Runge-Kutta schemes:

$$\sum_i b_i a_{ij} = b_j(1 - c_j),$$

$j = 1, 2, 3, 4$.

References

[1] J. C. BUTCHER, *The Numerical Analysis of Ordinary Differential Equations*, John Wiley, New York, 1987.

[2] A. L. DONTCHEV AND W. W. HAGER, *The Euler approximation in state constrained optimal control*, Mathematics of Computation, 2000, 31 pages.

[3] A. L. DONTCHEV AND W. W. HAGER, *Lipschitzian stability for state constrained nonlinear optimal control*, SIAM Journal on Control and Optimization, **36** (1998), 696–718.

[4] A. DONTCHEV AND W. W. HAGER, *An inverse mapping theorem for set-valued maps*, Proceedings of the American Mathematical Society, **121** (1994), 481–489.

[5] A. DONTCHEV AND W. W. HAGER, *Lipschitzian stability in nonlinear control and optimization*, SIAM Journal on Control and Optimization, **31** (1993), 569–603.

[6] A. L. DONTCHEV, W. W. HAGER, AND K MALANOWSKI, *Error bounds for Euler approximation of a state and control constrained optimal control problem*, Numerical Functional Analysis and Optimization, **21** (2000), 653–682.

[7] A. DONTCHEV, W. W. HAGER, A. POORE, AND B. YANG, *Optimality, stability, and convergence in nonlinear control*, Applied Mathematics and Optimization, **31** (1995), 297–326.

[8] A. L. DONTCHEV, W. W. HAGER, AND V. M. VELIOV, *Second-order Runge-Kutta approximations in constrained optimal control*, SIAM Journal on Numerical Analysis, **38** (2000), 202–226.

[9] W. W. HAGER, *Runge-Kutta methods in optimal control and the transformed adjoint system*, Numerische Mathematik, **87** (2000), pp. 247–282.

[10] W. W. HAGER, *Multiplier methods for nonlinear optimal control*, SIAM Journal on Numerical Analysis, **27** (1990), 1061–1080.

[11] W. W. HAGER, *Approximations to the multiplier method*, SIAM Journal on Numerical Analysis, **22** (1985), 16–46.

[12] W. W. HAGER, *Rates of convergence for discrete approximations to unconstrained control problems*, SIAM Journal on Numerical Analysis, **13** (1976), 449–472.

[13] W. W. HAGER AND R. ROSTAMIAN, *Optimal coatings, bang-bang controls, and gradient techniques*, Optimal Control: Applications and Methods, **8** (1987), 1–20.

[14] E. HAIRER, S. P. NØRSETT, AND G. WANNER, *Solving Ordinary Differential Equations I*, second revised edition, Springer-Verlag, Berlin, 1993.

[15] P. RENTROP, *Partitioned Runge-Kutta methods with stiffness detection and stepsize control*, Numerische Mathematik, **47** (1985), 545–564.

Department of Mathematics
University of Florida
358 Little Hall
Gainesville, FL 32611 USA
Web: `http://www.math.ufl.edu/~hager`
E-mail address: `hager@math.ufl.edu`

International Series of Numerical Mathematics, Vol. 139, 95–106

Optimal Control of the Drift Diffusion Model for Semiconductor Devices

Michael Hinze and René Pinnau

Abstract. The design problem for semiconductor devices is studied via an optimal control approach for the standard drift diffusion model. The solvability of the minimization problem is proved. The first–order optimality system is derived and the existence of Lagrange–multipliers is established. Further, estimates on the sensitivities are given. Numerical results concerning a symmetric n–p–diode are presented.

1. Introduction

Due to the rapidly increasing demand for semiconductor technology lots of effort has been spend on the development of new semiconductor devices. Especially, the ongoing miniaturization revealed several challenging problems for electrical engineers and applied mathematicians, too. Numerical simulations proved to be the main tool for reducing the time of a design cycle. For this purpose a hierarchy of models is employed, which ranges from microscopic, like the the Boltzmann–Poisson or the Wigner–Poisson model, to macroscopic models, like the energy transport, the hydrodynamic and the drift diffusion model (DD) [10]. Most popular and widely used in commercial simulation packages is the DD, which allows for a very efficient numerical study of the charge transport in many cases of practical relevance.

Many performance properties of semiconductor devices can be derived from the so–called current–voltage characteristics (IVC), which relates the applied biasing voltage and the current density. Typically, such an ideal IVC is given and the engineer meets the following *design problem*: Adjust physical and/or geometrical parameters of a semiconductor device such that the given ideal IVC is matched optimally with respect to certain performance criteria.

1991 *Mathematics Subject Classification.* 35J50,49J20,49K20.
Key words and phrases. semiconductor design, drift diffusion, optimal control, existence, first–order necessary condition, Lagrange–multipliers, sensitivity, numerics.
The second author acknowledges financial support from the TMR Project 'Asymptotic Methods in Kinetic Theory', grant number ERB FMRX CT97 0157.

In most applications one changes the geometry and the doping profile, which describes the density of charged background ions. In the conventional design cycle simulation tools are employed to compute the IVC for a certain set of parameters and then, the parameters are adjusted empirically. Thus, the total design time depends crucially on the knowledge and experience of the electrical engineer.

Although this problem can clearly be tackled by an optimization approach, only little efforts were yet made in semiconductor industry to solve the design problem via optimization techniques. As far as the authors know, up to now only one attempt has been made to address semiconductor design with mathematical optimization techniques. In [8] *Lee et al.* present a finite–dimensional least-squares approach for adjusting the parameters of a semiconductor to fit a given, ideal IVC. Their work has its focus on testing different approaches to solve numerically the least-squares problem.

In standard applications a working point, i.e. a certain voltage–current pair, for the device is fixed [14]. Thus, in this paper we consider the *modified design question*:

> Is it possible to gain an amplified current at the working point only by a slight change of the doping profile?

We give an positive answer to this question by means of an optimal control problem for the DD. The focus on the DD is due to the fact that this model is today most widely used in simulation codes, since it allows for an accurate description of the underlying physics in combination with low computational costs. There exists a large amount of literature on this model, which covers questions of the mathematical analysis [11, 12] as well as of the numerical discretization and simulation [4, 7]. For an excellent overview see [9, 10]

The standard drift diffusion model for semiconductor devices stated on a bounded domain $\Omega \subset \mathbb{R}^d$, $d = 1, 2$ and 3 reads in its scaled form

$$J_n = \mu_n \left(\nabla n + n \nabla V\right), \tag{1.1a}$$

$$J_p = \mu_p \left(\nabla p - p \nabla V\right), \tag{1.1b}$$

$$\operatorname{div} J_n = 0, \tag{1.1c}$$

$$\operatorname{div} J_p = 0, \tag{1.1d}$$

$$-\lambda^2 \Delta V = n - p - C. \tag{1.1e}$$

The variables are the densities of electrons $n(x)$ and holes $p(x)$, the current densities of electrons $J_n(x)$ and holes $J_p(x)$, respectively, and the electrostatic potential $V(x)$. The doping profile is denoted by $C(x)$. The parameter λ is the scaled Debye length of the device and μ_n, μ_p denote the carrier mobilities. The total current density is given by

$$J = J_n + J_p. \tag{1.1f}$$

Note that for the sake of simplicity we assume constant mobilities and that no generation–recombination processes occur [14].

Inserting the relations for the current densities into the continuity equations yields the system

$$\Delta n + \operatorname{div}(n\,\nabla V) = 0, \tag{1.2a}$$

$$\Delta p - \operatorname{div}(p\,\nabla V) = 0, \tag{1.2b}$$

$$-\lambda^2 \Delta V = n - p - C, \tag{1.2c}$$

which will be considered in the following.

To get a well-posed problem, system (1.2) has to be supplemented with appropriate boundary conditions. We assume that the boundary $\partial\Omega$ of the domain Ω splits into two disjoint parts Γ_D and Γ_N, where Γ_D models the Ohmic contacts of the device and Γ_N represents the insulating parts of the boundary. Let ν denote the unit outward normal vector along the boundary. Firstly, assuming charge neutrality and thermal equilibrium at the Ohmic contacts Γ_D and, secondly, zero current flow and vanishing electric field and the insulating part Γ_N yields the following set of boundary data

$$n = n_D, \quad p = p_D, \quad V = V_D \quad \text{on } \Gamma_D, \tag{1.2d}$$

$$\nabla n \cdot \nu = \nabla p \cdot \nu = \nabla V \cdot \nu = 0 \quad \text{on } \Gamma_N, \tag{1.2e}$$

where n_D, p_D, V_D are the $H^2(\Omega)$–extensions of

$$n_D = \frac{C + \sqrt{C^2 + 4\,\delta^2}}{2},$$

$$p_D = \frac{-C + \sqrt{C^2 + 4\,\delta^2}}{2},$$

$$V_D = -\log\left(\frac{C + \sqrt{C^2 + 4\,\delta^2}}{2\,\delta}\right) + U, \qquad \text{on } \Gamma_D.$$

Here, δ is the scaled intrinsic density and U is an applied voltage.

To solve the modified design question we start from given reference doping profile $\bar{C}$ and specify the working point $(\bar{U}, \bar{J})$. Let Γ_O be a portion of the Ohmic contacts Γ_D at which we can measure the total current J. At this contact we prescribe a gained current density J_g and allow for deviations of the doping profile from $\bar{C}$ in some suitable norm to gain this current flow.

Especially, we intend to minimize cost functionals of the form

$$Q(n, p, V, C) = \frac{1}{2}\int_{\Gamma_O} |J \cdot \nu - J_g \cdot \nu|^2 \, ds + \frac{\gamma}{2}\int_\Omega |C - \bar{C}|^2 \, dx, \tag{1.3a}$$

and

$$Q(n, p, V, C) = \frac{1}{2}\int_{\Gamma_O} |J \cdot \nu - J_g \cdot \nu|^2 \, ds + \frac{\gamma}{2}\int_\Omega |\nabla(C - \bar{C})|^2 \, dx, \tag{1.3b}$$

where the total current J is given by the solution of (1.1). Here, $\gamma > 0$ is a parameter which allows to balance the effective cost. Clearly, the proposed evaluation of the total current along the boundary poses some restrictions on the regularity of the solutions to (1.1), which will be addressed below.

The optimal control problem for the system (1.1) will be considered as a constrained optimization problem. The approach presented is closely related to that discussed by *Ito et al.* [6] for the control of nonlinear partial differential equations.

Further analytical results related to the work presented are given by *Fang et al.* in [2], where a mathematical model is developed for a non destructive optical testing technique for semiconductors called laser–beam–induced currents (LBIC), and by *Busenberg et al.* [1], where the identifiability of defects in a semiconductor from its LBIC–image is studied mathematically.

The paper is organized as follows. In Section 2 we specify the optimal control problem and its analytical setting. We present an existence result in Section 3. The first–order optimality system is studied in Section 4. After its derivation we establish the existence of Lagrange–multipliers, comment on their uniqueness and give estimates on the sensitivities. Lastly, numerical results for a symmetric n–p–diode are presented in Section 5.

2. Problem Formulation and Analytic Setting

We now introduce a functional analytic framework which allows to deal with general optimization problems that in particular include those introduced in Section 1. For the subsequent considerations we impose the following assumptions.

A.1 Let $\Omega \subset \mathbb{R}^d$, $d = 1, 2$ or 3 be a bounded domain with boundary $\partial\Omega \in C^{1,1}$. The boundary $\partial\Omega$ is piecewise regular and splits into two disjoint parts Γ_N and Γ_D. The set Γ_D has nonvanishing $(d-1)$–dimensional Lebesgue–measure. Γ_N is closed.

A.2 The boundary data fulfils $(n_D, p_D, V_D) \in H^2(\Omega)$. For the gained total current we require $J_g \in H^1(\Omega; \mathbb{R}^d)$ and $\bar{C} \in H^1(\Omega)$ for the reference doping profile.

A.3 There exists a constant $K = K(\Omega, \Gamma_D, \Gamma_N) > 0$ such that for $f \in L^2(\Omega)$ and $w_D \in H^2(\Omega)$ there exists a solution $w \in H^2(\Omega)$ of

$$\Delta w = f, \quad w - w_D \in H_0^1(\Omega \cup \Gamma_N),$$

which fulfils

$$\|w\|_{H^2(\Omega)} \le K \left(\|w_D\|_{H^2(\Omega)} + \|f\|_{L^2(\Omega)} \right).$$

Remark 2.1. Assumption **A.3** is essentially a restriction on the geometry of Ω. It is fulfilled in the case where the Dirichlet and the Neumann boundary do not meet, i.e. $\overline{\Gamma}_D \cap \Gamma_N = \emptyset$ [15]. In the two dimensional case it is fulfilled if Γ_D and Γ_N meet under angles larger than $\pi/2$ [3]. Let us recall that the space $H_0^1(\Omega \cup \Gamma_N)$ is the closure of $C_c^\infty(\Omega \cup \Gamma_N)$ with respect to the $H^1(\Omega)$–norm [15].

We introduce the space of states

$$X \stackrel{\text{def}}{=} x_D + X_0,$$

where $x_D \stackrel{\text{def}}{=} (n_D, p_D, V_D)$ denotes the boundary data introduced in (1.1) and $X_0 \stackrel{\text{def}}{=} \left(H^2(\Omega) \cap H_0^1(\Omega \cup \Gamma_N)\right)^3$. The set of admissible controls is given by

$$\mathcal{C} \stackrel{\text{def}}{=} \{C \in H^1(\Omega) : C = \bar{C} \text{ on } \Gamma_D\}. \tag{2.1}$$

Remark 2.2. At a first glance this imposes a further restriction on the doping profile, since we freeze the boundary values of C. Nevertheless, considering the cost functional (1.3a) and deriving *formally* the first order optimality system under the assumption $C \in H^1(\Omega)$ yields exactly the additional restriction on the doping profile given in (2.1).

According to the habits in control theory in the following doping profiles are denoted by the variable u. We abbreviate $x \stackrel{\text{def}}{=} (n, p, V)$ and rewrite the state equations (1.2) as $e(x, u) = 0$, where the nonlinear mapping $e : X \times \mathcal{C} \to (L^2(\Omega))^3$ is defined by

$$e(x,u) \stackrel{\text{def}}{=} \begin{pmatrix} \Delta n + \operatorname{div}(n\,\nabla V) \\ \Delta p - \operatorname{div}(p\,\nabla V) \\ -\lambda^2 \Delta V - n + p + u \end{pmatrix}. \tag{2.2}$$

Next we state the differentiability properties of the mapping e.

Theorem 2.3. *The mapping e defined in (2.2) is infinitely often Fréchet differentiable with respect to the topology of $H^1(\Omega)$. Further, the derivatives vanish for order greater than or equal to 2. The actions of the first derivatives at a point $(x,u) \in X \times \mathcal{C}$ are given by*

$$\begin{aligned} \langle e_x(x,u)\tilde{x}, z\rangle = {} & \langle \Delta\tilde{n} + \operatorname{div}(\tilde{n}\nabla V), z^n\rangle - \\ & \left\langle \tilde{n}, z^V \right\rangle + \langle \Delta\tilde{p} - \operatorname{div}(\tilde{p}\nabla V), z^p\rangle - \left\langle \tilde{p}, z^V \right\rangle + \\ & \left\langle \operatorname{div}(n\nabla\tilde{V}), z^n \right\rangle - \left\langle \operatorname{div}(p\nabla\tilde{V}), z^p \right\rangle - \left\langle \lambda^2 \Delta\tilde{V}, z^V \right\rangle \end{aligned} \tag{2.3}$$

for all $\tilde{x} = (\tilde{n}, \tilde{p}, \tilde{V})$, $z = (z^n, z^p, z^V) \in X_0$ and

$$\langle e_u(x,u)\tilde{u}, z\rangle = \left\langle \tilde{u}, z^V \right\rangle$$

for all $\tilde{u} \in \mathcal{C}$ and $z \in X_0$.

Here, $\langle \cdot, \cdot \rangle$ denotes the inner product in $L^2(\Omega)$. For a proof see [5].

3. Existence of Solutions

In this section we establish the existence of a solution to the optimal control problem. We require standard regularity properties of the cost functional Q.

A.4 Let $Q : X \times \mathcal{C} \to \mathbb{R}$ denote a cost functional which is assumed to be twice continuously Fréchet differentiable with Lipschitz continuous second derivatives. Further, let Q be radially unbounded w.r.t. $\mathcal{C}$, bounded from below and weakly lower semi-continuous.

Remark 3.1. Clearly, the cost functional (1.3b) fits into this setting.

We now consider the minimization problem

$$\min_{X\times\mathcal{C}} Q(n,p,V,u) \quad \text{s.t.} \quad e(n,p,V,u)=0. \tag{3.1}$$

The solvability of the state equations for every $u \in \mathcal{C}$ is a consequence of the following result.

Proposition 3.2. *Assume* **A.1**–**A.3**. *Then for each $C \in H^1(\Omega)$ there exists a solution $(n,p,V) \in (H^2(\Omega))^3$ of system* (1.2) *fulfilling $(n,p) \geq (\underline{n},\underline{p}) > 0$.*

For a proof see e.g. [9]. Note that uniqueness of solutions can only be expected for small potentials U, i.e. if the device is operated near thermal equilibrium.

Due to the $H^2(\Omega)$–regularity of (n,p,V) it holds $J \in H^1(\Omega;\mathbb{R}^d)$. Hence, the trace of the total current fulfils $trJ \in H^{1/2}(\partial\Omega)$. This implies that the functionals Q given by (1.3) are well defined.

The existence result reads

Theorem 3.3. *Assume* **A.1**–**A.4**. *Then, the minimization problem* (3.1) *admits a solution $(n^*,p^*,V^*,u^*) \in X \times \mathcal{C}$.*

Proof. Since Q is bounded from below, $Q_0 \stackrel{\text{def}}{=} \inf_{(n,p,V,u)\in X\times\mathcal{C}} Q(n,p,V,u)$ is finite. Consider a minimizing sequence $\{(x^n,u^n)\}_{n\in\mathbb{N}} \subset X \times \mathcal{C}$. From the radially unboundedness of Q we infer that $\{u_n\}_{n\in\mathbb{N}}$ is bounded in $\mathcal{C}$. Hence, there exists a weakly convergent subsequence, again denoted by $\{u^n\}_{n\in\mathbb{N}}$, such that

$$u^n \rightharpoonup u^*, \qquad \text{weakly in } \mathcal{C}.$$

Since $\mathcal{C}$ is weakly closed w.r.t. the $H^1(\Omega)$-norm, $u^* \in \mathcal{C}$. By the continuous embedding $H^1(\Omega) \hookrightarrow L^p(\Omega)$ ($p \in [1,6)$) the sequence $\{u^n\}_{n\in\mathbb{N}}$ is also bounded in $L^p(\Omega)$. Now one can employ Stampaccia's method [13] to derive the following estimates [12]

$$\|n^k\|_{H^1(\Omega)} + \|n^k\|_{L^\infty(\Omega)} \leq K\left(\|n_D\|_{L^\infty(\Gamma_D)} + \|u^k\|_{L^p(\Omega)}\right), \tag{3.2a}$$

$$\|p^k\|_{H^1(\Omega)} + \|p^k\|_{L^\infty(\Omega)} \leq K\left(\|p_D\|_{L^\infty(\Gamma_D)} + \|u^k\|_{L^p(\Omega)}\right), \tag{3.2b}$$

$$\|V^k\|_{H^1(\Omega)} + \|V^k\|_{L^\infty(\Omega)} \leq K\left(\|V_D\|_{L^\infty(\Gamma_D)} + \|u^k\|_{L^p(\Omega)}\right), \tag{3.2c}$$

for some constant $K = K(\Omega) > 0$. These are by far sufficient to pass to the limit in the state equations (1.2), which can be seen as follows. Every solution (n^k, p^k, V^k) of (1.2) associated to u^k satisfies the a priori estimates (3.2). Hence, there exists a subsequence, again denoted by $\{(n^k, p^k, V^k)\}_{k\in\mathbb{N}}$ such that

$$(n^k, p^k, V^k) \rightharpoonup (n^*, p^*, V^*) \quad \text{weakly in } (H^1(\Omega))^3,$$

which by Rellich's Theorem [16] implies strong convergence of $\{(n^k, p^k, V^k)\}_{k\in\mathbb{N}}$ in $(L^2(\Omega))^3$. Further, the uniform $L^\infty(\Omega)$–bounds imply

$$(n^k, p^k, V^k) \rightharpoonup (n^*, p^*, V^*) \quad \text{weak-* in } L^\infty(\Omega).$$

Utilizing these convergences one can pass to the limit in (1.2), which satisfies

$$\begin{aligned} \Delta n^* + \operatorname{div}(n^* \nabla V^*) &= 0, \\ \Delta p^* - \operatorname{div}(p^* \nabla V^*) &= 0, \\ -\lambda^2 \Delta V^* &= n^* - p^* - u^* \end{aligned}$$

together with the boundary conditions in (1.1).

Lastly, we have to ensure $(n^*, p^*, V^*) \in X$, which is an easy consequence of assumption **A.3**. This completes the proof of the existence of a minimizer. □

Remark 3.4. The existence proof also works for less regular states in $H^1(\Omega)$. In this case one has to modify the cost functional and to investigate carefully the evaluation of the total current along the boundary. These techniques will be presented by the authors in [5].

4. The First–order Optimality System

Next we derive the first–order optimality system and state a result on the existence of Lagrange–multipliers. Their uniqueness is commented and estimates on the sensitivities are given.

The Lagrangian $\mathcal{L} : X \times \mathcal{C} \times X_0 \to \mathbb{R}$ associated to the minimization problem (3.1) is defined by

$$\mathcal{L}(x, u, \xi) \stackrel{\text{def}}{=} Q(x, u) + \langle e(x, u), \xi \rangle . \tag{4.1}$$

By Theorem 2.3 and **A.4** the Lagrangian $\mathcal{L}$ is twice continuously Fréchet differentiable with Lipschitz continuous second derivatives. The first order optimality system corresponding to problem (3.1) is given by

$$\nabla_{(x,u,\xi)} \mathcal{L}(x, u, \xi) = 0. \tag{4.2}$$

All together we find that the first–order necessary optimality condition (4.2) is equivalent to the following nonlinear system of coupled equations for the state x, the control u and the adjoint variable ξ. It consists of the state equations

$$\Delta n + \operatorname{div}(n \nabla V) = 0, \tag{4.3a}$$

$$\Delta p - \operatorname{div}(p \nabla V) = 0, \tag{4.3b}$$

$$-\lambda^2 \Delta V = n - p - u \tag{4.3c}$$

with boundary data given in (1.2) and the adjoint system

$$\Delta \xi^n - \nabla V \, \nabla \xi^n = \xi^V, \tag{4.4a}$$

$$\Delta \xi^p + \nabla V \, \nabla \xi^p = -\xi^V, \tag{4.4b}$$

$$-\lambda^2 \Delta \xi^V + \operatorname{div}(n \nabla \xi^n) - \operatorname{div}(p \nabla \xi^p) = 0, \tag{4.4c}$$

supplemented with boundary data

$$\xi^n = \xi^p = \begin{cases} 0 & \text{on } \Gamma_D \setminus \Gamma_O, \\ (J_g - J) \cdot \nu & \text{on } \Gamma_O, \end{cases} \tag{4.5a}$$

$$\xi^V = 0 \text{ on } \Gamma_D, \tag{4.5b}$$

$$\nabla \xi^n \cdot \nu = \nabla \xi^p \cdot \nu = \nabla \xi^V \cdot \nu = 0 \quad \text{on } \Gamma_N. \tag{4.5c}$$

Moreover, the adjoint and the control are coupled via the equation

$$\xi^V = -\gamma (u - \bar{C}) \tag{4.6}$$

in the case of the cost functional (1.3a), and via

$$\begin{aligned} \gamma \Delta \left(u - \bar{C} \right) &= \xi^V && \text{in } \Omega, \\ u &= \bar{C} && \text{on } \Gamma_D, \\ \nabla u \cdot \nu &= \nabla \bar{C} \cdot \nu && \text{on } \Gamma_N. \end{aligned} \tag{4.7}$$

in the case of cost functional (1.3b).

4.1. Existence of Lagrange–multipliers

In Section 3 we proved that the minimizer satisfies the state equations. Now, we want to establish the existence of the Lagrange–multipliers.

For the following investigations it is most convenient to write system (4.4) in symmetric form. This is accomplished by multiplying the first two equations of (4.4) with e^{-V} and e^V, respectively.

$$\operatorname{div}\left(e^{-V} \nabla \xi^n \right) = e^{-V} \xi^V, \tag{4.8a}$$

$$\operatorname{div}\left(e^{V} \nabla \xi^p \right) = -e^{V} \xi^V, \tag{4.8b}$$

$$-\lambda^2 \Delta \xi^V + (n + p)\, \xi^V = -J_n \nabla \xi^n + J_p \nabla \xi^p. \tag{4.8c}$$

Note, that $J_n, J_p \in H^1(\Omega; \mathbb{R}^d)$ due to Proposition 3.2. Now we are in the position to state the existence theorem for the Lagrange–multipliers.

Theorem 4.1. *Assume* **A.1–A.3**. *Then there exists a constant* $j = j(\Omega, \lambda, V) > 0$ *such that for*

$$\left\| \frac{J_n^2}{n} \right\|_{L^\infty(\Omega)} + \left\| \frac{J_p^2}{p} \right\|_{L^\infty(\Omega)} \leq j$$

there exists a solution $(\xi^n, \xi^p, \xi^V) \in (H^1(\Omega))^3$ *of system* (4.8) *supplemented with the boundary data* (4.5).

The proof is done by means of Schauder's fixed point theorem by decoupling the equations appropriately and deriving a priori estimates ensuring the compactness of the fixed point mapping. The existence of Lagrange–multipliers also holds for less regular states $(n, p, V) \in \left[H^1(\Omega) \cap L^\infty(\Omega)\right]^3$, see [5].

4.2. Sensitivity

Testing (4.4) with appropriate test functions and using the following estimate on the extension operator

$$\|\xi_D^n\|_{H^1(\Omega)} \leq c \, \|(J_g - J) \cdot \nu\|_{H^{1/2}(\Gamma_O)},$$

for some constant $c = c(\Omega) > 0$, yields

$$\|\xi^n\|_{H^1(\Omega)}^2 \leq K \left\{ e^{2(V_{\max} - V_{\min})} \left\|\xi^V\right\|_{L^2(\Omega)}^2 + e^{V_{\max} - V_{\min}} \, \|(J_g - J) \cdot \nu\|_{H^{1/2}(\Gamma_O)}^2 \right\},$$

for some constant $K = K(\Omega) > 0$ and in analogy

$$\|\xi^p\|_{H^1(\Omega)}^2 \leq K \left\{ e^{2(V_{\max} - V_{\min})} \left\|\xi^V\right\|_{L^2(\Omega)}^2 + e^{V_{\max} - V_{\min}} \, \|(J_g - J) \cdot \nu\|_{H^{1/2}(\Gamma_O)}^2 \right\}.$$

Further, we get

$$\lambda^2 \left\|\xi^V\right\|_{H^1(\Omega)}^2 \leq \frac{K}{4} \left\{ \left\| \frac{J_n^2}{n} \right\|_{L^\infty(\Omega)} \|\xi^n\|_{H^1(\Omega)}^2 + \left\| \frac{J_p^2}{p} \right\|_{L^\infty(\Omega)} \|\xi^p\|_{H^1(\Omega)}^2 \right\}.$$

These are the key estimates for the proof of the next theorem, which states the bound on the sensitivity of the electrostatic potential.

Theorem 4.2. *Under the assumptions of Theorem 4.1 it holds*

$$\left\|\xi^V\right\|_{H^1(\Omega)}^2 \leq \\ e^{-(V_{\max} - V_{\min})} \, j \left(\lambda^2 \, e^{-2(V_{\max} - V_{\min})} - \frac{K^2}{4} j \right)^{-1} \|(J_g - J) \cdot \nu\|_{H^{1/2}(\Gamma_O)}^2.$$

In analogy one proves the following sensitivity bounds.

Corollary 4.3. *Under the conditions of Theorem 4.1 it holds*

$$\|\xi^n\|^2_{H^1(\Omega)} + \|\xi^p\|^2_{H^1(\Omega)} \leq$$

$$e^{V_{\max}-V_{\min}} \left(j \left(\lambda^2 \, e^{-2(V_{\max}-V_{\min})} - \frac{K^2}{4} j \right)^{-1} + 1 \right) \|(J_g - J)\cdot \nu\|_{H^{1/2}(\Gamma_O)}.$$

Remark 4.4.

a) As expected the sensitivities are small in the case of small deviations from the desired state and for small current densities, i.e. if the device is operated near thermal equilibrium.

b) Under the conditions of Theorem 4.2 the Lagrange–multipliers prove to be unique. We cannot expect their uniqueness without less restrictions. Compare the uniqueness results for the DD [9].

5. Numerical Results

In this section we solve the 1–d first–order optimality system numerically on the domain $\Omega = (0,1)$ to gain an improved performance of a symmetric n–p–diode. The reference doping profile can be found in Figure 5.1 and the corresponding current–voltage characteristic is depicted in Figure 5.2. The Si–diode is 500nm long and operated at 300K. The maximum doping density is 10^{22}m^{-3}.

For the biasing voltage at the working point we choose $\bar{U} = 0.39$V and try to gain an amplification of the total current $\bar{J}$ by 50%, i.e. we set $J_g = \bar{J} \cdot 1.5$.

For the simulations we employ the cost functionals given by (1.3) and decouple the state system from the adjoint system, i.e. starting with some C_0 we compute the corresponding state via (4.3), solve the adjoint system (4.4) for the Lagrange–multipliers and update the doping profile via (4.6) and (4.7), respectively. Note that these updates do not require $\bar{C} \in H^1(\Omega)$. The iteration stops, if the difference of two consecutive doping profiles is below some specified threshold. The advantage of this approach is that during the iteration an existing fast solver for the state system can be used. To stabilize the numerics we use a continuation method in γ, starting with $\gamma = 16$ which is bisected in each step. Clearly, the convergence of the iteration depends crucially on the size γ as it slows down for decreasing γ.

The state system was discretized using an exponentially fitted mixed finite element method suitable for convection dominated equations [9], while the adjoint system was discretized in its symmetric form (4.8) using standard techniques. The nonlinear discrete state system was solved by a full Newton method, while for the linear adjoint system a GMRES–solver was employed.

The numerical results show that the system it not controllable into the desired state. Nevertheless, for $\gamma = 1/16$ the minimizer of (1.3b) yields a current gain of approximately 20%. The minimizing doping density can be found in Figure 5.1, where also minimizing doping density for (1.3a) is shown, which also yields an amplification of 20%. The corresponding current–voltage characteristics are depicted in Figure 5.2. Note, that the position of the junction is identical with the one of the reference profile and also the height of the junction did only change slightly.

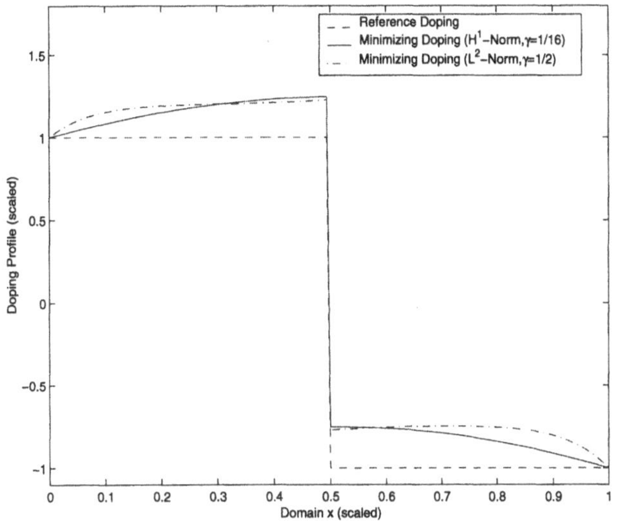

FIGURE 5.1. Doping Profiles

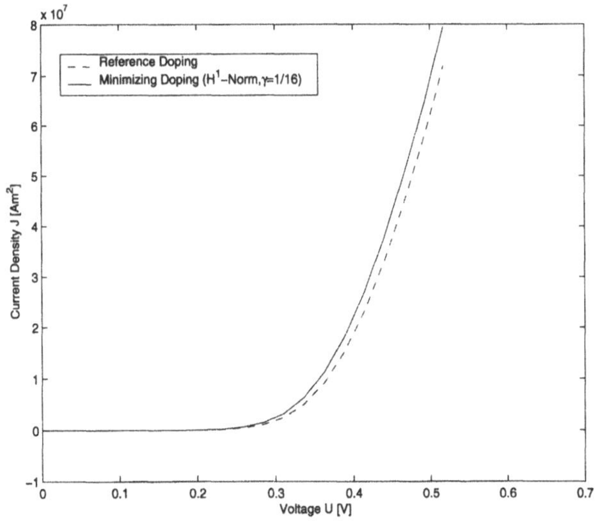

FIGURE 5.2. Current–Voltage Characteristics

References

[1] S. Busenberg, W. Fang, and K. Ito. Modeling and analysis of laser-beam-induced current images in semiconductors. *SIAM J. Appl. Math.*, 53:187–204, 1993.

[2] W. Fang and K. Ito. Identifiability of semiconductor defects from LBIC images. *SIAM J. Appl. Math.*, 52:1611–1625, 1992.

[3] P. Grisvard. *Elliptic Problems in Nonsmooth Domains.* Pitman, Boston, first edition, 1985.

[4] H. K. Gummel. A self–consistent iterative scheme for one–dimensional steady state transistor calculations. *IEEE Trans. Elec. Dev.*, ED–11:455–465, 1964.

[5] M. Hinze and R. Pinnau. An optimal control approach to semiconductor design. *Submitted for Publication.*

[6] K. Ito and K. Kunisch. Augmented Lagrangian-SQP-methods for nonlinear optimal control problems of tracking type. *SIAM J. Control and Optimization*, 34:874–891, 1996.

[7] T. Kerkhoven. A proof of convergence of Gummel's algorithm for realistic device geometries. *SIAM J. Numer. Anal.*, 23(6):1121–1137, December 1986.

[8] W.R. Lee, S. Wang, and K.L. Teo. An optimization approach to a finite dimensional parameter estimation problem in semiconductor device design. *Journal of Computational Physics*, 156:241–256, 1999.

[9] P. A. Markowich. *The Stationary Semiconductor Device Equations.* Springer–Verlag, Wien, first edition, 1986.

[10] P. A. Markowich, Ch. A. Ringhofer, and Ch. Schmeiser. *Semiconductor Equations.* Springer–Verlag, Wien, first edition, 1990.

[11] M. S. Mock. *Analysis of Mathematical Models of Semiconductor Devices.* Boole Press, Dublin, first edition, 1983.

[12] J. Naumann and M. Wolff. A uniqueness theorem for weak solutions of the stationary semiconductor equations. *Appl. Math. Optim.*, 24:223–232, 1991.

[13] G. Stampaccia. Contributi alla regolarizzazione delle soluzioni dei problemi al contorno per secondo ordine ellittiche. *Ann. Scuola Norm. Suo. Pisa*, 12:223–245, 1958.

[14] S. M. Sze. *Physics of Semiconductor Devices.* Wiley, New York, second edition, 1981.

[15] G. M. Troianiello. *Elliptic Differential Equations and Obstacle Problems.* Plenum Press, New York, first edition, 1987.

[16] E. Zeidler. *Nonlinear Functional Analysis and its Applications*, volume II/A and II/B. Springer–Verlag, Berlin, first edition, 1990.

(M. Hinze) Institut für Numerische Mathematik
Technische Universität Dresden
D–01069 Dresden, Germany

E-mail address: `hinze@math.tu-dresden.de`

(R. Pinnau) Fachbereich Mathematik
Technische Universität Darmstadt
D–64289 Darmstadt, Germany

E-mail address: `pinnau@mathematik.tu-darmstadt.de`

International Series of Numerical Mathematics, Vol. 139, 107–118

Fully Coupled Model of a Nonlinear Thin Plate Excited by Piezoelectric Actuators

K.-H. Hoffmann and N. D. Botkin

Abstract. A model describing oscillations of nonlinear thin plates excited by piezoelectric actuators is considered. The specific of the model is that the mutual coupling between elastic deformations and electric fields is taken into consideration. Partial differential equations describing the model are stated and their solvability is proved. The question of homogenization when the number of the piezoelectric patches goes to infinity whereas their dimension goes to zero is investigated.

1. Introduction

This paper is a continuation of the works [1] and [2] where the interface between electric fields and elastic deformations was assumed to be rather simple: electric fields generate elastic deformations but not vice versa. In this paper, a full coupling between elastic deformations and electric fields is assumed. This means that the generation of electric fields trough elastic deformations is taken into consideration. Therefore, the model contains additional variables, the potential functions, that describe electric fields arising both through the voltage applied to the piezoelectric patches and due to elastic deformations. The aim of this paper consists in deriving partial differential equations describing the phenomena, statement of the solvability and applying a homogenization procedure when the number of the piezoelectric patches goes to infinity.

It should be mentioned that the elasticity part of our model is the Kármán system (see [3] and [4]) with discontinuous coefficients and additional terms related to piezoelectric properties of the patches. Thus, the model is applicable to the case of "great" deformations: the bending must be much less then the longitudinal dimension of the plate (see [5]). Taking into account geometrical nonlinearities is motivated by the fact that linear models are applicable to the case of extremely small bending that must be much less then the thickness of the plate.

2. Mathematical Model

2.1. Piezoelectric media

Assume that deformations of a medium are sufficiently small and use the following conventional form of the strain tensor:

$$d_{lm} := 1/2\left(\frac{\partial u_l}{\partial x_m} + \frac{\partial u_m}{\partial x_l} + \frac{\partial u_k}{\partial x_l}\cdot\frac{\partial u_k}{\partial x_m}\right),$$

where $u_l = y_l(x_1, x_2, x_3, t) - x_l$ is the displacement. As usually, summation over repeating indices is assumed. Latin symbols run from 1 to 3 whenever Greek symbols run from 1 to 2.

Let σ_{ij} be the stress tensor. We consider linear material laws (see [7] and see [6] for more general models):

$$\begin{aligned} \sigma_{ij} &= C_{ijkl}d_{kl} - e_{kij}E_k, \\ D_i &= \varepsilon_{ij}E_j + e_{ikl}d_{kl}. \end{aligned} \tag{1}$$

The coefficients are such that

C_{ijkl}	is the stiffness tensor,
e_{ikl}	is the piezoelectric tensor,
ε_{ij}	is the permittivity tensor.

To cover all possible cases of piezoelectric ceramics, assume that all coefficients may be nonzero and different. This is really the case for the triclinic crystal systems (see [7]).

In conclusion of this section, we give the formula for the density of the energy of a piezoelectric medium (see e.g. [6]):

$$\chi = \frac{1}{2}(\sigma_{ij}d_{ij} - E_iD_i). \tag{2}$$

2.2. A plate with piezoelectric actuators

For simplicity, consider a thin plate supplied with two symmetric patches made of a piezoelectric ceramic. The plate itself consists of a metal, and the horizontal external surfaces of the patches are covered by a metal. Therefore, the voltage can be applied to the patches as shown in Figure 1.

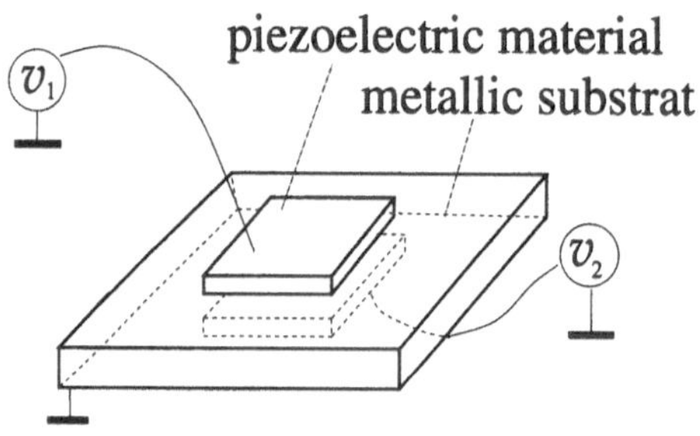

Figure 1.

As usually, we assume the existence of a neutral plane on which all deformations caused by the "pure bending" are equal to zero (see Figure 2). Note that the deformations caused by the stretching of the plate do not vanish on the neutral plane.

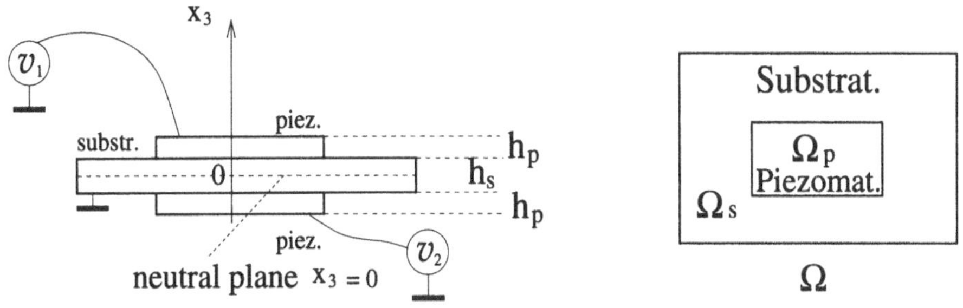

Figure 2.

Let Ω be the projection of the plate onto the plane $x_3 = 0$, Ω_p the projection of the piezopatches onto the plane $x_3 = 0$, and Ω_s the complement of Ω_p, i.e. $\Omega = \Omega_p \cup \Omega_s$ (see Figure 2). Let I_p and I_s denote indicator functions of Ω_p and Ω_s respectively.

2.3. Kirchhoff-Love-Koiter hypothesis

Let Q be a point of the plate, M orthogonal projection of Q onto the neutral plane, u_α, $\alpha = 1, 2$ longitudinal components of the displacement of M, and w is the transversal displacement of M. According to the Kirchhoff-Love-Koiter hypothesis (see e.g. [8]), the components of the strain tensor at the point Q are defined as follows:

$$
\begin{aligned}
&d_{\alpha\beta} = 1/2(u_{\alpha x_\beta} + u_{\beta x_\alpha} + w_{x_\alpha} w_{x_\beta}) - x_3 \cdot w_{x_\alpha x_\beta},\\
&d_{3\alpha} = d_{\alpha 3} = 0, \quad d_{33} = -\frac{1}{C_{3333}} C_{33\alpha\beta} d_{\alpha\beta} + \frac{1}{C_{3333}} e_{k33} E_k.
\end{aligned} \tag{3}
$$

Note that the assumption $d_{3\alpha} = d_{\alpha 3} = 0$, $\alpha = 1, 2$, means the absence of transversal shear strains, which implies the conservation of the normal. The component d_{33} is computed from the relation $\sigma_{33} = 0$, which expresses the in-plain stress condition. Therefore, d_{33} is material and field dependent.

Using (1), (2), (3) and taking into account that $\sigma_{33} = 0$ yield the energy density for the piezopatches:

$$
\chi^p = \frac{1}{2} c^p_{\alpha\beta\gamma\mu} d_{\alpha\beta} d_{\gamma\mu} - e_{i\alpha\beta} d_{\alpha\beta} E_i - \frac{1}{2} \epsilon_{ij} E_i E_j \tag{4}
$$

where

$$
c^p_{\alpha\beta\gamma\mu} = C_{\alpha\beta\gamma\mu} - \frac{C_{33\alpha\beta} C_{33\gamma\mu}}{C_{3333}}, \quad e_{i\alpha\beta} = e_{i\alpha\beta} - \frac{C_{33\alpha\beta} e_{i33}}{C_{3333}}, \quad \epsilon_{ij} = \varepsilon_{ij} + \frac{e_{i33} e_{j33}}{C_{3333}}.
$$

One can verify that the tensors $c^p_{\alpha\beta\gamma\mu}$ and ϵ_{ij} are positive definite for all realistic materials. This means that

$$c^p_{\alpha\beta\gamma\mu} r_{\alpha\beta} r_{\gamma\mu} \geq \nu r_{\alpha\beta} r_{\alpha\beta} \quad \text{and} \quad \epsilon_{ij} E_i E_j \geq \nu E_i E_i$$

for all symmetric $r_{\alpha\beta} \in R^{2\times2}$ and all $E_i \in R^3$, respectively.

Substituting relations (3) into (4), we obtain:

$$\chi^p = \frac{1}{2}c^p_{\alpha\beta\gamma\mu} s_{\alpha\beta} s_{\gamma\mu} + \frac{1}{2}x_3^2 c^p_{\alpha\beta\gamma\mu} w_{x_\alpha x_\beta} w_{x_\gamma x_\mu} - e_{i\alpha\beta} s_{\alpha\beta} E_i - \frac{1}{2}\epsilon_{ij} E_i E_j$$
$$+ x_3 c^p_{\alpha\beta\gamma\mu} s_{\alpha\beta} w_{x_\gamma x_\mu} - x_3 e_{i\alpha\beta} w_{x_\alpha x_\beta} E_i, \tag{5}$$

where

$$s_{\alpha\beta} = 1/2(u_{\alpha x_\beta} + u_{\beta x_\alpha} + w_{x_\alpha} \cdot w_{x_\beta}).$$

The energy density of the substrate is given by the formula

$$\chi^s = \frac{1}{2}c^s_{\alpha\beta\gamma\mu} s_{\alpha\beta} s_{\gamma\mu} + \frac{1}{2}x_3^2 c^s_{\alpha\beta\gamma\mu} w_{x_\alpha x_\beta} w_{x_\gamma x_\mu} + x_3 c^s_{\alpha\beta\gamma\mu} s_{\alpha\beta} w_{x_\gamma x_\mu}, \tag{6}$$

where

$$c^s_{\alpha\beta\gamma\mu} = C_{\alpha\beta\gamma\mu} - \frac{C_{33\alpha\beta} C_{33\gamma\mu}}{C_{3333}}$$

with $C_{\alpha\beta\gamma\mu}$ being the stiffness tensor of the substrate.

2.4. Electric fields in piezoelectric patches

Note that electric fields in the upper and lower patches are independent each from other because of the metal plate between them. The interface between the fields occurs due to deformation of the whole structure. Thus, we need two potential functions to describe electric fields above and below the plate.

$$\bar{E}_i = \frac{\partial\bar{\varphi}}{\partial x_i}, \quad \underline{E}_i = \frac{\partial\underline{\varphi}}{\partial x_i}$$

$$\bar{z} = x_3 - (h_p + h_s)/2$$

$$\underline{z} = x_3 + (h_p + h_s)/2$$

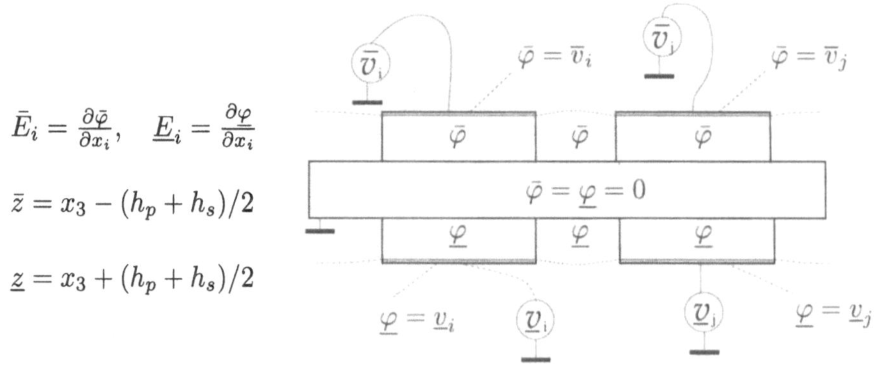

Figure 3. *Fragment of a plate with many piezoelectric patches. Here $\bar{\varphi}$ and $\underline{\varphi}$ are potential functions, $\bar{v}_i$ and $\underline{v}_i$ are applied voltages.*

Using assumption that the potential functions are quadratic in x_3 and taking into account boundary conditions, we obtain for piezoelectric patches:

$$\bar{\varphi}(x_1,x_2,x_3) = \left(\frac{1}{h_p}\bar{z} + \frac{2}{h_p^2}\bar{z}^2\right)\bar{v} + \left(1 - \frac{4}{h_p^2}\bar{z}^2\right)\bar{\phi}(x_1,x_2),$$

$$\underline{\varphi}(x_1,x_2,x_3) = \left(\frac{2}{h_p^2}\underline{z}^2 - \frac{1}{h_p}\underline{z}\right)\underline{v} + \left(1 - \frac{4}{h_p^2}\underline{z}^2\right)\underline{\phi}(x_1,x_2).$$

We have for the free space:

$$\bar{\varphi}(x_1,x_2,x_3) = \left(1 - \frac{4}{h_p^2}\bar{z}^2\right)\bar{\phi}(x_1,x_2),$$

$$\underline{\varphi}(x_1,x_2,x_3) = \left(1 - \frac{4}{h_p^2}\underline{z}^2\right)\underline{\phi}(x_1,x_2).$$

Here, $\bar{\phi}(x_1,x_2)$ and $\underline{\phi}(x_1,x_2)$ are unknown distributions of the electric field above and below the plate along x_1, x_2 plane. The densities $\bar{\chi}^p$ and $\underline{\chi}^p$ of the energy and the total energy F are of the following form:

$$\bar{\chi}^p = \frac{1}{2}c^p_{\alpha\beta\gamma\mu}s_{\alpha\beta}s_{\gamma\mu} + \frac{1}{2}x_3^2 c^p_{\alpha\beta\gamma\mu}w_{x_\alpha x_\beta}w_{x_\gamma x_\mu} - e_{i\alpha\beta}s_{\alpha\beta}\bar{E}_i$$
$$-\frac{1}{2}\epsilon_{ij}\bar{E}_i\bar{E}_j + +x_3 c^p_{\alpha\beta\gamma\mu}s_{\alpha\beta}w_{x_\gamma x_\mu} - x_3 e_{i\alpha\beta}w_{x_\alpha x_\beta}\bar{E}_i.$$

$$\underline{\chi}^p = \frac{1}{2}c^p_{\alpha\beta\gamma\mu}s_{\alpha\beta}s_{\gamma\mu} + \frac{1}{2}x_3^2 c^p_{\alpha\beta\gamma\mu}w_{x_\alpha x_\beta}w_{x_\gamma x_\mu} - e_{i\alpha\beta}s_{\alpha\beta}\underline{E}_i$$
$$-\frac{1}{2}\epsilon_{ij}\underline{E}_i\ \underline{E}_j + +x_3 c^p_{\alpha\beta\gamma\mu}s_{\alpha\beta}w_{x_\gamma x_\mu} - x_3 e_{i\alpha\beta}w_{x_\alpha x_\beta}\underline{E}_i.$$

$$\chi^s = \frac{1}{2}c^s_{\alpha\beta\gamma\mu}s_{\alpha\beta}s_{\gamma\mu} + \frac{1}{2}x_3^2 c^s_{\alpha\beta\gamma\mu}w_{\alpha x_\beta}w_{x_\gamma x_\mu} + x_3 c^s_{\alpha\beta\gamma\mu}s_{\alpha\beta}w_{x_\gamma x_\mu}.$$

$$F = \iint\limits_{\Omega_p}\int\limits_{h_s/2}^{h_s/2+h_p}\bar{\chi}^p dx_3 + \iint\limits_{\Omega_p}\int\limits_{h_s/2\ h_p}^{-h_s/2}\underline{\chi}^p dx_3 + \iint\limits_{\Omega}\int\limits_{-h_s/2}^{h_s/2}\chi^s dx_3$$
$$+\iint\limits_{\Omega_s}\int\limits_{h_s/2}^{h_s/2+h_p}\epsilon_0\bar{E}_i^2 dx_3 + \iint\limits_{\Omega_s}\int\limits_{-h_s/2-h_p}^{-h_s/2}\epsilon_0\ \underline{E}_i^2 dx_3.$$

Here $s_{\alpha\beta} = 1/2(u_{\alpha x_\beta} + u_{\beta x_\alpha} + w_{x_\alpha}w_{x_\alpha})$ is the in-plain strain tensor.

The equations assume the form:

$$\rho w_{tt} + \frac{\partial}{\partial x_\beta}(\tau_{\alpha\beta} w_{x_\alpha}) + \frac{\partial^2}{\partial x_\gamma \partial x_\mu}(\gamma_{\alpha\beta\gamma\mu} w_{x_\alpha x_\beta}) - \theta \frac{\partial}{\partial x_\beta}(e_{\gamma\alpha\beta} w_{x_\alpha}(\bar{\phi}_{x_\gamma} + \underline{\phi}_{x_\gamma}))$$
$$+\theta \frac{\partial^2}{\partial x_\alpha \partial x_\beta}(e_{3\alpha\beta}(\bar{\phi} + \underline{\phi})) - \sigma \frac{\partial^2}{\partial x_\alpha \partial x_\beta}(e_{\gamma\alpha\beta}(\bar{\phi}_{x_\gamma} - \underline{\phi}_{x_\gamma}))$$
$$-h_s^{-1} \frac{\partial}{\partial x_\beta}(e_{3\alpha\beta} w_{x_\alpha}(\bar{v} - \underline{v})) - \ell \frac{\partial^2}{\partial x_\alpha \partial x_\beta}(e_{3\alpha\beta}(\bar{v} + \underline{v})) = 0.$$

$$\rho u_{1tt} - \frac{\partial}{\partial x_\beta}\tau_{1\beta} + \theta \frac{\partial}{\partial x_\beta}(e_{\gamma 1\beta}(\bar{\phi}_{x_\gamma} + \underline{\phi}_{x_\gamma})) + h_s^{-1} \frac{\partial}{\partial x_\beta} e_{31\beta}(\bar{v} - \underline{v}) = 0. \quad (7)$$
$$\rho u_{2tt} - \frac{\partial}{\partial x_\beta}\tau_{2\beta} + \theta \frac{\partial}{\partial x_\beta}(e_{\gamma 2\beta}(\bar{\phi}_{x_\gamma} + \underline{\phi}_{x_\gamma})) + h_s^{-1} \frac{\partial}{\partial x_\beta} e_{32\beta}(\bar{v} - \underline{v}) = 0.$$

$$a \frac{\partial}{\partial x_\alpha}(\epsilon_{\alpha\beta} \frac{\partial}{\partial x_\beta}\bar{\phi}) + b\epsilon_{33}\bar{\phi} + \theta \frac{\partial}{\partial x_\gamma}(e_{\gamma\alpha\beta} s_{\alpha\beta}) + \sigma \frac{\partial}{\partial x_\gamma} e_{\gamma\alpha\beta} w_{x_\alpha x_\beta}$$
$$-\theta \frac{\partial}{\partial x_\alpha} e_{3\alpha\beta} w_{x_\beta} - g\epsilon_{33}\bar{v} = 0.$$

$$a \frac{\partial}{\partial x_\alpha}(\epsilon_{\alpha\beta} \frac{\partial}{\partial x_\beta}\underline{\phi}) + b\epsilon_{33}\underline{\phi} + \theta \frac{\partial}{\partial x_\gamma}(e_{\gamma\alpha\beta} s_{\alpha\beta}) - \sigma e_{\gamma\alpha\beta} \frac{\partial}{\partial x_\gamma} w_{x_\alpha x_\beta}$$
$$-\theta \frac{\partial}{\partial x_\alpha} e_{3\alpha\beta} w_{x_\beta} - g\epsilon_{33}\underline{v} = 0.$$

Here,

$s_{\alpha\beta} = \frac{1}{2}(u_{\alpha x_\beta} + u_{\beta x_\alpha} + w_{x_\alpha} w_{x_\beta})$ is the in-plane strain tensor,

$\tau_{\alpha\beta} = c_{\alpha\beta\lambda\nu}\frac{1}{2}(u_{\alpha x_\beta} + u_{\beta x_\alpha} + w_{x_\alpha} w_{x_\beta})$ is the in-plane stress tensor,

$c_{\alpha\beta\lambda\nu}, \gamma_{\alpha\beta\lambda\nu}, \rho, e_{i\alpha\beta}, \epsilon_{\alpha\beta}$ are discontinuous functions of the form:

$$c_{\alpha\beta\lambda\nu}(x_1, x_2) = I_p(x_1, x_2) c^p_{\alpha\beta\lambda\nu} + I_s(x_1, x_2) c^s_{\alpha\beta\lambda\nu}.$$

The quantities $a, b, \theta, \sigma, \ell$ are some constants that can be easily computed from the problem data.

2.5. Weak formulation

Introduce the notation:

$$\vec{u} = (u_1, u_2, w), \quad \vec{\psi} = (\psi_1, \psi_2, \psi_3), \quad \vec{\phi} = (\bar{\phi}, \underline{\phi}), \quad \vec{\eta} = (\bar{\eta}, \underline{\eta}), \quad \vec{v} = (\bar{v}, \underline{v}),$$

$$H = (L_2(\Omega))^3, \quad V = \left(H_0^1(\Omega)\right)^2 \times H_0^2(\Omega), \quad \Phi = H^1(\Omega) \times H^1(\Omega).$$

Then the weak formulation looks like this:

$$(\rho \vec{u}_{tt}, \vec{\psi}) + \omega(\vec{u}, \vec{\psi}) - b(\vec{u}, \vec{\phi}, \vec{\psi}) - e(\vec{u}, \vec{\psi}, \vec{v}) = 0,$$
$$a(\vec{\phi}, \vec{\eta}) + \hat{b}(\vec{u}, \vec{\eta}, \vec{u}) - g(\vec{\eta}, \vec{v}) = 0,$$
$$\vec{u}(0) = \vec{u}_0, \ \vec{u}_t(0) = \vec{u}_0',$$

where $\omega, b, e, \hat{b}, g$ are some forms that arise from the variation of the total energy.

Functions

$$\vec{u} \in L_2(0,T;V) \cap H^1(0,T;H), \quad \vec{\phi} \in L^2(0,T;\Phi), \quad \vec{u}(0) = \vec{u}_0.$$

are called energy solution, if

$$\int_0^T \Big\{ -(\rho\vec{u}_t, \vec{\psi}_t) + \omega(\vec{u}, \vec{\psi}) - b(\vec{u}, \vec{\phi}, \vec{\psi}) - e(\vec{u}, \vec{\psi}, \vec{v}) \Big\} dt - (\rho\vec{u}'_0, \vec{\psi}(0)) = 0, \quad (8)$$

$$\int_0^T \Big\{ a(\vec{\phi}, \vec{\eta}) + \hat{b}(\vec{u}, \vec{\eta}, \vec{u}) - g(\vec{\eta}, \vec{v}) \Big\} dt = 0, \quad (9)$$

for all $\vec{\psi} \in L_2(0,T;V) \cap H^1_T(0,T;H)$, $\vec{\eta} \in L^2(0,T;\Phi)$ (the index "T" denotes that functions vanish at $t = T$).

Theorem 2.1. (Solvability) *If* $\vec{u}_0 \in V$, $\vec{u}'_0 \in H$, $\vec{v} \in H^1(0,T;R^2)$. *Then the system has an energy solution such that:*

$$\vec{u} \in L_\infty(0,T;V), \quad \vec{u}_t \in L_\infty(0,T;H), \quad \vec{\phi} \in L_\infty(0,T;\Phi). \qquad \blacksquare$$

Proof (sketch). Let $\vec{u}^m$ are Galerkin approximations. Then

$$(\rho\vec{u}^m_t, \vec{u}^m_t) + \hat{\omega}(\vec{u}^m, \vec{u}^m) + a(\vec{\phi}^m, \vec{\phi}^m) \le 2\int_0^t \Big\{ |g(\vec{\phi}^m, \vec{v}_t)|$$

$$+ |\hat{e}(\vec{u}^m, \vec{u}^m, \vec{v}_t)| \Big\} dt + 2|\hat{e}(\vec{u}^m, \vec{u}^m, \vec{v})| + C,$$

where $\hat{\omega}$ and $\hat{e}$ are some slight modifications of the forms ω and e. Using properties of the forms $\hat{\omega}, a, g, \hat{e}$ yields:

$$\|\vec{u}^m_t\|^2_H + \|s^m_{\alpha\beta}\|^2_{L_2(\Omega)} + \|w^m_{x_\alpha x_\beta}\|^2_{L_2(\Omega)} + \|\vec{\phi}^m\|^2_\Phi \le C.$$

The use of Korn's inequality yields:

$$\|\vec{u}^m_t\|^2_H + \|\vec{u}^m\|^2_V + \|\vec{\phi}^m\|^2_\Phi \le C. \quad (10)$$

Therefore:

$$\{\vec{u}^m\} \to \vec{u}^0 \quad \text{weak* in} \quad L_\infty(0,T;V),$$

$$\{\vec{u}^m_t\} \to \vec{u}^0_t \quad \text{weak* in} \quad L_\infty(0,T;H),$$

$$\{\vec{\phi}^m\} \to \vec{\phi}^0 \quad \text{weak* in} \quad L_\infty(0,T;\Phi),$$

$$\{w^m\} \to w^0 \quad \text{in} \quad C([0,T];W^{1,q}(\Omega)\forall q > 1.$$

This is sufficient to prove the passage to the limit in equations (8), (9) (see [1] for technical details).

3. Homogenization

Assume that the piezoelectric patches form a self-similar structure that can be refined by varying a parameter ε.

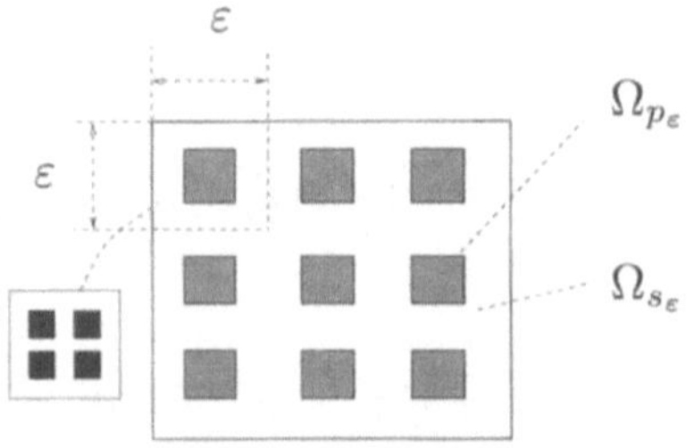

Figure 4. *Refinement of each cell if* $\varepsilon := \varepsilon/2$.

We assume that $\underline{v} = 0$ and $\underline{\phi} = 0$ to obtain a simplified system from (7):

$$\rho w^\varepsilon_{tt} + \frac{\partial}{\partial x_\beta}(\tau^\varepsilon_{\alpha\beta} w^\varepsilon_{x_\alpha}) + \frac{\partial^2}{\partial x_\gamma \partial x_\mu}(\gamma_{\alpha\beta\gamma\mu} w^\varepsilon_{x_\alpha x_\beta}) - \theta \frac{\partial}{\partial x_\beta}(e_{\gamma\alpha\beta} w^\varepsilon_{x_\alpha} \phi^\varepsilon_{x_\gamma}) +$$

$$\theta \frac{\partial^2}{\partial x_\alpha \partial x_\beta}(e_{3\alpha\beta}\phi^\varepsilon) - \sigma \frac{\partial^2}{\partial x_\alpha \partial x_\beta}(e_{\gamma\alpha\beta}\phi^\varepsilon_{x_\gamma}) - h_s^{-1} \frac{\partial}{\partial x_\beta}(e_{3\alpha\beta} w^\varepsilon_{x_\alpha} v^\varepsilon) - \ell \frac{\partial^2}{\partial x_\alpha \partial x_\beta}(e_{3\alpha\beta} v^\varepsilon) = 0.$$

$$\rho u^\varepsilon_{\alpha tt} - \frac{\partial}{\partial x_\beta}\tau^\varepsilon_{\alpha\beta} + \theta \frac{\partial}{\partial x_\beta}(e_{\gamma\alpha\beta}\phi^\varepsilon_{x_\gamma}) + h_s^{-1}\frac{\partial}{\partial x_\beta}(e_{3\alpha\beta} v^\varepsilon) = 0, \quad \alpha = 1, 2.$$

$$a \frac{\partial}{\partial x_\alpha}(\epsilon_{\alpha\beta}\frac{\partial}{\partial x_\beta}\phi^\varepsilon) + b\epsilon_{33}\phi^\varepsilon + \theta \frac{\partial}{\partial x_\gamma}(e_{\gamma\alpha\beta} s^\varepsilon_{\alpha\beta}) + \sigma \frac{\partial}{\partial x_\gamma} e_{\gamma\alpha\beta} w^\varepsilon_{x_\alpha x_\beta}$$
$$-\theta \frac{\partial}{\partial x_\alpha} e_{3\alpha\beta} w^\varepsilon_{x_\beta} - g\epsilon_{33} v^\varepsilon = 0.$$

Let $Y = [0,1] \times [0,1]$ be the unit square. Assume that the equations are rewritten in the form such that the coefficients are Y-periodic functions of $\frac{x}{\varepsilon}$. For example:

$$\tau^\varepsilon_{\alpha\beta} = C_{\alpha\beta\lambda\nu}\left(\tfrac{x}{\varepsilon}\right)\left(u^\varepsilon_{\lambda\, x_\nu} + u^\varepsilon_{\nu\, x_\lambda} + w^\varepsilon_{x_\lambda} w^\varepsilon_{x_\nu}\right).$$

3.1. Two-scale convergence

The tool for derivation of limiting equations is two-scale convergence. For pioneering works on two-scale convergence for time-independent problems we refer to [10] and [11]. Two-scale convergence for time-dependent problems was considered in [12]. These results were generalized in [13]. Let us remember the definition of two-scale convergence.

Definition 3.1. **(Definition 6.8 of Haller** [13]**)** *Let* $v^\varepsilon \in L_2\big((0,T)\times\Omega\big)$, $v^0 \in L_2\big((0,T)\times\Omega\times Y\big)$. *We say that* $v^\varepsilon \overset{2-scale}{\longrightarrow} v^0$, *if*

$$\lim_{\varepsilon\to 0}\int_0^T \int_\Omega v^\varepsilon(t,x)\psi(t,x,\tfrac{x}{\varepsilon})\,dxdt = \int_0^T \int_\Omega \int_Y v^0(t,x,y)\psi(t,x,y)\,dydxdt$$

for all $\psi \in \overset{\circ}{C}{}^{\infty}_{0,T}((0,T)\times\Omega; C^{\infty}_{\#}(Y))$. *The last notation introduces the space of infinitely differentiable functions from* $(0,T)\times\Omega$ *into* $C^{\infty}_{\#}(Y)$ *which vanish on* $\partial\Omega$, *at* $t=0$, *and* $t=T$ *along with all derivatives, where* $C^{\infty}_{\#}(Y)\subset C^{\infty}(R^2)$ *is the subspace of periodic functions that have equal values on opposite edges of* Y.

It is proved (Theorem 6.15 of Haller [13]) that all properties of two-scale convergence hold, if the test functions in the above definition are replaced by more general functions of the form: $\psi(t,x,y)=\alpha(t,x)\beta(y)\sigma(t,x,y)$, where $\alpha\in L_{\infty}(Q)$, $\beta(y)\in L_{\infty\#}(Y)$, and $\sigma\in C^{\infty}((0,T)\times\Omega; C^{\infty}_{\#}(Y))$ (not necessary vanishes).

Proposition 3.2. (Theorem 6.12 of Haller [13]). *Under condition* $\|\vec{u}^{\varepsilon}_t\|^2_H+\|\vec{u}^{\varepsilon}\|^2_V+\|\vec{\phi}^{\varepsilon}\|^2_{\Phi}\le C$, *there exist auxiliary functions* $\tilde{w}(t,x,y)$, $\tilde{u}_m(t,x,y)$ *and* $\tilde{\phi}(t,x,y)$ *such that*

$$w^{\varepsilon}_{x_\alpha x_\beta}(t,x)\overset{2-scale}{\longrightarrow} w_{x_\alpha x_\beta}(t,x)+\tilde{w}_{y_\alpha y_\beta}(t,x,y),$$

$$u^{\varepsilon}_{m\,x_\alpha}(t,x)\overset{2-scale}{\longrightarrow} u_{m\,x_\alpha}(t,x)+\tilde{u}_{m\,y_\alpha}(t,x,y),\quad m=1,2,$$

$$\phi^{\varepsilon}_{x_\alpha}(t,x)\overset{2-scale}{\longrightarrow} \phi_{x_\alpha}(t,x)+\tilde{\phi}^{\varepsilon}_{y_\alpha}(t,x,y),$$

where $w(t,x)$, $u_m(t,x)$, *and* $\phi(t,x)$ *are* L_2*-limits of* $w^{\varepsilon}(t,x)$, $u^{\varepsilon}_m(t,x)$, *and* $\phi^{\varepsilon}(t,x)$.

Some analysis shows that the auxiliary functions can be found in the form:

$$\tilde{w}=P_{\alpha\beta}(y)w_{x_\alpha x_\beta}(t,x)+Q_\alpha(y)\phi_{x_\alpha}(t,x)+D(y)\phi(t,x)+R(y)v(t,x),$$

$$\tilde{u}_m=O_{m\alpha\beta}s_{\alpha\beta}+G_{m\,\alpha\beta}(y)w_{x_\alpha x_\beta}(t,x)+H_{m\,\alpha}(y)\phi_{x_\alpha}(t,x)+J_{m\,\alpha}(y)v(t,x),$$

$$\tilde{\phi}=S_\alpha(y)\phi_{x_\alpha}(t,x)+T_{\alpha\beta}(y)w_{x_\alpha x_\beta}(t,x)+V_\alpha(y)w_{x_\alpha}(t,x)+W_{\alpha\beta}(y)s_{\alpha\beta}(t,x),$$

where $P_{\alpha\beta}, Q_\alpha, D, R, G_{m\,\alpha\beta}, H_{m\,\alpha}, J_{m\,\alpha}, T_{\alpha\beta}, V_\alpha$ and $W_{\alpha\beta}$ are new unknown functions. To find these functions and obtain limiting equations defining $w(t,x)$, $u_m(t,x)$, and $\phi(t,x)$, the following test functions are being used for the weak formulation of the equations of section 3:

$$\psi(t,x)=\psi_1(t,x)+\varepsilon^2\psi_2(t,x,x/\varepsilon)$$

for the first equation and

$$\eta(t,x)=\eta_1(t,x)+\varepsilon\eta_2(t,x,x/\varepsilon)$$

for the second, third and forth equations.

3.2. Cell equations

Using just the same arguments as in [2], we obtain the so-called cell equations defining the auxiliary functions.

$$\begin{cases}
\dfrac{\partial^2}{\partial y_\mu\partial y_\nu}\left[\gamma_{ij\mu\nu}\left(\delta_{in}\delta_{jm}+\dfrac{\partial^2 P_{nm}}{\partial y_i\partial y_j}\right)-\sigma e_{i\mu\nu}\dfrac{\partial T_{nm}}{\partial y_i}\right]=0,\\[2ex]
\dfrac{\partial}{\partial y_\mu}\left[a\epsilon_{i\mu}\dfrac{\partial T_{nm}}{\partial y_i}+\dfrac{1}{2}\theta e_{\mu ij}\left(\dfrac{\partial G_{inm}}{\partial y_j}+\dfrac{\partial G_{jnm}}{\partial y_i}\right)+\sigma e_{\mu ij}\left(\delta_{in}\delta_{jm}+\dfrac{\partial^2 P_{nm}}{\partial y_i\partial y_j}\right)\right]=0,\\[2ex]
\dfrac{\partial}{\partial y_\mu}\left[\dfrac{1}{2}C_{ijk\mu}\left(\dfrac{\partial G_{inm}}{\partial y_j}+\dfrac{\partial G_{jnm}}{\partial y_i}\right)+\theta e_{ik\mu}\dfrac{\partial T_{nm}}{\partial y_i}\right]=0.
\end{cases}$$

$$\begin{cases} \dfrac{\partial^2}{\partial y_\mu \partial y_\nu}\left[\gamma_{ij\mu\nu}\dfrac{\partial^2 Q_n}{\partial y_i \partial y_j} - \sigma e_{i\mu\nu}\left(\delta_{in} + \frac{\partial S_n}{\partial y_i}\right)\right] = 0, \\ \dfrac{\partial}{\partial y_\mu}\left[a\epsilon_{i\mu}\left(\delta_{in} + \frac{\partial S_n}{\partial y_i}\right) + \sigma e_{\mu ij}\dfrac{\partial^2 Q_n}{\partial y_i \partial y_j} + \frac{1}{2}\theta e_{\mu ij}\left(\frac{\partial H_{in}}{\partial y_j} + \frac{\partial H_{jn}}{\partial y_i}\right)\right] = 0, \\ \dfrac{\partial}{\partial y_\mu}\left[\frac{1}{2}C_{ijk\mu}\left(\frac{\partial H_{in}}{\partial y_j} + \frac{\partial H_{jn}}{\partial y_i}\right) + \theta e_{ik\mu}\delta_{in}\frac{\partial S_n}{\partial y_i}\right] = 0. \end{cases}$$

$$\begin{cases} \dfrac{\partial}{\partial y_\mu}\left[C_{ijk\mu}\left(\delta_{in}\delta_{jm} + \frac{1}{2}\left(\frac{\partial O_{inm}}{\partial y_j} + \frac{\partial O_{jnm}}{\partial y_i}\right)\right) + \theta e_{ik\mu}\frac{\partial W_{nm}}{\partial y_i}\right] = 0, \\ \dfrac{\partial}{\partial y_\mu}\left[a\epsilon_{\mu i}\frac{\partial W_{nm}}{\partial y_i} + \theta e_{\mu ij}\left(\delta_{in}\delta_{jm} + \frac{1}{2}\left(\frac{\partial O_{inm}}{\partial y_j} + \frac{\partial O_{jnm}}{\partial y_i}\right)\right)\right] = 0. \end{cases}$$

Here, $n, m = 1, 2; \quad m \le n; \quad k = 1, 2.$

$$\begin{cases} \dfrac{\partial^2}{\partial y_\mu \partial y_\nu}\left[\gamma_{ij\mu\nu}\dfrac{\partial^2 R}{\partial y_i \partial y_j} - \ell e_{3\mu\nu}\right] = 0, \\ \dfrac{\partial}{\partial y_\mu}\left[a\epsilon_{\mu i}\frac{\partial V_n}{\partial y_i} - \theta e_{3\mu n}\right] = 0, \\ \dfrac{\partial}{\partial y_\mu}\left[\frac{1}{2}C_{ijk\mu}\left(\frac{\partial J_i}{\partial y_j} + \frac{\partial J_j}{\partial y_i}\right) + h_s^{-1} e_{3k\mu}\right] = 0, \\ \dfrac{\partial^2}{\partial y_\mu \partial y_\nu}\left[\gamma_{ij\mu\nu}\dfrac{\partial^2 D}{\partial y_i \partial y_j} + \theta e_{3\mu\nu}\right] = 0. \end{cases}$$

Here, $n = 1, 2, \quad k = 1, 2$. The equations should be solved on the unit square Y with periodic boundary conditions on the opposite edges. Note that on the contrary to the case of [2], each cell equation contains several auxiliary functions.

3.3. Limiting equations

The derivation of the limiting equations is being done using the technique of [2].

$$\hat{\rho} w_{tt} + \frac{\partial}{\partial x_\beta}(\hat{\tau}_{\alpha\beta} w_{x_\alpha}) + \frac{\partial^2}{\partial x_\gamma \partial x_\mu}(\hat{\gamma}_{\alpha\beta\gamma\mu} w_{x_\alpha x_\beta}) - \frac{\partial}{\partial x_\beta}(\hat{m}_{\gamma\alpha\beta} w_{x_\alpha}\phi_{x_\gamma}) -$$

$$-\underline{\frac{\partial}{\partial x_\beta}(\hat{p}_{\mu\nu\alpha\beta} w_{x_\alpha} w_{x_\mu x_\nu})} - \underline{\frac{\partial}{\partial x_\beta}(\hat{t}_{\mu\alpha\beta} w_{x_\alpha} w_{x_\mu})} - \underline{\frac{\partial}{\partial x_\beta}(\hat{f}_{\mu\nu\alpha\beta} w_{x_\alpha} s_{\mu\nu})} +$$

$$+ \frac{\partial^2}{\partial x_\alpha \partial x_\beta}(\hat{g}_{\alpha\beta}\phi) - \frac{\partial^2}{\partial x_\alpha \partial x_\beta}(\hat{e}_{\gamma\alpha\beta}\phi_{x_\gamma}) - \frac{\partial}{\partial x_\beta}(\hat{q}_{\alpha\beta} w_{x_\alpha} v) - \frac{\partial^2}{\partial x_\alpha \partial x_\beta}(\hat{r}_{\alpha\beta} v) = 0.$$

$$\hat{\rho} u_{\alpha tt} - \frac{\partial}{\partial x_\beta}\hat{\tau}_{\alpha\beta} + \frac{\partial}{\partial x_\beta}(\hat{\ell}_{\gamma\alpha\beta}\phi_{x_\gamma}) + \underline{\frac{\partial}{\partial x_\beta}(\hat{d}_{\mu\nu\alpha\beta} w_{x_\mu x_\nu})} + \underline{\frac{\partial}{\partial x_\beta}(\hat{h}_{\mu\alpha\beta} w_{x_\mu})} + \frac{\partial}{\partial x_\beta}(\hat{q}_{\alpha\beta} v) = 0.$$

$$\frac{\partial}{\partial x_\alpha}(\hat{\epsilon}_{\alpha\beta}\frac{\partial}{\partial x_\beta}\phi) + \hat{b}\phi + \frac{\partial}{\partial x_\gamma}(\hat{k}_{\gamma\alpha\beta} s_{\alpha\beta}) + \frac{\partial}{\partial x_\gamma}(\hat{n}_{\gamma\alpha\beta} w_{x_\alpha x_\beta}) - \frac{\partial}{\partial x_\alpha}\hat{o}_{\alpha\beta} w_{x_\beta} - \hat{c} v = 0.$$

All of the coefficients of the limiting equations are constant and can be easily computed from the auxiliary functions. For example:

$$\hat{\tau}_{\alpha\beta} = \hat{C}_{\alpha\beta\lambda\nu}\left(u_{\lambda x_\nu} + u_{\nu x_\lambda} + w_{x_\lambda} w_{x_\nu}\right),$$

where

$$\hat{C}_{nmk\mu} = \left\langle C_{ijk\mu}\left(\delta_{in}\delta_{jm} + \frac{1}{2}\left(\frac{\partial O_{inm}}{\partial y_j} + \frac{\partial O_{jnm}}{\partial y_i}\right)\right)\right\rangle.$$

Moreover,

$$\hat{\gamma}_{nm\mu\nu} = \left\langle \gamma_{ij\mu\nu}\left(\delta_{in}\delta_{jm} + \frac{\partial^2 P_{nm}}{\partial y_i \partial y_j}\right) - \sigma e_{i\mu\nu}\frac{\partial T_{nm}}{\partial y_i}\right\rangle,$$

$$\hat{p}_{nmk\mu} = \left\langle \frac{1}{2} C_{ijk\mu}\left(\frac{\partial G_{inm}}{\partial y_j} + \frac{\partial G_{jnm}}{\partial y_i}\right) + \theta e_{ik\mu}\frac{\partial T_{nm}}{\partial y_i}\right\rangle,$$

$$\hat{d}_{nmk\mu} = \left\langle \theta e_{ik\mu}\frac{\partial T_{nm}}{\partial y_i}\right\rangle, \quad \hat{\epsilon}_{k\mu} = \left\langle \epsilon_{kj}\left(\delta_{j\mu} + \frac{\partial S_\mu}{\partial y_j}\right)\right\rangle,$$

$$\hat{q}_{k\mu} = \left\langle \frac{1}{2} C_{ijk\mu}\left(\frac{\partial J_i}{\partial y_j} + \frac{\partial J_j}{\partial y_i}\right) + \theta e_{3k\mu}\right\rangle,$$

and so on. Here, $\langle g \rangle := \int\int_Y g(y)dy$ is the mean value of a function on Y.

4. Conclusion

On the contrary to [2], some new nonlinear terms (they are underlined) have appeared. The concept of two-scale convergence provides the existence of a weak solution where all time derivatives are shifted to the test function. If we add to the first equation a term of the type $-\Delta w_{tt}$ that expresses the inertia of mechanical moments in Kármán systems, we can prove the existence of energy-solutions. Now the authors try to extend the results of [2] (unique strong solvability of the limiting equations and a strong convergence of all solution of the original system to this unique solution) to the fully coupled model.

References

[1] K. -H. Hoffmann and N. D. Botkin, *Oscillations of nonlinear thin plates excited by piezoelectric patches,* ZAMM: Z. angew. Math. Mech., **78** (1998), 495–503.

[2] K. -H. Hoffmann and N. D. Botkin, *Homogenization of von Kármán plates excited by piezoelectric patches,* ZAMM: Z. angew. Math. Mech., **133** (1999), 191–200.

[3] J-P. Puel and M. Tucsnak, *Global Existence for the Full Kármán System,* Appl. Math. Optimiz., **34** (1996), 139–160.

[4] I.Lasiecka and R. Trigiani, *Control theory for partial equations: continuous and approximation theories.II. Abstract hyperbolic-like systems over a finite time horizon.* Encyclopedia of mathematics and its applications **75**, Edited by G.-C.Rota, Cambridge University Press **2000**.

[5] L. D. Landau and E. M. Lifschitz, *Elastizitätstheorie.* Akademie–Verlag (Berlin) **1975**.

[6] G. A. Maugin, *Continuum mechanics of electromagnetic solids.* North-Holland series in Applied Mathematics and mechanics. **33**, North-Holland **1987**.

[7] J. Zelenka, *Piezoelectric Resonators and their Applications.* Studies in Electrical and Electronic Engineering. **24**, Elsevier **1986**.

[8] M. Bernadou, *Finite Element Methods for Thin Shell Problems.* Wiley (Chichester) **1996**.

[9] J. Simon, *Compact sets in the space $L^p(0,T;b)$,* Ann. Mat. Pura Appl., **IV (146)** (1987), 65–96.

[10] G. Nguetseng, *A general convergence result for a functional related to the theory of homogenization,* SIAM J. Math. Anal., **20 (3)** (1989), 608–623.

[11] G. Allaire, *Homogenization and two-scale Convergence,* SIAM J. Math. Anal., **23 (6)** (1992), 1482–1518.

[12] G. Allaire, *Homogenization of the unsteady Stokes equations in porous media,* in "Progress in partial differential equations: calculus of variations, applications" Pitman Research Notes in Mathematics Series. C. Bandle et al. eds, Longman Higher Education, New York. **267**, (1992), 109–123.

[13] H. Haller, *Verbundwerkstoffe mit Formgedächtnislegierung - Mikromechanische Modellierung und Homogenisierung,* Dissertation, TU-München **1997**.

[14] J. L. Lions and E. Magenes, *Non-homogeneous Boundary Value Problems and applications I,* Springer-Verlag (Berlin - Heidelberg - New York) **1972**.

Stiftung CAESAR
center of advanced european studies and research
Friedensplatz 16
D-53111 Bonn
E-mail address: hoffmann@caesar.de, botkin@caesar.de

International Series of Numerical Mathematics, Vol. 139, 119–131

Topology Optimization of High Power Electronic Devices

Ronald H.W. Hoppe, Svetozara Petrova, and Volker Schulz

Abstract. Modern high power electronic devices such as converter modules used as electric drives for high power electromotors are characterized by extremely high switched currents and very fast switching times. The avoidance of significant losses in the power transmission due to parasitic inductivities requires a subtle layout of the devices. Using the electric conductivity as a design parameter and electromagnetic potentials associated with the eddy currents equations as the state variables, the design issue gives rise to a topology optimization with both equality and inequality constraints where the design objective is to distribute the material in such a way that the electromagnetic energy dissipation is minimized. Based on appropriate finite element approximations of the eddy currents equations, for the numerical solution of the discretized optimization problem we suggest a primal-dual Newton interior-point method with a hierarchy of two merit functions and a watchdog strategy for convergence monitoring.

1. Introduction

The optimal design of mechanical structures described by continuum mechanical models is by now a well established discipline both with regard to mathematical theory, numerical simulation, and engineering applications. It includes shape and topology optimization as well as the design of composites by homogenization approaches (cf., e.g., the textbooks [5, 17, 28, 31], and the references therein). On the other hand, the use of modern discretization and numerical solution techniques such as multigrid and domain decomposition methods in an optimization framework, in particular their appropriate combination with advanced optimization approaches, is still in its infancy (cf., e.g., [7, 10, 18, 22, 23, 25], and [30]).

As far as the optimal design and layout of electronic devices and systems are concerned, a lot of work has been done in electric circuitry with emphasis on the application of discrete optimization techniques (cf., e.g., [1] and [11]), but considerably less work has been devoted to the optimization of devices and systems whose operational behavior is strongly dictated by Maxwell's equations.

Key words and phrases. topology optimization, high power electronic devices, eddy currents, edge elements, primal-dual interior-point methods, watchdog strategy.

In this contribution, we consider an optimal design problem arising in high power electronics, namely the layout of converter modules that are used as electric drives for high power electromotors. The objective is to minimize power losses due to parasitic inductivities by an optimal distribution of the material. From a mathematical point of view, this leads to a topology optimization problem where the design variable, the electric conductivity, and the state variables, the generated electromagnetic fields resp. the associated potentials, are subject to equality and inequality constraints. In particular, the scalar electric potential and the magnetic vector potential are required to satisfy the potential formulation of the eddy currents equations given by the quasistationary limit of Maxwell's equations. As far as the numerical solution is concerned, we use finite element methods based on curl-conforming edge elements for the magnetic vector potential and nonconforming P1 elements (Crouzeix-Raviart elements) for the scalar electric potential. The discretized optimization problem is then solved by a primal-dual Newton interior-point method featuring logarithmic barrier functions to take care of the inequality constraints and a simultaneous sequential quadratic programming approach for the resulting equality constrained minimization subproblems. The convergence to a local minimizer is monitored by a hierarchy of two merit functions used within an appropriate watchdog strategy.

2. The Topology Optimization Problem

A typical example for a high power electric device is a converter module designed to convert dc into ac or vice versa and to be used in electric drives for high power electromotors. As shown in Figure 1, a converter module consists of modern semiconductor devices such as IGBTs (Insulated Gate Bipolar Transistors) or GTOs (Gate Turn-Off Thyristors) interconnected and linked with the high power voltage source and load by copper made bus bars. The IGBTs and GTOs that can be viewed as valves for the electric currents admit switching times of less than 100 nanoseconds and switched currents up to five kiloamperes. Figure 2 shows the typical 3D geometry of a bus bar with several ports where the semiconductor devices can be attached.

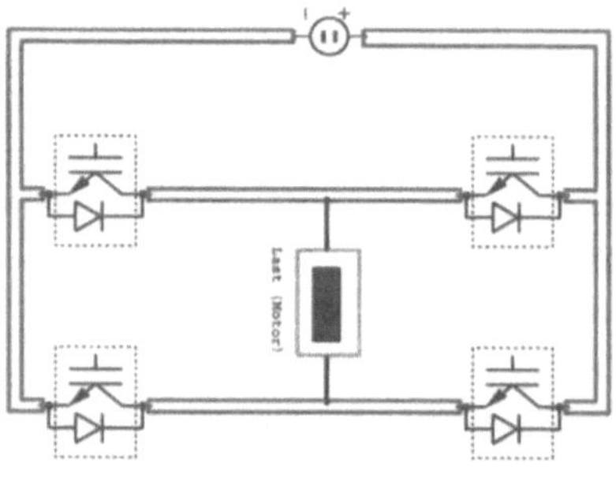

Fig. 1: Converter Module

Fig. 2: Geometry of a bus bar

The problem that occurs is that due the fast switching times and steep current ramps, eddy currents build up inside the bus bars causing parasitic inductivities that lead to a considerable loss in the power transmission (cf. [12]). Therefore, the primary goal is to design the bus bars in such a way that the energy dissipation is minimized. It is known that the topology of the bus bars plays a prominent role in so far as it has a significant impact on the distribution and size of the generated eddy currents. Consequently, the task is to distribute the material in an optimal way. From a mathematical point of view, the problem will be stated as a topology optimization problem with constraints on the state and design variables.

The eddy currents are described by the quasistationary limit of Maxwell's equations

$$\frac{\partial \mathbf{B}}{\partial t} + \operatorname{curl} \mathbf{E} = \mathbf{0} \ , \quad \operatorname{div} \mathbf{B} = 0 \ , \quad \operatorname{curl} \mathbf{H} = \mathbf{J} \ , \tag{1}$$

$$\mathbf{B} = \mu \mathbf{H} \ , \quad \mathbf{J} = \sigma \mathbf{E} \ , \tag{2}$$

where $\mathbf{E}, \mathbf{H}$ denote the electric and the magnetic field, $\mathbf{B}$ is the magnetic induction, $\mathbf{J}$ stands for the current density, and the material parameters μ, σ refer to the magnetic permeability and the electric conductivity, respectively. Following [6], we resort to a potential formulation by introducing a scalar electric potential φ and a magnetic vector potential $\mathbf{A}$ according to

$$\mathbf{E} = -\operatorname{grad} \varphi - \frac{\partial \mathbf{A}}{\partial t} \ , \quad \mathbf{B} = \operatorname{curl} \mathbf{A} \ . \tag{3}$$

Considering a module $\Omega = \cup_{\nu=1}^{N} \Omega_\nu$ with N bars Ω_ν, $1 \le \nu \le N$, each bar containing N_ν ports $\Gamma_{\nu\alpha}$, $1 \le \alpha \le N_\nu$, and denoting by $I_{\nu\alpha}$ the flux at the port $\Gamma_{\nu\alpha}$, we are thus led to the following coupled system of PDEs

$$\operatorname{div} (\sigma \operatorname{grad} \varphi) = 0 \quad \text{in} \quad \Omega \ , \tag{4}$$

$$\sigma\, \mathbf{n} \cdot \operatorname{grad} \varphi = \begin{cases} -I_{\nu\alpha}(t) & \text{on } \Gamma_{\nu\alpha} \\ 0 & \text{elsewhere} \end{cases} \tag{5}$$

$$\sigma \frac{\partial \mathbf{A}}{\partial t} + \operatorname{curl} \mu^{-1} \operatorname{curl} \mathbf{A} = \begin{cases} -\sigma \operatorname{grad} \varphi & \text{in } \Omega \\ 0 & \text{in } \mathbf{R}^3 \setminus \bar{\Omega} \end{cases} \tag{6}$$

with appropriate initial and boundary conditions. The energy dissipation on $[0, T]$ is then given by

$$L = \int_0^T \int_\Omega \mathbf{J} \cdot \mathbf{E} \, dx \, dt \ . \tag{7}$$

Taking (1),(2) and (3) into account, we may view L as a functional depending on the conductivity σ which will serve as the design parameter and on the potentials $\varphi, \mathbf{A}$ which are chosen as the state variables. Prescribing the total amount of material in terms of the conductivity and allowing σ to vary between a maximum value $\sigma_{\max}$ (conductivity of copper) and a minimum value $0 < \sigma_{\min} := \varepsilon \ll 1$ which is chosen small but positive in order to keep the ellipticity of the problem, we are faced with the topology optimization problem

$$\inf_{\sigma,\varphi,\mathbf{A}} L(\sigma,\varphi,\mathbf{A}) \tag{8}$$

subject to the equality constraints

$$\varphi \text{ and } \mathbf{A} \text{ satisfy the state equations (4),(5),(6),} \tag{9}$$

$$\int_\Omega \sigma\, dx \;=\; C \tag{10}$$

and the inequality constraints

$$\sigma_{\min} \;\leq\; \sigma \;\leq\; \sigma_{\max} \;. \tag{11}$$

In order to enforce the extreme values $\sigma_{\max}$ and $\sigma_{\min}$, we use the SIMP-approach (Simple Isotropic Material with Penalization) known from structural mechanics. In the present context, it means that we "replace" the conductivity σ by

$$\eta(\sigma) \;=\; \Big(\frac{\sigma - \sigma_{\min} + \varepsilon}{\sigma_{\max} - \sigma_{\min}}\Big)^q \tag{12}$$

with an appropriately chosen penalty parameter $q \geq 1$.

3. The Primal-dual Newton Interior-point Method

In this section, we will present a primal-dual Newton interior-point method for the numerical solution of the discretized optimization problem. Realizing the exterior domain by an artificial exterior boundary and using simplicial triangulations $\mathcal{T}_h^{(I)}, \mathcal{T}_h^{(E)}$ of the interior and exterior domain, the discretization is performed by applying the implicit Euler scheme in time and curl-conforming edge elements of lowest order [26] in space to (6) whereas nonconforming Crouzeix-Raviart elements are used for (4),(5). The conductivity is approximated by elementwise constants, i.e., $\vec{\sigma} = (\sigma_1, \ldots, \sigma_{m_h})^T$, $m_h = \operatorname{card} \mathcal{T}_h^{(I)}$. The discretized state variables are denoted by $\vec{\varphi} = (\varphi_1, \ldots, \varphi_{n_h})^T$ and $\vec{A} = (A_1, \ldots, A_{p_h})^T$ where n_h, p_h are the dimensions of the associated nonconforming resp. edge element spaces. For notational convenience, we comprise them to a vector $\vec{u} = (\vec{\varphi}, \vec{A})$ and refer to

$$A(\vec{\sigma})\,\vec{u} \;=\; \vec{b} \tag{13}$$

as the discretized state equations with $A(\vec{\sigma})$ denoting the matrix of the associated system of equations. The constraints (10) and (11) take the form

$$g(\vec{\sigma}) \;:=\; \sum_{i=1}^{m_h} |\, K_i \,|\; \sigma_i \;=\; C\,, \tag{14}$$

$$\sigma_{\min}\,\vec{e} \;\leq\; \vec{\sigma} \;\leq\; \sigma_{\max}\,\vec{e}\,, \tag{15}$$

where $K_i \in \mathcal{T}_h^{(I)}$, $1 \leq i \leq m_h$, and $\vec{e} := (1, \ldots, 1)^T$.

Denoting the discretized objective functional by $L_h(\vec{u},\vec{\sigma})$, the discrete optimization problem reads as follows:

$$\min_{\vec{u},\vec{\sigma}} L_h(\vec{u},\vec{\sigma}) \tag{16}$$

subject to the constraints (13),(14), and (15).

We note that the discretization can be performed within a multilevel and/or domain decomposition framework by means of multigrid iterative solvers based on edge element discretizations of the implicitly in time discretized equation for the magnetic vector potential (6) and nonconforming P1 approximations of the equation for the scalar electric potential (5) (cf., e.g., [2, 24]) or domain decomposition methods on nonmatching grids with respect to a nonoverlapping geometrically conforming partition of the computational domain dictated by the geometry of the bus bars (cf., e.g., [4, 8, 19, 20, 29]). For adaptive grid refinement/coarsening relying on efficient and reliable a posteriori error estimators we refer to [2, 3].

In contrast to traditional design strategies where the optimization loop consists of the numerical solution of the field equations for the current design followed by a Newton-type procedure for the computation of the increments for the design parameters, we will use an integrated approach by means of a primal-dual Newton interior-point method where the convergence is monitored by a hierarchy of merit functions combined with an appropriate watchdog strategy. Such techniques have been recently developed and tested for nonlinear programming problems (cf., e.g., [13, 14, 16]). Typically, the inequality constraints are taken care of by classical logarithmic barrier functions with a barrier parameter resulting in a parametrized family of minimization subproblems which is then solved by a simultaneous sequential quadratic programming technique.

The first step in the primal-dual interior-point approach is to introduce the logarithmic barrier functions

$$B(\vec{u},\vec{\sigma},p) \; := \; L_h(\vec{u},\vec{\sigma}) \; - \; p\left[\log\left(\vec{\sigma} \; - \; \sigma_{\min}\,\vec{e}\right) \; + \; \log\left(\sigma_{\max}\,\vec{e} \; - \; \vec{\sigma}\right)\right] , \tag{17}$$

where $p > 0$ is a suitably chosen barrier parameter. We consider the family of minimization subproblems

$$\min_{\vec{u},\vec{\sigma}} B(\vec{u},\vec{\sigma},p) \tag{18}$$

subject to the equality constraints

$$A(\vec{\sigma})\,\vec{u} \; = \; \vec{b} \quad , \quad g(\vec{\sigma}) \; = \; C \; . \tag{19}$$

For an isolated local minimum $(\vec{u}^*,\vec{\sigma}^*)$ of (16) it can be shown that for a null sequence $(p_n)_{\mathbf{N}}$ of sufficiently small barrier parameters the minimization problems (18) have solutions $(\vec{u}_n,\vec{\sigma}_n)$ converging to $(\vec{u}^*,\vec{\sigma}^*)$.

The second step is to invoke a simultaneous SQP approach for the solution of (18). To be more specific, the equality constraints (19) are coupled by Lagrangian multipliers leading to the saddle point problem

$$\min_{\vec{u},\vec{\sigma}} \max_{\vec{\lambda},\eta} \mathcal{L}^{(p)}(\vec{u},\vec{\sigma},\vec{\lambda},\eta) \tag{20}$$

for the Lagrangian

$$\mathcal{L}^{(p)}(\vec{u},\vec{\sigma},\vec{\lambda},\eta) \;:=\; B(\vec{u},\vec{\sigma},p) \;+\; \vec{\lambda}^T\,(A(\vec{\sigma})\,\vec{u} - \vec{b}) \;+\; \eta\,(g(\vec{\sigma}) - C)\,. \tag{21}$$

The Karush-Kuhn-Tucker conditions are given by

$$\mathcal{F}^{(p)}(\vec{u},\vec{\sigma},\vec{\lambda},\eta) \;=\; 0\,, \tag{22}$$

where

$$\begin{aligned}
\mathcal{F}_1^{(p)} &= \nabla_{\vec{u}}\,\mathcal{L}^{(p)} &&= \nabla_{\vec{u}}\,L \,+\, A(\vec{\sigma})^T\,\vec{\lambda}\,,\\
\mathcal{F}_2^{(p)} &= \nabla_{\vec{\sigma}}\,\mathcal{L}^{(p)} &&= \partial_{\vec{\sigma}}\,(\vec{\lambda}^T A(\vec{\sigma})\vec{u}) \,+\, \eta\,\nabla_{\vec{\sigma}}\,g(\vec{\sigma}) \,-\, pD_1^{-1}\vec{e} \,+\, pD_2^{-1}\vec{e}\,,\\
\mathcal{F}_3^{(p)} &= \nabla_{\vec{\lambda}}\,\mathcal{L}^{(p)} &&= A(\vec{\sigma})\vec{u} \,-\, \vec{b}\,,\\
\mathcal{F}_4^{(p)} &= \nabla_{\eta}\,\mathcal{L}^{(p)} &&= g(\vec{\sigma}) \,-\, C\,,
\end{aligned}$$

and $D_1 := \operatorname{diag}(\sigma_i - \sigma_{\min})$ and $D_2 := \operatorname{diag}(\sigma_{\max} - \sigma_i)$.

Since for $p \to 0$ the expressions $pD_1^{-1}\vec{e}$ and $pD_2^{-1}\vec{e}$ approximate the complementarity conditions associated with (16), it is standard to introduce $\vec{z} := pD_1^{-1}\vec{e}$ and $\vec{w} := pD_2^{-1}\vec{e}$ as some kind of approximate complementarity. Then, Newton's method is applied to three sets of equations

- primal feasibility $(\vec{u},\vec{\sigma})$,
- dual feasibility $(\vec{\lambda},\eta)$,
- perturbed complementarity $(\vec{z},\vec{w})$

resulting in the linear algebraic system

$$\begin{pmatrix}
0 & \mathcal{L}_{\vec{u}\vec{\sigma}} & \mathcal{L}_{\vec{u}\vec{\lambda}} & 0 & 0 & 0\\
\mathcal{L}_{\vec{\sigma}\vec{u}} & \mathcal{L}_{\vec{\sigma}\vec{\sigma}} & \mathcal{L}_{\vec{\sigma}\vec{\lambda}} & \mathcal{L}_{\vec{\sigma}\eta} & -I & I\\
\mathcal{L}_{\vec{\lambda}\vec{u}} & \mathcal{L}_{\vec{\lambda}\vec{\sigma}} & 0 & 0 & 0 & 0\\
0 & \mathcal{L}_{\eta\vec{\sigma}} & 0 & 0 & 0 & 0\\
0 & Z & 0 & 0 & D_1 & 0\\
0 & -W & 0 & 0 & 0 & D_2
\end{pmatrix}
\begin{pmatrix}
\Delta\vec{u}\\ \Delta\vec{\sigma}\\ \Delta\vec{\lambda}\\ \Delta\eta\\ \Delta\vec{z}\\ \Delta\vec{w}
\end{pmatrix}
= -
\begin{pmatrix}
\nabla_{\vec{u}}\,\mathcal{L}\\ \nabla_{\vec{\sigma}}\,\mathcal{L}\\ \nabla_{\vec{\lambda}}\,\mathcal{L}\\ \nabla_{\eta}\,\mathcal{L}\\ \nabla_{\vec{z}}\,\mathcal{L}\\ \nabla_{\vec{w}}\,\mathcal{L}
\end{pmatrix}. \tag{23}$$

Note that the coefficient matrix is usually referred to as the primal-dual Hessian. Obviously, it is not symmetric but can be easily symmetrized, since the matrices Z and W are diagonal (cf., e.g., [15]). We do not adapt this approach here, but instead perform a block elimination of the increments $\Delta\,\vec{z}$ and $\Delta\,\vec{w}$ yielding the condensed system

$$\begin{pmatrix}
0 & \mathcal{L}_{\vec{u}\vec{\sigma}} & \mathcal{L}_{\vec{u}\vec{\lambda}} & 0\\
\mathcal{L}_{\vec{\sigma}\vec{u}} & \tilde{\mathcal{L}}_{\vec{\sigma}\vec{\sigma}} & \mathcal{L}_{\vec{\sigma}\vec{\lambda}} & \mathcal{L}_{\vec{\sigma}\eta}\\
\mathcal{L}_{\vec{\lambda}\vec{u}} & \mathcal{L}_{\vec{\lambda}\vec{\sigma}} & 0 & 0\\
0 & \mathcal{L}_{\eta\vec{\sigma}} & 0 & 0
\end{pmatrix}
\begin{pmatrix}
\Delta\vec{u}\\ \Delta\vec{\sigma}\\ \Delta\vec{\lambda}\\ \Delta\eta
\end{pmatrix}
= -
\begin{pmatrix}
\nabla_{\vec{u}}\,\mathcal{L}\\ \tilde{\nabla}_{\vec{\sigma}}\,\mathcal{L}\\ \nabla_{\vec{\lambda}}\,\mathcal{L}\\ \nabla_{\eta}\,\mathcal{L}
\end{pmatrix}, \tag{24}$$

where

$$\tilde{\mathcal{L}}_{\vec{\sigma}\vec{\sigma}} \;:=\; \mathcal{L}_{\vec{\sigma}\vec{\sigma}} + D_1^{-1}Z + D_2^{-1}W\;,\quad \tilde{\nabla}_{\vec{\sigma}}\mathcal{L} \;:=\; \nabla_{\vec{\sigma}}\mathcal{L} - D_1^{-1}\nabla_{\vec{z}}\mathcal{L} + D_2^{-1}\nabla_{\vec{w}}\mathcal{L}\,.$$

Following [25], we consider a null space decomposition of the condensed primal-dual Hessian, i.e., we interchange the second and third rows and columns and partition the resulting matrix according to

$$\mathcal{K} = \begin{pmatrix} \mathcal{A} & \mathcal{B}^T \\ \mathcal{B} & \mathcal{D} \end{pmatrix} = \left(\begin{array}{cc|cc} 0 & \mathcal{L}_{\vec{u}\vec{\lambda}} & \mathcal{L}_{\vec{u}\vec{\sigma}} & 0 \\ \mathcal{L}_{\vec{\lambda}\vec{u}} & 0 & \mathcal{L}_{\vec{\lambda}\vec{\sigma}} & 0 \\ \hline \mathcal{L}_{\vec{\sigma}\vec{u}} & \mathcal{L}_{\vec{\sigma}\vec{\lambda}} & \tilde{\mathcal{L}}_{\vec{\sigma}\vec{\sigma}} & \mathcal{L}_{\vec{\sigma}\eta} \\ 0 & 0 & \mathcal{L}_{\eta\vec{\sigma}} & 0 \end{array}\right) . \tag{25}$$

We remark that the first diagonal block

$$\mathcal{A} = \begin{pmatrix} 0 & \mathcal{L}_{\vec{u}\vec{\lambda}} \\ \mathcal{L}_{\vec{\lambda}\vec{u}} & 0 \end{pmatrix}$$

is indefinite, but nonsingular with $\mathcal{L}_{\vec{\lambda}\vec{u}}$ being the stiffness matrix $A(\vec{\sigma})$ associated with the discretized potential equations.

We choose $\tilde{A}(\vec{\sigma})$ as an appropriate approximation of $A(\vec{\sigma})$ realized, for instance, by an SSOR iteration. Then, for

$$\mathcal{K}^R = \begin{pmatrix} I & -\tilde{\mathcal{A}}^{-1}\mathcal{B}^T \\ 0 & I \end{pmatrix} = \left(\begin{array}{cc|cc} I & 0 & -\tilde{A}^{-1}(\vec{\sigma})\mathcal{L}_{\vec{\lambda}\vec{\sigma}} & 0 \\ 0 & I & -\tilde{A}^{-1}(\vec{\sigma})\mathcal{L}_{\vec{u}\vec{\sigma}} & 0 \\ \hline 0 & 0 & I & 0 \\ 0 & 0 & 0 & I \end{array}\right) \tag{26}$$

and taking advantage of the regular splitting

$$\mathcal{K}\mathcal{K}^R = \underbrace{\left(\begin{array}{cc|cc} 0 & A(\vec{\sigma}) & 0 & 0 \\ A(\vec{\sigma}) & 0 & 0 & 0 \\ \hline \mathcal{L}_{\vec{\sigma}\vec{u}} & \mathcal{L}_{\vec{\sigma}\vec{\lambda}} & \tilde{S} & \mathcal{L}_{\vec{\sigma}\eta} \\ 0 & 0 & \mathcal{L}_{\eta\vec{\sigma}} & 0 \end{array}\right)}_{=:\mathcal{M}_1} - \underbrace{\left(\begin{array}{cc|cc} 0 & 0 & \mathcal{L}_{\vec{u}\vec{\sigma}} - A(\vec{\sigma})\tilde{A}^{-1}(\vec{\sigma})\mathcal{L}_{\vec{u}\vec{\sigma}} & 0 \\ 0 & 0 & \mathcal{L}_{\vec{\lambda}\vec{\sigma}} - A(\vec{\sigma})\tilde{A}^{-1}(\vec{\sigma})\mathcal{L}_{\vec{\lambda}\vec{\sigma}} & 0 \\ \hline 0 & 0 & 0 & 0 \\ 0 & 0 & 0 & 0 \end{array}\right)}_{=:\mathcal{M}_2}$$

where

$$\tilde{S} := \tilde{\mathcal{L}}_{\vec{\sigma}\vec{\sigma}} - \mathcal{L}_{\vec{\sigma}\vec{u}}\,\tilde{A}^{-1}(\vec{\sigma})\,\mathcal{L}_{\vec{\lambda}\vec{\sigma}} - \mathcal{L}_{\vec{\sigma}\vec{\lambda}}\,\tilde{A}^{-1}(\vec{\sigma})\,\mathcal{L}_{\vec{u}\vec{\sigma}} ,$$

we perform the transforming iterations

$$\Delta\Psi^{\nu+1} = \Delta\Psi^{\nu} + \mathcal{K}^R\mathcal{M}_1^{-1}(d - \mathcal{K}\Delta\Psi^{\nu}) , \tag{27}$$

where $\Delta\Psi := (\Delta\vec{u}, \Delta\vec{\lambda}, \Delta\vec{\sigma}, \Delta\eta)^T$. The new iterate $\Psi^{(\mathrm{new})} := (\vec{u}^{(\mathrm{new})}, \vec{\lambda}^{(\mathrm{new})}, \vec{\sigma}^{(\mathrm{new})}, \eta^{(\mathrm{new})})^T$ is then obtained by a line search in the direction $\Delta\Psi$:

$$\Psi_i^{(\mathrm{new})} = \Psi_i^{(\mathrm{old})} + s_i\,(\Delta\Psi)_i \quad , \quad 1 \le i \le 4 . \tag{28}$$

A standard convergence monitor in nonlinear programming is to choose the Euclidean norm $\|\mathcal{F}^{(p)}(\vec{u}, \vec{\sigma}, \vec{\lambda}, \eta)\|$ of the residual with respect to the KKT-conditions (22) as a merit function. However, in the situation under consideration this is an inappropriate tool, since it does not allow to tell the difference between a local minimizer and a stationary nonminimizing point. Indeed, the computations reveal that using the residual as a merit function one often gets stuck with a saddle point. A better approach is to rely on a hierarchy of two merit functions (cf., e.g., [16, 27]). In particular, the primary merit function is chosen as a modified augmented Lagrangian incorporating the logarithmic barrier functions according to

$$M_1(\vec{x},\vec{y},p,p_A) \;:=\; L_h(\vec{x}) \;-\; p\sum_{i=1}^{2}\log d_i(\vec{x}) \;+\; \vec{y}^T\,\vec{c}(\vec{x}) \;+\; \frac{1}{2}\,p_A\,\vec{c}(\vec{x})^T\,\vec{c}(\vec{x}) \quad (29)$$

where $\vec{x} := (\vec{u},\vec{\sigma})^T$, $\vec{y} := (\vec{\lambda},\eta)^T$, $\vec{c}(\vec{x}) \;:=\; (c_1(\vec{x}),c_2(\vec{x})^T$, and

$$\begin{aligned} c_1(\vec{x}) \;&:=\; A(\vec{\sigma})\vec{u} \;-\; \vec{b} \quad , \quad c_2(\vec{x}) \;:=\; g(\vec{\sigma}) \;-\; C \;, \\ d_1(\vec{x}) \;&:=\; \vec{\sigma} \;-\; \sigma_{\min}\,\vec{e} \quad , \quad d_2(\vec{x}) \;:=\; \sigma_{\max}\,\vec{e} \;-\; \vec{\sigma} \;. \end{aligned}$$

Note that p_A is a positive penalty parameter. For p_A sufficiently large it is always possible to realize a decrease in M_1.

The residual with respect to the KKT-conditions is chosen as the secondary merit function

$$M_2(\vec{x},\vec{y}) \;:=\; \|\mathcal{F}_p(\vec{u},\vec{\sigma},\vec{\lambda},\eta)\| \;. \quad (30)$$

In practice, the hierarchy of merit functions is used by means of the following strategy: If the steplengths s_i, $1 \le i \le 4$, lead to a decrease in M_1, they are accepted. If M_1 does not decrease, M_2 is checked and the steplengths are accepted in case it has decreased. However, if there is no reduction of M_1 after at most $N_{\rm wd}$ iterations, the penalty parameter p_A is chosen sufficiently large in order to guarantee a decrease in M_1. Note that in our computations $N_{\rm wd} = 4$ turned out to be a suitable choice.

4. Numerical Results

The simulation results obtained by the application of the primal-dual Newton interior-point method can be displayed by a grey-scale representing the range of the computed material distribution from dark ($\sigma = \sigma_{\max}$) to light ($\sigma = \sigma_{\min}$). Figures 3 and 4 show such a scale for a 2D computation where the bar contains 2 ports (Fig. 3) resp. 6 ports (Fig. 4) with a current inflow at the upper port(s) and an equal amount of current outflow at the lower port, and the design objective is to minimize the electric energy dissipation. In particular, Figure 3 displays the influence of the penalty parameter q in (12) on the material distribution whereas Figure 4 reflects the impact of the granularity of the triangulation. Table 1 contains the convergence history of the optimization algorithm where N_c stands for the number of ports, N_x resp. N_y are the numbers of nodal points in x- resp. y-direction, "iter" is the number of required iterations with $\|\mathcal{F}_p^k\| \le 10^{-6}\,\|\mathcal{F}_p^0\|$ as stopping criterion ($\mathcal{F}_p^k$ denotes the k-th residual), "p" is the last value of the barrier parameter, "M_1 " and $\|\mathcal{F}_p\|_2$ are the final values of the primary and secondary merit functions, and $\|v\|_2$ is the ℓ_2-norm related to the compatibility conditions at the last iteration. In all experiments the watchdog never "barked", i.e., we achieved a reduction of the primary merit function within the prescribed maximal number of watchdog iterations. For more details concerning the performance of the primal-dual Newton interior-point method and the watchdog strategy we refer to [22]. We remark that the numerical results are usually postprocessed in order to obtain a strict material/no material distribution (cf., e.g., [5]).

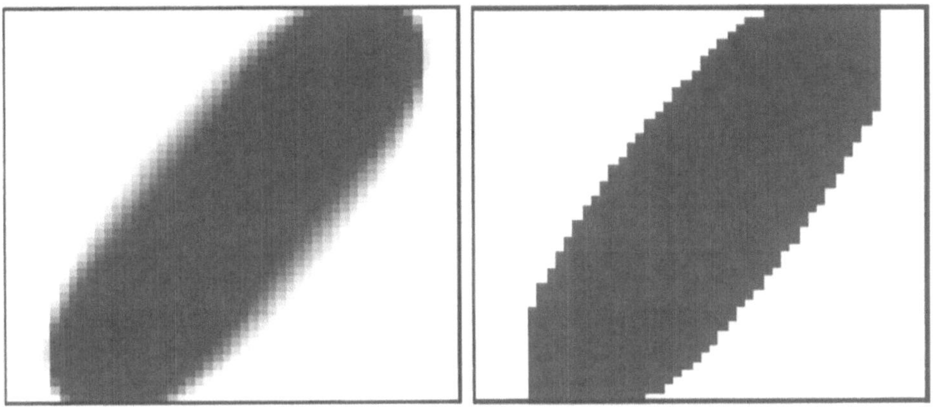

Fig. 3: Material distribution (2 contacts, q=1 (left) and q=2 (right)

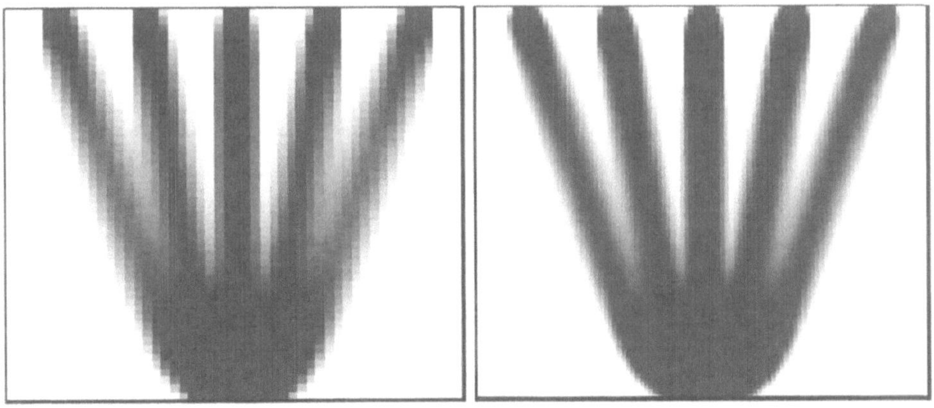

Fig. 4: Material distribution (6 contacts)
50 × 50 mesh (left) and 100 × 100 mesh (right)

For an individual optimized 3D bus bar with prescribed fluxes through the ports, Figure 4 displays the computed magnetic induction between two ports illustrating the effect of the holes.

Finally, we note that the primal-dual techniques described in the previous section lead to considerable savings in computational time compared to traditional approaches and allow to determine local minima representing improved designs by a margin between 10% and 20% depending on the specific operating conditions.

N_c	N_x	N_y	q	iter	p	M_1	$\|\mathcal{F}_p\|_2$	$\|\mathbf{v}\|_2$
2	25	25	1	17	4.92e-17	4.69	9.64e-4	e-9
2	25	25	2	19	4.29e-18	4.83	2.85e-5	e-9
2	50	50	1	19	1.28e-18	5.10	2.99e-4	e-10
2	50	50	2	90	1.18e-9	5.23	5.70e-3	e-6
3	50	50	1	30	6.44e-19	3.78	2.40e-4	e-11
3	50	50	2	75	1.07e-6	4.33	9.67e-3	e-4
3	100	100	1	84	3.68e-17	4.05	3.35e-4	e-9
3	100	100	2	24	1.78e-7	4.20	9.69e-3	e-5
6	50	50	1	20	9.85e-17	85.99	1.27e-3	e-9
6	50	50	2	45	9.64e-7	97.46	1.61e-2	e-4
6	100	100	1	24	3.46e-17	84.30	7.62e-4	e-9
6	100	100	2	43	4.18e-7	89.12	1.64e-2	e-5

TABLE 1. Convergence history

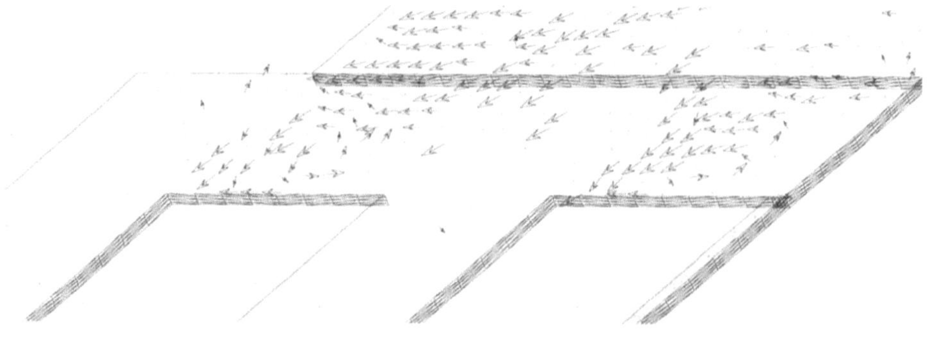

Fig. 5: Magnetic induction in converter module (zoom)

Acknowledgments. The work of the first author has been supported by a grant from the Federal Ministry for Education and Research (BMBF) under Grant No. 03HO7AU1-8. The second author is indebted to the Alexander-von-Humboldt Foundation for an AvH-Fellowship. The first and the second author are also grateful for grants from the German National Science Foundation (DFG; Grant No. HO877/4-1 and HO877/5-1).

References

[1] Chr. Albrecht, B. Korte, J. Schietke, and J. Vygen, "Cycle time and slack optimization for VLSI-chips", Rep. No. 99881, Forschungsinstitut für Diskrete Mathematik, Bonn, (1999).

[2] R. Beck, P. Deuflhard, R. Hiptmair, R.H.W. Hoppe, and B. Wohlmuth, "Adaptive multilevel methods for edge element discretizations of Maxwell's equations", Surveys of Math. in Industry, Vol. **8**, pp. 271–312, (1999).

[3] R. Beck, R. Hiptmair, R.H.W. Hoppe, and B. Wohlmuth, "Residual based a posteriori error estimators for eddy current computation", M^2AN Mathematical Modelling and Numerical Analysis **34**, 159–182, (2000).

[4] F. Ben Belgacem, A. Buffa, and Y. Maday, "The mortar finite element method for 3D Maxwell equations: First results", Report 99023, Laboratoire d'Analyse Numérique, Université Pierre et Marie Curie, Paris, (1999).

[5] M.P. Bendsøe, "Optimization of Sructural Topology, Shape, and Material", Springer, Berlin-Heidelberg-New York, (1995).

[6] O. Biro and K. Preis, "Various FEM formulations for the calculation of transient 3D eddy currents in nonlinear media", IEEE Trans. Magn. **31**, 1307–1312 ,(1995).

[7] G. Biros and O. Ghattas, "Parallel preconditioners for KKT systems arising in optimal control of viscous incompressible flows". to appear in: Proc. Parallel CFD '99, Williamsburg, Virginia, May 23–26, 1999, North Holland, Amsterdam, London, New York, (2000).

[8] P. Böhm, R.H.W. Hoppe, G. Mazurkevitch, S. Petrova, G. Wachutka, and E. Wolfgang, "Optimal design of high power electronic devices by topology optimization", to appear in: Mathematik – Schlüsseltechnologie für die Zukunft. Verbundprojekte zwischen Mathematik und Industrie, Springer, Berlin-Heidelberg-New York, (2001).

[9] Th. Borrvall and J. Petersson, "Topology optimization using regularized intermediate density control", Techn. Rep. LiTH-IKP-R-1086, Department of Mechanical Engineering, University of Linköping, (2000).

[10] A. Borzi and K. Kunisch, "A multigrid method for optimal control of time-dependent reaction diffusion processes". to appear in: Proc. Workshop "Fast Solution of Discretized Optimization Problems", Weierstrass Institute Berlin, May 08–12, 2000 (K.-H. Hoffmann, R.H.W. Hoppe, V. Schulz; eds.), Birkhäuser, Basel, (2000).

[11] K. Doll, F.M. Johannes, and K.J. Antreich, "Iterative placement improvement by network flow methods", IEEE Trans. Comp.-Aided Design Integr. Circuits and Systems **CAD-13**, 1190–1200, (1994).

[12] St. Dürndorfer, V. Gradinaru, R.H.W. Hoppe, E.-R. König, G. Schrag, and G. Wachutka, "Numerical simulation of microstructured semiconductor devices, transducers, and systems", In: High Performance Scientific and Engineering Computing. Proc. "Int. FORTWIHR-Symposium", Munich, March 1998 (Bungartz, H., Durst, F.,and Zenger, Chr.; eds.), pp. 309–323, Lecture Notes in Computational Science and Engineering, Vol. **8**, Springer, Berlin-Heidelberg-New York, (1999).

[13] A.S. El-Bakry, R.A. Tapia, T. Tsuchiya, and Y. Zhang, "On the formulation of the Newton interior-point method for nonlinear programming", Journal of Optimization Theory and Applications **89**, 507–541 (1996).

[14] A. Forsgen and Ph. Gill, "Primal-dual interior methods for nonconvex nonlinear programming", SIAM J. Optimization **8**, 1132–1152 (1998).

[15] A. Forsgen, Ph. Gill, and J.R. Shinnerl, "Stability of symmetric ill-conditioned systems arising in interior methods for constrained optimizatiom", SIAM J. Matrix Anal. Appl. **17**, 187–211, (1996).

[16] D.M. Gay, M.L. Overton, and M.H. Wright, "A primal-dual interior method for nonconvex nonlinear programming", Techn. Rep. 97-4-08, Computing Science Research Center, Bell Laboratories, Murray Hill, New Jersey, 1997.

[17] J. Haslinger and P. Neittaanmäki, "Finite Element Approximation for Optimal Shape Design: Theory and Applications", J. Wiley & Sons, Chichester, 1988.

[18] M. Heinkenschloss, "Time-domain decomposition iterative methods for the solution of parabolic linear-quadratic optimal control problems", Techn. Rep., Department of Comput. and Appl. Math., Rice University, Houston, (2000).

[19] R.H.W. Hoppe, "Mortar edge elements in $\mathbf{R}^3$", East-West J. Numer. Anal., Vol. **7**, 159–173, (1999).

[20] R.H.W. Hoppe, "Adaptive mortar edge elements in the computation of eddy currents", In: Proc. Conf. "Analysis and Approximation of Boundary Value Problems", Jyväskylä, Finland, October 1998 (P. Neittaanmäki et al.; eds.), pp. 83–96, Dept. Math. Inf. Techn., No. A 2/2000, Jyväskylä, 2000.

[21] R.H.W. Hoppe, Y. Iliash, and G. Mazurkevitch, "Domain decomposition methods in the design of high power electronic devices", In: Proc. Conf. on Multifield Problems, Stuttgart, October 1999 (M. Sändig and W. Wendland, eds.), pp. 169–182, Springer, Berlin-Heidelberg-New York, (2000).

[22] R.H.W. Hoppe, S. Petrova, and V. Schulz, "A primal-dual Newton-type interior-point method for topology optimization", Preprint, Institute of Mathematics, University of Augsburg, (2000).

[23] R.H.W. Hoppe, S. Petrova, and V. Schulz, "Topology optimization of conductive media described by Maxwell's equations", to appear in Proc. 2nd Conf. Numer. Anal. Appl., Rousse, Bulgaria, Lect. Notes in Comp. Sci., Springer,(2001).

[24] R.H.W. Hoppe and B. I. Wohlmuth, "Adaptive multilevel iterative techniques for nonconforming finite element discretizations", East-West J. Numer. Math. **3**, 179–197, (1995).

[25] B. Maar and V. Schulz, "Interior point multigrid methods for topology optimization", Structural Optimization **19**, 214–224, (2000).

[26] J.-C. Nédélec, "Mixed finite elements in $\mathbf{R}^3$", Numer. Math. **35**, 315–341, (1980).

[27] Z. Parada and R.A. Tapia, "Computational experience with a modified augmented Lagrangian merit function in a primal-dual interior-point method", Techn. Rep., Department of Comput. and Appl. Math., Rice University, Houston, (1995).

[28] O. Pironneau, "Optimal Shape Design for Elliptic Systems", Springer, Berlin-Heidelberg-New York, (1984).

[29] F. Rapetti, Y. Maday, and F. Bouillaut, "The mortar edge element method in three dimensions: application to magnetostatics", Preprint, ASCI-UPR 9029 CNRS,Paris Sud University, Orsay, (2000).

[30] V. Schulz, "Mehrgittermethoden für Optimierungsprobleme bei PDE", Habilitationsschrift, Universität Heidelberg, (2000).
[31] J. Sokolowski and J.P. Zolesio, "Introduction to Shape Optimization", Springer, Berlin-Heidelberg-New York, (1992).

Institute of Mathematics
University of Augsburg
D-86159 Augsburg, Germany
E-mail address: hoppe@math.uni-augsburg.de
E-mail address: petrova@math.uni-augsburg.de

Weierstrass Institute for Applied Analysis and Stochastics
D-10117 Berlin, Germany
E-mail address: vschulz@na-net.ornl.gov

International Series of Numerical Mathematics, Vol. 139, 133–144

Effects of Thickness on Sharp Trace Regularity for a Kirchhoff Plate with Free Boundary Conditions

Mary Ann Horn

Abstract. Sharp trace regularity estimates for a Kirchhoff plate with free boundary conditions are established with the primary goal of tracking the effects of thickness in the estimates. Microlocal analysis is used in the proof with an alternative localization to the one seen in the earlier work of Lasiecka and Triggiani. Knowledge of how thickness appears in the estimates has important implications in uniform stability for more complex systems which involve the Kirchhoff plate equation.

1. Introduction

As attention narrows on the effects of individual parameters upon desired stability properties, thickness has played an important role when the focus is on elastic structures. In many models, including thin plates and spherical and cylindrical shells, thickness plays a crucial role. Incorporating the effects of thickness into the model is important in accurately deriving a mathematical system describing the motion of the elastic structure. However, because thickness is assumed to be an extremely small parameter, terms which correspond to it are often neglected to simplify the analysis. As has been seen in the past, this can lead to models with radically different mathematical properties, a typical example being the case of the Kirchhoff plate equation.

In the recent work of Horn and McMillan [3], the challenges became abundantly clear. The model for the cylindrical shell involves a system that couples a Kirchhoff plate equation with a two-dimensional system of elasticity. To illustrate the need for careful analysis in tracking the effects of thickness, this paper focuses on a single Kirchhoff plate equation. Consider the following Kirchhoff plate equation defined in a bounded domain, Ω, with sufficiently smooth boundary,

The author gratefully acknowledges the support of the Alexander von Humboldt-Stiftung and the Technische Universität Darmstadt during her sabbatical.
Partially supported by National Science Foundation Grant NSF DMS-9803547.

$\partial\Omega = \partial\Omega_0 \cup \partial\Omega_1$:

$$\begin{aligned}
w_{tt} - \gamma\Delta w_{tt} + \Delta^2 w = f &\quad \text{in } \Omega \times (0,T), &(1.a)\\
w(0,\cdot) = w_0 &\quad \text{in } \Omega, &(1.b)\\
w_t(0,\cdot) = w_1 &\quad \text{in } \Omega, &(1.c)\\
w \equiv \frac{\partial w}{\partial \nu} \equiv 0 &\quad \text{on } \partial\Omega_0 \times (0,T), &(1.d)\\
\Delta w + \beta_1 w = u_1 &\quad \text{on } \partial\Omega_1 \times (0,T), &(1.e)\\
\frac{\partial \Delta w}{\partial \nu} + \beta_2 w - \gamma \frac{\partial w_{tt}}{\partial \nu} = u_2 &\quad \text{on } \partial\Omega_1 \times (0,T), &(1.f)
\end{aligned}$$

where the constant $\gamma > 0$ is directly proportional to the square of the thickness of the plate and the boundary operators β_1 and β_2 are defined by

$$\begin{aligned}
\beta_1 &= -(1-\mu)\left(\tfrac{\partial^2}{\partial\tau^2} + k\tfrac{\partial}{\partial\nu}\right),\\
\beta_2 &= (1-\mu)\left[\tfrac{\partial}{\partial\nu}\tfrac{\partial^2}{\partial\tau^2} - \tfrac{\partial}{\partial\tau}\left(k\tfrac{\partial}{\partial\tau}\right)\right],
\end{aligned} \tag{2}$$

with $0 < \mu < 1$ and curvature k [4].

While this system is hyperbolic with finite speed of propagation when $\gamma > 0$, if one uses the assumption that the thickness is very small and neglects the terms corresponding to γ, the result is a system of Petrovskii type with infinite speed of propagation. Thus, γ plays a critical role in determining the mathematical behavior of the system.

To uniformly stabilize system (1), feedback controls, u_1 and u_2, defined by

$$\begin{aligned}
u_1(t,x) &= -\frac{\partial w_t}{\partial \nu} \quad \text{on } \partial\Omega_1 \times (0,T) &(3.a)\\
u_2(t,x) &= w_t - \sqrt{\gamma}\frac{\partial}{\partial\tau}\left(\frac{\partial}{\partial\tau}w_t\right) + k_0 w \quad \text{on } \partial\Omega_1 \times (0,T), &(3.b)
\end{aligned}$$

are implemented, where the constant $k_0 \geq 0$ is assumed to satisfy

$$\begin{aligned}
&\textit{i.)}\ k_0 \geq 0 \text{ if } \partial\Omega_0 \text{ is nonempty},\\
&\textit{ii.)}\ k_0 > 0 \text{ if } \partial\Omega_0 \text{ is empty}.
\end{aligned} \tag{4}$$

With these controllers, which act as moments and torques along the boundary of the domain, the associated energy function for the system is defined by

$$E_\gamma(t) \equiv a(w,w) + \int_\Omega (w_t^2 + \gamma|\nabla w_t|^2)\,dx + k_0 \int_{\partial\Omega_1} w^2 dx \tag{5}$$

where

$$a(w,w) = \int_\Omega \left[|\Delta w|^2 + (1-\mu)(2w_{xy}^2 - w_{xx}^2 - w_{yy}^2)\right] dx. \tag{6}$$

The norm induced by the energy function is equivalent to $H^2(\Omega) \times H^1(\Omega)$ when $\gamma > 0$ and to $H^2(\Omega) \times L_1(\Omega)$ when $\gamma = 0$. With the preceeding definition, the following theorem may be stated.

Theorem 1.1. *Consider the feedback system defined by (1) with controllers as defined by (3). Assume either*

i.) If $\partial\Omega_0 \neq$ and $k_0 \geq 0$, there exists a point $x_0 \in \mathbb{R}^2$ such that

$$(x - x_0) \cdot \nu \leq 0 \qquad \text{on } \partial\Omega_0, \tag{7}$$

where ν is the unit outward normal;

ii.) $\partial\Omega_0 =$ and $k_0 > 0$.

Then, in either case, there exist constants C, $\omega > 0$, such that the energy associated with the system satisfies

$$E_\gamma(t) \leq Ce^{-\omega t} E_\gamma(0) \tag{8}$$

Moreover, the constants C and ω are independent of γ.

While this theorem is much like the one proven in [6], our goal is to avoid any possible dependence of the constants upon the thickness of the plate. Once this theorem is established, uniform stability for the limit problem then follows, with boundary controls which are appropriate and physically realistic for the limiting system.

In establishing uniform stability for both the original problem and the limiting system when γ is taken to be identically zero, the role of thickness becomes very important. An illustration of how results for the limiting system can be derived from estimates when $\gamma > 0$ can be found in [2] for the Kirchhoff plate with simply supported boundary conditions. Lasiecka and Triggiani considered free boundary conditions as in system (1) [6], but performed separate analyses for the two systems rather than deriving results for the limit problem from the estimates when $\gamma > 0$. In their work, sharp trace regularity estimates permit uniform stability, as well as corresponding exact controllability results, to be proven while eliminating unnatural geometrical conditions on the domain Ω which had been frequently seen in earlier literature (see, e.g., [4, 5]). However, in their analysis, no effort was made to track the effects of thickness and the constants in their estimates depend upon γ. In addition, formal application of the trace estimates derived in [6] would give rise to uniform stability results for the limiting system with *stronger boundary controls than necessary.* To derive stability results for both systems while using the controls which naturally arise when considering the systems independently, a more delicate analysis is needed.

Critical to this analysis is the need to follow the effects of γ through the derivation of the sharp trace estimates. While various answers to the questions of uniform stability and exact controllability for the Kirchhoff plate equation have been found by many researchers, the need to understand the role of thickness comes into play in far more complex models, such as the system describing the motion of a cylindrical shell. Thus, the goal of this paper is to illustrate the effects of thickness on such sharp trace regularity estimates in a single Kirchhoff plate equation, one of the simplest cases where this issue arises.

2. Sharp Trace Regularity

Imposition of geometric conditions in earlier work (see, e.g., [4]) was due to the need to eliminate second order traces which arise when multiplier techniques are used. To avoid this, an estimate on the second order traces of the solution, w, as stated in the following theorem, is required.

Theorem 2.1. *Let $0 < \alpha < T$ and $\epsilon > 0$ be arbitrary. Let $s_0 < \frac{1}{2}$. Then the solution, w, of the Kirchhoff plate equation (1) with $\gamma > 0$ satisfies:*

$$\begin{aligned}\int_\alpha^{T-\alpha} \int_{\partial\Omega_1} & \left\{ \left(\frac{\partial^2 w}{\partial\tau^2}\right)^2 + \left(\frac{\partial^2 w}{\partial\nu^2}\right)^2 + \left(\frac{\partial^2 w}{\partial\tau\partial\nu}\right)^2 \right\} dx dt \\ & \leq C_{T,\alpha,\epsilon} \left\{ \|f\|^2_{H^{-s_0}(0,T;\Omega)} + \|g_1\|^2_{L_2(0,T;\partial\Omega_1)} + \|g_2\|^2_{L_2(0,T;H^{-1}(\partial\Omega_1))} \right. \\ & \quad + \|\tfrac{\partial}{\partial\nu} w_t\|^2_{L_2(0,T;\partial\Omega_1)} + \gamma\|\tfrac{\partial}{\partial\tau} w_t\|^2_{L_2(0,T;\partial\Omega_1)} \\ & \quad \left. + (1+\gamma)\|w\|^2_{L_2(0,T;H^{3/2+\epsilon}(\Omega))} + \|w_t\|^2_{L_2(0,T;\partial\Omega_1)} \right\}\end{aligned} \tag{9}$$

2.1. Microlocal analysis

Via a partition of unity and a flattening of the boundary procedure, it is sufficient to establish the estimate of Theorem 2.1 for the following half-space problem. Notation will follow that of [2, 6], restricted to two-dimensions. Let $\Omega = \{(x,y)|x \geq 0\}$ and $\partial\Omega = \Omega|_{x=0}$ and consider the general problem related to the original system.

$$\mathcal{P}w = f \quad \text{in } \Omega \times (0,T), \tag{10.a}$$

$$\mathcal{B}_1 w = g_1 \quad \text{on } \partial\Omega \times (0,T), \tag{10.b}$$

$$\mathcal{B}_2 w = g_2 \quad \text{on } \partial\Omega \times (0,T), \tag{10.c}$$

where the operator $\mathcal{P}$ (modulo lower order terms) is defined by

$$\mathcal{P}(x,y;D_t,D_x,D_y) = -aD_t^2 - \gamma a D_t^2 \left(\tilde{D}_x^2 + \tilde{D}_y^2\right) + \left(\tilde{D}_x^2 + \tilde{D}_y^2\right)^2. \tag{11}$$

Here, we have adopted the notation $D_t = -\sqrt{-1}(\partial/\partial t)$, etc., and

$$\tilde{D}_x = D_x + a_2(x,y)D_y, \qquad \tilde{D}_y^2 = a_1(x,y)D_y^2 - a_2^2 D_y^2. \tag{12}$$

In the above definition, it is assumed that

$$\begin{aligned} &i.)\ a(x,y) > 0,\ a_j(x,y) \text{ depend smoothly on } (x,y) \text{ for } j=1,2; \\ &ii.)\ a_1(x,y) > \rho > 0 \qquad \forall (x,y) \in \Omega \\ &iii.)\ a_1(x,y) - a_2^2(x,y) > \rho > 0. \end{aligned} \tag{13}$$

Boundary operators become

$$\begin{aligned} \mathcal{B}_1 &= \left(\tilde{D}_x^2 + \tilde{D}_y^2\right) + \beta_1 && \text{at } x = 0, \\ \mathcal{B}_2 &= \tilde{D}_x \left(\tilde{D}_x^2 + \tilde{D}_y^2\right) + \beta_2 + \gamma a \tilde{D}_x D_t^2 && \text{at } x = 0, \end{aligned} \tag{14}$$

where

$$\begin{aligned} \beta_1 &= b_{11}D_y^2 + b_{12}D_y\tilde{D}_x + l.o.t. \\ \beta_2 &= b_{21}D_y^3 + b_{22}D_y^2\tilde{D}_x + b_{23}D_y\tilde{D}_x^2 + l.o.t. \end{aligned} \tag{15}$$

Corresponding the the above operators are the symbols:

$$p(x,y;s,\xi,\eta) = symb\{\mathcal{P}\} = -as^2 - \gamma as^2(\tilde{\xi}^2+\tilde{\eta}^2) + (\tilde{\xi}^2+\tilde{\eta}^2)^2, \tag{16.a}$$

$$\tilde{\xi} = symb\{\tilde{D}_x\} = \xi + a_2\eta, \tag{16.b}$$

$$\tilde{\eta}^2 = symb\{\tilde{D}_y^2\} = (a_1 - a_2^2)\eta^2 \tag{16.c}$$

$$symb\{\beta_1\} = \mathcal{O}\left(|\eta|^2 + |\eta|\tilde{\xi}\right). \tag{16.d}$$

2.2. Localization in time

Let $\psi(t) \in C_0^\infty(\mathbb{R})$ be a cutoff function defined such that $0 \le \psi(t) \le 1 \forall t$ and

$$\psi(t) = \begin{cases} 1 & \text{in } [\alpha, T-\alpha] \\ 0 & \text{outside } (\alpha/2, T-\alpha/2) \end{cases} \tag{17}$$

Defining $w_c(t,\cdot) \equiv \psi(t)w(t,\cdot)$ and recalling (10), $w_c(t,\cdot)$ satisfies

$$\begin{aligned} \mathcal{P}w_c &= [\mathcal{P},\psi]w + \psi f \quad && \text{in } \Omega \times (-\infty,\infty), \\ \mathcal{B}_1 w_c &= \psi g_1 && \text{on } \partial\Omega \times (-\infty,\infty), \\ \mathcal{B}_2 w_c &= \psi g_2 + [\mathcal{B}_2,\psi]w && \text{on } \partial\Omega \times (-\infty,\infty), \end{aligned} \tag{18}$$

where $[A,B]$ denotes the commutator of two operators A and B. In our case,

$$[\mathcal{P},\psi] = i2\gamma a(\tilde{D}_x^2 + \tilde{D}_y^2)D_t(\psi' w) + l.o.t., \tag{19}$$

$$[\mathcal{B}_2,\psi]w = i2\gamma a\tilde{D}_x D_t(\psi' w) + l.o.t.. \tag{20}$$

With this localization, the properties established in [6] remain true. For completeness, we note those that will be used in this analysis.

$$\|[\mathcal{B}_2,\psi]w\|_{H^{-1}(-\infty,\infty;L_2(\partial\Omega))} \le C_\epsilon\gamma\|w\|_{L_2(\alpha/2,T-\alpha/2;H^{3/2}(\Omega))} \tag{21}$$

$$symb\{\beta_2\}w_c = \mathcal{O}\left([|\eta|^3 + |\eta|^2\tilde{\xi}]w_c\right) + \mathcal{O}\left(|\eta|g_1\right) \tag{22}$$

$$symb\{[\mathcal{P},\psi]\}w = \gamma\mathcal{O}\left(s\tilde{\xi}^2 + |\eta|^2 s\right)(\psi' w). \tag{23}$$

As noted in [6], since w_c has compact support in $(\alpha/2, T-\alpha/2)$, we may view $s = \sigma - ir_0$ as the Laplace transform variable of D_t and η is the Fourier variable corresponding to D_y.

2.3. Localization in the dual variables

The goal of localization in the dual variable is to study of the behavior of the Laplace and Fourier variables, σ and η, with respect to one another. To this end, it is sufficient to consider only the quadrant where $\sigma > 0$ and $\eta > 0$ and we define the following mutually disjoint regions:

$$\mathcal{R}_1 \equiv \{(x,y;\sigma,\eta) \in \Omega \times \mathbb{R}^2 |\ \sigma^2 + \gamma\sigma^2|\eta|^2 \le c_0|\eta|^4\} \tag{24.a}$$

$$\mathcal{R}_{tr} \equiv \{(x,y;\sigma,\eta) \in \Omega \times \mathbb{R}^2 |\ c_0|\eta|^4 \le \sigma^2 + \gamma\sigma^2|\eta|^2 \le 2c_0|\eta|^4\} \tag{24.b}$$

$$\mathcal{R}_2 \equiv \{(x,y;\sigma,\eta) \in \Omega \times \mathbb{R}^2 |\ 2c_0|\eta|^4 \le \sigma^2 + \gamma\sigma^2|\eta|^2\} \tag{24.c}$$

In $\mathcal{E}_1 \equiv \mathcal{R}_1 \cup \mathcal{R}_{tr}$, it can be readily seen by noting that

$$\sigma^2(1+\gamma|\eta|^2) \le 2c_0|\eta|^4 \Longrightarrow \sigma^2 \le 2c_0\left(\frac{|\eta|^4}{1+\gamma|\eta|^2}\right) \le \frac{2c_0}{\gamma}|\eta|^2, \tag{25}$$

that the symbol $p(x,y;\sigma,\xi,\eta)$ is elliptic of order four uniformly with respect to γ, i.e., there exists a constant $\delta > 0$, where δ is independent of γ, such that

$$p(x,y;\sigma,\xi,\eta) \ge \delta\left(\xi^4+\sigma^4+|\eta|^4\right) \qquad \text{in } \mathcal{E}_1. \tag{26}$$

To take advantage of this microelliptic behavior, let $\lambda(x,y;\sigma,\eta) \in C^\infty$ be a cutoff function defined such that $0 \le \lambda(x,y;\sigma,\eta) \le 1$, λ is homogeneous of order zero in σ and η, and

$$\lambda(x,y;\sigma,\eta) = \begin{cases} 1 & \text{in } \mathcal{R}_1, \\ 0 & \text{in } \mathcal{R}_2. \end{cases} \tag{27}$$

The corresponding zero-order pseudodifferential operator, $\Lambda \in \mathcal{S}^0(\mathbb{R}^2_{ty})$ may then be applied to (18) which yields

$$\begin{aligned} \mathcal{P}(\Lambda w_c) = \Lambda\psi f + \Lambda[\mathcal{P},\psi]w + [\mathcal{P},\Lambda]w_c \equiv \Lambda\psi f + \mathcal{F} &\quad \text{in } \Omega\times(-\infty,\infty), \\ \mathcal{B}_1 w_c = \Lambda\psi g_1 + [\mathcal{B}_1,\Lambda]w_c \equiv \hat{g}_1 &\quad \text{on } \partial\Omega\times(-\infty,\infty), \\ \mathcal{B}_2 w_c = \Lambda\psi g_2 + \Lambda[\mathcal{B}_2,\psi]w + [\mathcal{B}_2,\Lambda]w_c \equiv \hat{g}_2 &\quad \text{on } \partial\Omega\times(-\infty,\infty), \end{aligned} \tag{28}$$

3. Elliptic Estimates in $\mathcal{E}_1$

Because the symbol of the operator $\mathcal{P}$ is microelliptic of order four in the region $\mathcal{E}_1$, problem (28) is elliptic and satisfies elliptic estimates in all variables. Thus, from [1, 7], the following lemma holds.

Lemma 3.1. *There exists a constant C such that for $s_0 < \frac{1}{2}$, the solution, Λw_c, of (28) satisfies the inequality*

$$\begin{aligned} &\|\Lambda w_c\|_{H^{5/2}(-\infty,\infty;\Omega)} + \|\Lambda w_c\|_{H^2(-\infty,\infty;\partial\Omega)} + \|\tilde{D}_x(\Lambda w_c)\|_{H^1(-\infty,\infty;\partial\Omega)} \\ &\quad \le C\Big\{\|\Lambda\psi f\|_{H^{-s_0}(-\infty,\infty;\Omega)} + \|\mathcal{F}\|_{L_2(\mathbb{R}^1_{x+};H^{-3/2}(\mathbb{R}^2_{ty}))} \\ &\qquad + \|\hat{g}_1\|_{L_2(-\infty,\infty;\partial\Omega)} + \|\hat{g}_2\|_{H^{-1}(-\infty,\infty;\partial\Omega)} \\ &\qquad + \|w_c\|_{L_2(-\infty,\infty;\Omega)}\Big\}. \end{aligned} \tag{29}$$

Further estimates for the boundary terms that do not involve γ are identical to those derived in [6] and can be summarized as follows:

$$\|[\mathcal{B}_1,\Lambda]w_c\|_{L_2(-\infty,\infty;\partial\Omega)} \le C_\epsilon\|w_c\|_{L_2(0,T;H^{3/2+\epsilon}(\Omega))} \tag{30}$$

$$\|\hat{g}_1\|_{L_2(-\infty,\infty;\partial\Omega)} \le C_\epsilon\left\{\|w_c\|_{L_2(0,T;H^{3/2+\epsilon}(\Omega))} + \|g_1\|_{L_2(0,T;\partial\Omega)}\right\} \tag{31}$$

$$\begin{aligned} \|\hat{g}_2\|_{H^{-1}(-\infty,\infty;\partial\Omega)} \le C_\epsilon\Big\{&\|w_c\|_{L_2(0,T;H^{3/2+\epsilon}(\Omega))} + \gamma\|w\|_{L_2(\alpha,T-\alpha;H^{3/2}(\Omega))} \\ &+\|g_1\|_{H^{-1}(0,T;\partial\Omega)} + \|g_2\|_{H^{-1}(0,T;\partial\Omega)}\Big\} \end{aligned} \tag{32}$$

Thus, our concentration will be on the three commutators which involve γ, $[\mathcal{B}_2,\Lambda]$, $[\mathcal{P},\Lambda]$ and $\Lambda[\mathcal{P},\psi]$.

3.1. Estimates for the boundary term $[\mathcal{B}_2, \Lambda]$

Using the asymptotic expansions of symbols [1],

$$\begin{aligned} symb\{[\mathcal{B}_2,\Lambda]\} &= \phi_{00}(y;\sigma,\eta)\tilde{\xi}^2 + \phi_{01}(y;\sigma,\eta)\tilde{\xi} + \phi_{02}(y;\sigma,\eta) \\ &\quad +\gamma\sigma^2\phi_{00}(y;\sigma,\eta) + l.o.t., \end{aligned} \tag{33}$$

where $\phi_{ij}(\cdot;\sigma,\eta)$ has support in $\mathcal{E}_1$ and represents the symbol of a generic pseudodifferential operator, Φ_{ij} of order i in σ and j in η. Therefore,

$$[\mathcal{B}_2,\Lambda] = \Phi_{00}\tilde{D}_x^2 + \Phi_{00}D_y\tilde{D}_x + \Phi_{00}D_y^2 + \gamma\Phi_{00}D_t^2 + l.o.t. \tag{34}$$

Lemma 3.2. *For all $\epsilon > 0$, the boundary operator, $[\mathcal{B}_1,\Lambda]$, arising from system (28), satisfies*

$$\begin{aligned} \|[\mathcal{B}_2,\Lambda]w_c\|_{H^{-1}(-\infty,\infty;\partial\Omega)} \leq C_\epsilon \big\{ &\|w_c\|_{L_2(-\infty,\infty;H^{3/2}(\Omega))} \\ &+ \|g_1\|_{H^{-1}(-\infty,\infty;\partial\Omega)} \big\} \end{aligned} \tag{35}$$

Proof. From the definitions of $\mathcal{B}_1$ and β_1, $\tilde{D}_x^2 w_c$ cand be written as

$$\tilde{D}_x^2 w_c = \psi g_1 - \tilde{D}_y^2 - \beta_1 w_c. \tag{36}$$

By trace theory,

$$\begin{aligned} &\|(\tilde{D}_y^2 + \beta_1)w_c\|_{H^{-1}(-\infty,\infty;\partial\Omega)} \\ &\quad \leq C\left\{ \|D_y^2 w_c\|_{H^{-1}(-\infty,\infty;\partial\Omega)} + \|D_y\tilde{D}_x w_c\|_{H^{-1}(-\infty,\infty;\partial\Omega)} \right\} \\ &\quad \leq C\left\{ \|w_c\|_{L_2(-\infty,\infty;H^1(\partial\Omega))} + \|\tilde{D}_x w_c\|_{L_2(-\infty,\infty;\partial\Omega)} \right\} \\ &\quad \leq C_\epsilon \|w_c\|_{L_2(-\infty,\infty;H^{3/2+\epsilon}(\partial\Omega))}, \end{aligned} \tag{37}$$

and, therefore,

$$\begin{aligned} \|\tilde{D}_x^2 w_c\|_{H^{-1}(-\infty,\infty;\partial\Omega)} \leq C_\epsilon \big\{ &\|g_1\|_{H^{-1}(-\infty,\infty;\partial\Omega)} \\ &+ \|w_c\|_{L_2(-\infty,\infty;H^{3/2+\epsilon}(\partial\Omega))} \big\} \end{aligned} \tag{38}$$

Finally, since the symbol ϕ_{00} is supported in $\mathcal{E}_1$,

$$\begin{aligned} \|\gamma\phi_{00}D_t^2 w_c\|^2_{H^{-1}(-\infty,\infty;\partial\Omega)} &\leq C\gamma^2 \int_{\mathcal{E}_2} \left| \frac{\sigma^2}{|\eta|} w_c \right|^2 d\sigma d\eta \\ &\leq C \int_{\mathcal{E}_2} ||\eta| w_c|^2 \, d\sigma d\eta \\ &\leq C\|D_y w_c\|^2_{L_2(-\infty,\infty;\partial\Omega)}, \end{aligned} \tag{39}$$

Combining all of the above estimates and applying trace theory yields the desired result. □

3.2. Estimates for the interior term $[\mathcal{P}, \Lambda]$

We recall Lemma 4.6 of [6], which applies again in our case.

Lemma 3.3. ([6], Lemma 4.6.) *$[\mathcal{P},\Lambda]$ can be written as*

$$[\mathcal{P},\Lambda] = \Phi_{00}\tilde{D}_x^3 + \Phi_{00}D_y\tilde{D}_x^2 + \mathcal{G}, \tag{40}$$

where the symbol of the operator $\mathcal{G}$ is supported in $\mathcal{E}_1$ and $\mathcal{G}$ satisfies

$$\|\mathcal{G}w_c\|_{L_2(\mathbb{R}^1_{x+};H^{-3/2}(\mathbb{R}^2_{ty}))} \leq C_\epsilon \|w_c\|_{L_2(-\infty,\infty;H^{3/2}(\Omega))} \tag{41}$$

Thus, to complete the estimate for this term, we must determine if estimates analagous to those derived in [6] (see equation (4.7.7)). In this case, the following proposition is required.

Proposition 3.4. *With reference to the first two terms of (40), the following estimate holds for all* $s_0 < \frac{1}{2}$*:*

$$\begin{aligned} &\|\Phi_{00}\tilde{D}_x^3 w_c\|_{L_2(\mathbb{R}^1_{x+};H^{-3/2}(\mathbb{R}^2_{ty}))} + \|\Phi_{00}D_y\tilde{D}_x^2 w_c\|_{L_2(\mathbb{R}^1_{x+};H^{-3/2}(\mathbb{R}^2_{ty}))} \\ &\quad \le C_\epsilon \left\{ \|g_1\|_{L_2(0,T;\partial\Omega)} + \|g_2\|_{L_2(0,T;H^{-1}(\partial\Omega))} + \|f\|_{H^{-s_0}(0,T;\Omega)} \right. \\ &\qquad + \|w_c\|_{L_2(-\infty,\infty;H^{3/2+\epsilon}(\Omega))} + (1+\gamma)\|w\|_{L_2(0,T;H^{3/2+\epsilon}(\Omega))} \\ &\qquad \left. + \gamma\|\Phi_{00}\tilde{D}_x^2(\phi' w)\|_{L_2(\mathbb{R}^1_{x+};H^{-3/2}(\mathbb{R}^2_{ty}))} \right\}, \end{aligned} \tag{42}$$

where Φ_{00} *is, as before, a pseudodifferential operator with symbol* ϕ_{00} *of order zero in* σ *and* η *which is supported in* $\mathcal{E}_1$.

Proof. A sketch of the proof, closely following that in [6] will be given with notes as to where changes must be made to account for γ. Rewriting the system for w_c, (18), as

$$\tilde{D}_x^4 w_c + \left[\tilde{D}_y^2 - \gamma a D_t^2\right]\tilde{D}_x^2 w_c = \psi f + \mathcal{U}w_c + [\mathcal{P},\psi]w \quad \text{in} \quad \Omega\times(-\infty,\infty), \tag{43.a}$$

$$\tilde{D}_x^2 w_c|_{x=0} = \psi g_1 - \tilde{D}_y^2 w_c - \beta_1 w_c \quad \text{on} \quad \partial\Omega\times(-\infty,\infty), \tag{43.b}$$

$$\begin{aligned} \tilde{D}_x^3 w_c|_{x=0} = \psi g_2 - \tilde{D}_x\tilde{D}_y^2 w_c - \beta_2 w_c \\ -\gamma a\tilde{D}_x D_t^2 w_c + [\mathcal{B}_2,\psi]w \quad \text{on} \quad \partial\Omega\times(-\infty,\infty), \end{aligned} \tag{43.c}$$

where

$$\mathcal{U} = aD_t^2 - \tilde{D}_y^4 - [\tilde{D}_x^2, \tilde{D}_y^2] + \gamma a D_t^2\tilde{D}_y^2 \tag{44}$$

and

$$symb\{\mathcal{U}\} = \mathcal{O}(|\eta|^4) + \mathcal{O}(|\eta|^2\tilde{\xi}) \text{ in } \mathcal{E}_1 \tag{45}$$

Properties of system (43) which account for the alternative definition of $\mathcal{E}_1$ in (24), include:

i.) Letting $\mathcal{Q} \equiv \tilde{D}_y^2 - \gamma a D_t^2$, its symbol satisfies

$$q(x,y;\sigma,\eta) \equiv |\tilde{\eta}|^2 - \gamma a\sigma^2 \ge \delta|\eta|^2 \text{ in } \mathcal{E}_1 \text{ for some } \delta > 0, \tag{46}$$

provided c_0 is sufficiently small.

ii.) In the first boundary condition,

$$symb\{\tilde{D}_y^2 + \beta_1\} = \mathcal{O}(|\eta|^2 + |\eta|\xi) \tag{47}$$

iii.) In the second boundary condition,

$$symb\{[\tilde{D}_x\tilde{D}_y^2 + \beta_2 + \gamma a\tilde{D}_x D_t^2]w_c\} = \mathcal{O}([|\eta|^3 + |\eta|^2\xi]w_c) + \mathcal{O}(|\eta|g_1) \tag{48}$$

Introducing the change of variable

$$z(x;\sigma,\eta) \equiv \tilde{D}_x^2 w_c(x;\sigma,\eta), \qquad (\sigma,\eta)\in\mathcal{E}_1, \tag{49}$$

a symbolic problem can be derived for $x > 0$ in $\mathcal{E}_1$,

$$\begin{aligned}&\tilde{D}_x^2 z + qz = r,\\ &\quad z|_{x=0} = \psi g_1 + \mathcal{O}([|\eta|^2 + |\eta|\xi]w_c),\\ &\quad \tilde{D}_x z|_{x=0} = \psi g_2 + \mathcal{O}([|\eta|^3 + |\eta|^2\xi]w_c) + \mathcal{O}(|\eta| g_1) + symb\{[\mathcal{B}_2, \psi]w\},\end{aligned} \tag{50}$$

where $(\sigma, \eta) \in \mathcal{E}_1$ and

$$r = \psi f + \mathcal{O}(|\eta|^4 w_c) + \mathcal{O}(|\eta|^2 \tilde{\xi} w_c) + symb\{[\mathcal{P}, \psi]w\} + l.o.t. \tag{51}$$

To establish the result of Proposition 3.4, the following two lemmas are required.

Lemma 3.5. *For all $\epsilon_1 > 0$ and $s_0 < \frac{1}{2}$,*

i.)
$$\begin{aligned}\int_0^\infty rz dx \ &\le \epsilon_1 \|z\|^2_{H^{s_0}(\mathbb{R}^1_{x+})} + \frac{C}{\epsilon_1}\|f\|^2_{H^{-s_0}(\mathbb{R}^1_{x+})} + \epsilon_1 \||\eta| z\|^2_{L_2(\mathbb{R}^1_{x+})}\\ &+\epsilon_1 \mathcal{O}\left(\||\eta|^3 w_c\|^2_{L_2(\mathbb{R}^1_{x+})} + \||\eta|\tilde{D}_x w_c\|^2_{L_2(\mathbb{R}^1_{x+})}\right) \qquad \text{in } \mathcal{E}_1\\ &+\gamma \||\eta|^2 (\psi' w)\|^2_{L_2(\mathbb{R}^1_{x+})} + \gamma \|\tilde{D}_x^2(\psi' w)\|^2_{L_2(\mathbb{R}^1_{x+})}\end{aligned} \tag{52}$$

ii.)
$$\begin{aligned}&\int_{\mathcal{E}_1} \frac{(1-\epsilon_1)\|\tilde{D}_x z\|^2_{L_2(\mathbb{R}^1_{x+})} + (\delta - \epsilon_1)\||\eta| z\|^2_{L_2(\mathbb{R}^1_{x+})}}{\sigma^3 + |\eta|^3} d\sigma d\eta\\ &\le C_{\epsilon_1}\Big\{\|f\|^2_{H^{-s_0}(\mathbb{R}^1_{x+})} + \|w_c\|^2_{L_2(-\infty,\infty;H^{3/2}(\Omega))}\\ &\quad + \|w\|^2_{L_2(0,T;H^{3/2}(\Omega))} + \gamma \|\tilde{D}_x^2(\psi' w)\|^2_{L_2(\mathbb{R}^1_{x+};H^{-3/2}(\mathcal{E}_1))}\Big\}\\ &\quad + \int_{\mathcal{E}_1} \frac{|(\tilde{D}_x z|_{x=0})(z|_{x=0})|}{\sigma^3 + |\eta|^3} d\sigma d\eta.\end{aligned} \tag{53}$$

Lemma 3.6. *The solution of (50) satisfies*

$$\begin{aligned}&\int_{\mathcal{E}_1} \frac{|(\tilde{D}_x z|_{x=0})(z|_{x=0})|}{|\eta|^3} d\sigma d\eta\\ &\le C_{\epsilon_1}\Big\{\|g_1\|^2_{L_2(0,T;\partial\Omega)} + \|g_2\|^2_{L_2(0,T;H^{-1}(\partial\Omega))}\\ &\quad + \|w_c\|^2_{L_2(-\infty,\infty;H^{3/2+\epsilon}(\Omega))} + (1+\gamma)\|w_c\|^2_{L_2(-\infty,\infty;H^{3/2+\epsilon}(\Omega))}\Big\}.\end{aligned} \tag{54}$$

Notice the changes in how γ plays a role in the estimates, as compared with analagous lemmas in [6]. This is due both to the change in how $\mathcal{E}_1$ is defined, as well as to the fact that γ was neglected in portions of the proof in [6]. In particular, the role of γ in the terms involving $\psi' w$ plays a role. To illustrate, a brief proof of part *i.)* of Lemma 3.5 follows.

Proof of Lemma 3.5 i. Multiplying the definition of r in (51) by z and integrating in x over $\mathbb{R}^1_{x+}$ yields

$$\begin{aligned}\int_0^\infty rz dx \ &= (\psi f, z)_{L_2(\mathbb{R}^1_{x+})} + \mathcal{O}\left((|\eta|^3 w_c, |\eta| z)_{L_2(\mathbb{R}^1_{x+})}\right)\\ &+\mathcal{O}\left((|\eta|\tilde{D}_x w_c, |\eta| z)_{L_2(\mathbb{R}^1_{x+})}\right)\\ &+\sqrt{\gamma}\mathcal{O}\left((\tilde{D}_x^2(\psi' w), |\eta| z)_{L_2(\mathbb{R}^1_{x+})}\right)\\ &+\sqrt{\gamma}\mathcal{O}\left((|\eta|^2 \psi' w, |\eta| z)_{L_2(\mathbb{R}^1_{x+})}\right),\end{aligned} \tag{55}$$

since

$$symb\{[\mathcal{P},\psi]w\} = \sqrt{\gamma}\mathcal{O}\left(|\eta|\tilde{\xi}^2 + |\eta|^3\right)(\psi' w). \qquad \text{in } \mathcal{E}_1 \tag{56}$$

The conclusion of Lemma 3.5 is then found directly by applying the inequality $2ab \leq \epsilon_1 a^2 + (1/\epsilon_1)b^2$.

Conclusion of the Proof of Proposition 3.4. With Lemmas 3.5 and 3.6, the conclusion of Proposition 3.4 follows directly by recalling that Φ_{00} represents a pseudo-differential operator of zero order with support in $\mathcal{E}_1$. □

3.3. Estimates for the interior term $\Lambda[\mathcal{P},\psi]$

Adaptations to account for γ in estimates for this commutator take advantage of (56) and the alternative definition of $\mathcal{E}_1$. Since the proof is similar to that of the estimates for the earlier commutator, $[\mathcal{P},\Lambda]$, the result is simply stated at this point.

Lemma 3.7. *For $\epsilon > 0$ and $s_0 < \frac{1}{2}$, the operator, $\Lambda[\mathcal{P},\psi]$, arising from system (28), satisfies*

$$\begin{aligned} &\|\Lambda[\mathcal{P},\psi]w\|^2_{L_2(\mathbb{R}^1_{x+};H^{-3/2}(\mathbb{R}^2_{ty}))} \\ &\quad \leq C_\epsilon \Big\{\|f\|^2_{H^{-s_0}(0,T;\Omega)} + \|g_1\|^2_{L_2(0,T;\partial\Omega)} \\ &\qquad + \|g_2\|^2_{L_2(0,T;H^{-1}(\partial\Omega))} + (1+\gamma)\|w\|^2_{L_2(0,T;H^{3/2}(\Omega))}\Big\} \end{aligned} \tag{57}$$

3.4. Final estimates for Λw_c

Combining estimates (30–32) for the boundary terms and the estimates for the commutators which involve γ derived in the previous three sections with the elliptic estimate in Lemma 3.1, the following result is obtained.

Theorem 3.8. *There exists a constant C such that, for $\epsilon > 0$ and $s_0 < \frac{1}{2}$, the solution, Λw_c, of (28) satisfies the inequality*

$$\begin{aligned} &\|\Lambda w_c\|_{H^{5/2}(-\infty,\infty;\Omega)} + \|\Lambda w_c\|_{H^2(-\infty,\infty;\partial\Omega)} \\ &\qquad + \|\tilde{D}_x(\Lambda w_c)\|_{H^1(-\infty,\infty;\partial\Omega)} + \|\tilde{D}_x^2(\Lambda w_c)\|_{L_2(-\infty,\infty;\partial\Omega)} \\ &\quad \leq C\big\{\|f\|_{H^{-s_0}(-\infty,\infty;\Omega)} + \|g_1\|_{L_2(-\infty,\infty;\partial\Omega)} \\ &\qquad + \|g_2\|_{L_2(0,T;H^{-1}(\partial\Omega))} + (1+\gamma)\|w\|_{L_2(-\infty,\infty;\Omega)}\big\}. \end{aligned} \tag{58}$$

4. Estimates in the Nonelliptic Sector

In the nonelliptic sector, $\mathcal{E}_2 \equiv \mathcal{R}_{tr} \cup \mathcal{R}_2$, the analysis relies on the properties of the symbols.

Lemma 4.1. *There exists a constant C_ϵ, independent of γ, such that $(1-\Lambda)w_c$ satisfies*

i.)

$$\begin{aligned} &\|D_y^2(1-\Lambda)w_c\|^2_{L_2(-\infty,\infty;\partial\Omega)} \\ &\quad \leq C_\epsilon\left(\|w_t\|^2_{L_2(0,T;\partial\Omega)} + \gamma\|D_y w_t\|^2_{L_2(0,T;\partial\Omega)} + \|w\|^2_{L_2(0,T;H^{3/2}(\Omega))}\right), \end{aligned} \tag{59}$$

ii.)

$$\begin{aligned}&\|D_y\tilde{D}_x(1-\Lambda)w_c\|^2_{L_2(-\infty,\infty;\partial\Omega)}\\&\quad\le C_\epsilon\left(\|\tilde{D}_xD_tw_c\|^2_{L_2(-\infty,\infty;H^{-1}(\partial\Omega))}+\gamma\|\tilde{D}_xD_tw_c\|^2_{L_2(-\infty,\infty;\partial\Omega)}\right.\\&\qquad\left.+\|w\|^2_{L_2(0,T;H^{3/2}(\Omega))}\right),\end{aligned}\tag{60}$$

iii.)

$$\begin{aligned}&\|\tilde{D}_x^2(1-\Lambda)w_c\|^2_{L_2(-\infty,\infty;\partial\Omega)}\\&\quad\le C_\epsilon\left(\|g_1\|^2_{L_2(0,T;\partial\Omega)}+\|w_t\|^2_{L_2(0,T;\partial\Omega)}\right.\\&\qquad+\gamma\|D_yw_t\|^2_{L_2(0,T;\partial\Omega)}+\gamma\|\tilde{D}_xw_t\|^2_{L_2(0,T;\partial\Omega)}\\&\qquad\left.+\|\tilde{D}_xw_t\|^2_{L_2(0,T;H^{-1}(\partial\Omega))}+\|w\|^2_{L_2(0,T;H^{3/2}(\Omega))}\right),\end{aligned}\tag{61}$$

Proof. i.) In $\mathcal{E}_2$, $c_0|\eta|^4\le\sigma^2+\gamma\sigma^2|\eta|^2$. Using this, the following estimate follows directly:

$$\begin{aligned}&\|(1-\Lambda)D_y^2w_c\|^2_{L_2(-\infty,\infty;\partial\Omega)}\\&\quad\le C\int_{\mathcal{E}_2}\left||\eta|^2w_c\right|^2d\sigma d\eta\le\frac{C}{c_0}\int_{\mathcal{E}_2}\left(\sigma^2+\gamma\sigma^2|\eta|^2\right)|w_c|^2\,d\sigma d\eta\\&\quad\le C\left(\|w_t\|^2_{L_2(0,T;\partial\Omega)}+\gamma\|D_yw_t\|^2_{L_2(0,T;\partial\Omega)}\right),\end{aligned}\tag{62}$$

Combining this with the properties of commutators to account for the lower order commutator, $[D_y^2,1-\Lambda]$, which arises and applying trace theory yields (59).

ii.) Noting that $c_0|\eta|^2\le\frac{\sigma^2}{|\eta|^2}+\gamma\sigma^2$ in $\mathcal{E}_2$, it can be seen that

$$\begin{aligned}&\|(1-\Lambda)D_y\tilde{D}_xw_c\|^2_{L_2(-\infty,\infty;\partial\Omega)}\\&\quad\le C\int_{\mathcal{E}_2}\left||\eta|\tilde{D}_xw_c\right|^2d\sigma d\eta\\&\quad\le\frac{C}{c_0}\left(\int_{\mathcal{E}_2}\left|\frac{\sigma}{|\eta|}\tilde{D}_xw_c\right|^2d\sigma d\eta+\int_{\mathcal{E}_2}\left|\sigma\tilde{D}_xw_c\right|^2d\sigma d\eta\right)\\&\quad\le C\left(\|\tilde{D}_xD_tw_c\|^2_{L_2(-\infty,\infty;H^{-1}(\partial\Omega))}+\gamma\|\tilde{D}_xD_tw_c\|^2_{L_2(-\infty,\infty;\partial\Omega)}\right),\end{aligned}\tag{63}$$

As before, application of this inequality together with trace theory yields (60).

iii.) Recalling the alternative expression for β_2 in [6],

$$\beta_2w_c=b_{21}D_y^3w_c+b_{22}D_y^2\tilde{D}_xw_c-b_{23}D_y\tilde{D}_y^2w_c-b_{23}D_y\beta_1w_c+b_{23}D_y\psi g_1,\tag{64}$$

we obtain

$$(1-\Lambda)\tilde{D}_x^2w_c=(1-\Lambda)\psi g_1-(1-\Lambda)\tilde{D}_y^2w_c-(1-\Lambda)\beta_1w_c.\tag{65}$$

Combining this with (59) and (60), noting the definition of β_1, and again applying trace theory yields the desired inequality (61). □

5. Completion of the Proof of Theorem 2.1

Writing $w_c = \Lambda w_c + (1 - \Lambda)w_c$, the estimates of Theorem 3.8 and Lemma 4.1 may be applied. Recalling that $w_c(\cdot, t) = \psi(t)w(\cdot, t)$ and, therefore, $w \equiv w_c$ on $(\alpha, T - \alpha)$, then allows us to obtain the conclusion of Theorem 2.1.

References

[1] L. Hörmander, *The Analysis of Linear Partial Differential Operators III,* Springer-Verlag, New York, 1985.

[2] M. A. Horn and I. Lasiecka, *Asymptotic behavior with respect to thickness of boundary stabilizing feedback for the Kirchhoff plate,* Journal of Differential Equations, **114** (1994), 396–433.

[3] M. A. Horn and C. A. McMillan, *Uniform stability and asymptotic behavior with respect to thickness for a cylindrical shell.* Manuscript.

[4] J. Lagnese, *Boundary Stabilization of Thin Plates,* SIAM Studies in Applied Mathematics, Society for Industrial and Applied Mathematics, Philadelphia, 1989.

[5] J. E. Lagnese and J.-L. Lions, *Modelling, Analysis, and Control of Thin Plates,* Masson, Paris, 1988.

[6] I. Lasiecka and R. Triggiani, *Sharp trace estimates of solutions to Kirchhoff and Euler-Bernoulli equations,* Applied Mathematics and Optimization, **28** (1993), 277–306.

[7] J.-L. Lions and E. Magenes, *Non-Homogeneous Boundary Value Problems and Applications,* Springer-Verlag, Berlin, 1972.

Department of Mathematics
Vanderbilt University
1326 Stevenson Center
Nashville, Tennessee 37240
U.S.A.
E-mail address: horn@math.vanderbilt.edu

International Series of Numerical Mathematics, Vol. 139, 145–156

On the Solvability of Trigonometric Moment Problems Arising in the Problem of Controllability of Rotating Beams

V. I. Korobov, W. Krabs, and G. M. Sklyar

Abstract. In this paper the problem of controllability of a slowly rotating Timoshenko beam in a horizontal plane from the position of rest into an arbitrary position at some given time is investigated. It is solved with the aid of a theorem by D. Ullrich on a trigonometric moment problem which generalizes a classical result of Paley and Wiener.

1. Introduction: Ullrich's Theorem and Consequences

Generalizing a classical theorem of Paley and Wiener [3] D. Ullrich proved in [4] the following

Theorem 1.1. *Suppose that, for every $n \in \mathbb{Z}$, there are given distinct complex numbers $w_{n,0}, w_{n,1}, \ldots, w_{n,K}$ such that*

$$\lim_{n\to\pm\infty} |n - w_{n,j}| = 0 \quad \text{for every} \quad j = 0, \ldots, K.$$

Then, for given complex numbers $c_{n,j}$ $(n \in \mathbb{Z},\ j = 0, \ldots, K)$, the system of moment equations

$$\int_{-(K+1)\pi}^{(K+1)\pi} f(t)e^{itw_{n,j}}\,dt = c_{n,j} \tag{1.1}$$

has a solution $f \in L^2(-(K+1)\pi, (K+1)\pi)$, if and only if

$$\sum_{n=-\infty}^{\infty} \left(|c_0^n|^2 + |c_{0,1}^n|^2 + \ldots + |c_{0,1,\ldots,K}^n|^2\right) < \infty \tag{1.2}$$

where, for every $n \in \mathbb{Z}$ and $0 \le j < m \le K$,

$$c_j^n = c_{n,j}, \qquad c_{j,\ldots,m}^n = \frac{c_{j,\ldots,m-1}^n - c_{j+1,\ldots,m}^n}{w_{n,j} - w_{n,m}}. \tag{1.3}$$

If (1.1) has a solution, then it is unique.

This research was supported by NATO Linkage Grant No. CRG-LG 972980.

Remark 1.2. *Note that in (1.1) the interval of integration* $[-(K+1)\pi,(K+1)\pi]$ *can be replaced by an arbitrary interval of length* $2(K+1)\pi$.

In fact, if instead of (1.1) we consider the system

$$\int_a^{a+2(K+1)\pi} g(\tau)e^{i\tau w_{n,j}}d\tau = c'_{n,j}, \tag{1.1'}$$

then we obtain (1.1) by the substitutions $t=\tau-a-(K+1)\pi$, $f(t)=g(\tau)$, $c_{n,j}=c'_{n,j}e^{-i(a+(K+1)\pi)w_{n,j}}$. By Theorem 1.1 the system (1.1′) has a solution $g\in L^2(a,a+2(K+1)\pi)$, if and only if (1.2) holds true. This, however, is equivalent to

$$\sum_{n=1}^{\infty}\left(|c'^n_0|^2+|c'^n_{0,1}|^2+\ldots+|c'^n_{0,1,\ldots,K}|^2\right)<\infty. \tag{1.2'}$$

In order to see this we at first conclude from (1.3) that

$$c'^n_{0,1,\ldots,m}=\sum_{l=0}^{m}c^n_{0,1,\ldots,l}e^n_{l,\ldots,m} \tag{1.4}$$

where

$$e^n_j=e^{i(a+(K+1)\pi)w_{n,j}},\qquad e^n_{j,\ldots,m}=\frac{e^n_{j,\ldots,m-1}-e^n_{j+1,\ldots,m}}{w_{n,j}-w_{n,m}} \tag{1.5}$$

for $n\in\mathbb{Z},\ 0\le j<m\le K$.

If we define

$$\widetilde{e}^n_j=e^{i(a+(K+1)\pi)(w_{n,j}-n)}$$

and

$$\widetilde{e}^n_{j,\ldots,m}=\frac{\widetilde{e}^n_{j,\ldots,m-1}-\widetilde{e}^n_{j+1,\ldots,m}}{w_{n,j}-w_{n,m}},$$

then it follows from the statement (8) in [4] that

$$\lim_{n\to\pm\infty}\widetilde{e}^n_{j,\ldots,m}=\frac{[i(a+(K+1)\pi)]^{m-j}}{(m-j)!}$$

for $0\le j<m\le K$.

This implies that the sequence $\left(\widetilde{e}^n_{j,\ldots,m}\right)_{n\in\mathbb{Z}}$ is bounded for $0\le j<m\le K$. Hence there exists a constant $C>0$ such that

$$\left|\widetilde{e}^n_{j,\ldots,m}\right|\le C\quad\text{for all}\quad n\in\mathbb{Z}\quad\text{and}\quad 0\le j<m\le K.$$

As a further consequence taking into account that $\left|\widetilde{e}^n_{l,\ldots,m}\right|=\left|e^n_{l,\ldots,m}\right|$ for all $n\in\mathbb{N}$ and $0\le l<m\le K$ we obtain from (1.4)

$$\begin{aligned}\left|c'^n_{0,1,\ldots,m}\right|^2&\le\sum_{l=0}^{m}|c^n_{0,1,\ldots,l}|^2\sum_{l=0}^{m}|\widetilde{e}^n_{l,\ldots,m}|^2\\&\le C^2(K+1)\sum_{l=0}^{m}\left|c^n_{0,1,\ldots,l}\right|^2.\end{aligned}$$

From this we infer that

$$\sum_{n=-\infty}^{\infty}\left(\left|c'^{n}_{0}\right|^2+\left|c'^{n}_{0,1}\right|^2+\ldots+\left|c'^{n}_{0,1,\ldots,K}\right|^2\right)$$

$$\leq C^2(K+1)\sum_{n=-\infty}^{\infty}|c^n_0|^2+\left(|c^n_0|^2+\left|c^n_{0,1}\right|^2\right)+\ldots$$

$$+\left(|c^n_0|^2+\left|c^n_{0,1}\right|^2+\ldots+\left|c^n_{0,1,\ldots,K}\right|^2\right)$$

$$\leq C^2(K+1)^2\sum_{n=-\infty}^{\infty}|c^n_0|^2+\left|c^n_{0,1}\right|^2+\ldots+\left|c^n_{0,1,\ldots,K}\right|^2<\infty.$$

The inverse implication (1.2′) $\Longrightarrow$ (1.2) follows by similar arguments, if we use

$$c^n_{0,1,\ldots,m}=\sum_{l=0}^{m}c'^{n}_{0,1,\ldots,l}e'^{n}_{l,\ldots,m},$$

where

$$e'^{n}_{j}=\frac{1}{e^n_j}$$

and

$$e'^{n}_{j,\ldots,m}=\frac{e'^{n}_{j,\ldots,m-1}-e'^{n}_{j+1,\ldots,m}}{w_{n,j}-w_{n,m}}$$

for $n\in\mathbb{Z}$, $0\leq j<m\leq K$.

Due to Remark 1.2 we will replace in Theorem 1.1 the interval $[-(K+1)\pi,(K+1)\pi]$ by $[0,2(K+1)\pi]$. The following remark allows to apply Theorem 1.1 also on intervals $[0,T]$ where $T>2(K+1)\pi$.

Remark 1.3. *Condition (1.2) remains necessary and sufficient for the solvability (1.1) if we replace the interval of integration $[-(K+1)\pi,(K+1)\pi]$ by any interval $[0,T]$ for $T>2(K+1)\pi$. However in this case the solution of (1.1) is not unique.*

In fact, let (1.2) hold. Then, due to Remark 1.2, the problem (1.1) has a solution $f(t)$ on $[0,2(K+1)\pi]$ and, as a consequence, also on $[0,T]$ (one needs simply to extend $f(t)$ by $f(t)\equiv 0$ for $t\in(2(K+1)\pi,T]$). Nonuniqueness of this solution also easily follows from Remark 1.2.

Conversely, let there exist $f\in L^2(0,T)$ such that

$$\int_0^T f(t)e^{itw_{n,j}}dt=c_{n,j},\quad n\in\mathbb{Z},\quad 0\leq j\leq K.$$

Then we choose $p\in\mathbb{N}$ satisfying

$$2p(K+1)\pi<T\leq 2(p+1)(K+1)\pi$$

and denote

$$h_j^{(l)n} = \int\limits_{2l(K+1)\pi}^{2(l+1)(K+1)\pi} f(t)e^{itw_{n,j}}dt, \quad l = 0,\ldots,p-1,$$

and

$$h_j^{(p)n} = \int\limits_{2p(K+1)\pi}^{T} f(t)e^{itw_{n,j}}dt,$$

for $n \in \mathbb{Z}$, $0 \le j \le K$. Considering for any given $l = 0,1,\ldots,p$ the sequence $\left\{h_j^{(l)n}\right\}_{\substack{n\in\mathbb{Z}\\ 0\le j\le K}}$ as a moment one and applying Remark 1.2 on the interval $[2l(K+1)\pi, 2(l+1)(K+1)\pi]$ to this sequence we conclude that

$$\sum_{n=-\infty}^{\infty}\left(\left|h_0^{(l)n}\right|^2 + \left|h_{0,1}^{(l)n}\right|^2 + \ldots + \left|h_{0,1,\ldots,K}^{(l)n}\right|^2\right) < \infty,$$

where $h_{j,\ldots,m}^{(l)n}$ are divided differences for $h_j^{(l)n}$. Since for any $n \in \mathbb{Z}$, $0 \le j < m \le K$

$$c_{j,\ldots,m}^n = \sum_{l=0}^{p} h_{j,\ldots,m}^{(l)n}$$

we have

$$\sum_{n=-\infty}^{\infty}\left(|c_0^n|^2 + \left|c_{0,1}^n\right|^2 + \left|c_{0,1,\ldots,K}^n\right|^2\right) \le$$

$$\le \sum_{l=0}^{p}(p+1)\sum_{n=-\infty}^{\infty}\left(\left|h_0^{(l)n}\right|^2 + \left|h_{0,1}^{(l)n}\right|^2 + \ldots + \left|h_{0,1,\ldots,K}^{(l)n}\right|^2\right) < \infty,$$

thus (1.2) holds true.

In Section 2 of this paper we will show that the problem of controllability of a rotating Timoshenko beam leads to a trigonometric moment problem (see(2.14)) which turns out to be equivalent to the problem of finding a function $v \in L^2(0, 4\pi)$ such that

$$\int\limits_0^{4\pi} v(t)e^{itw_{n,j}}dt = c_{n,j}$$

holds true for $n \in \mathbb{Z}$ and $j = 0, 1$ where, for every $n \in \mathbb{Z}$, $w_{n,0}$ and $w_{n,1}$ are distinct real numbers satisfying

$$\lim_{n\to\pm\infty} |n - w_{n,j}| = 0 \quad \text{for} \quad j = 0, 1.$$

In Section 3 we show with the aid of Theorem 1.1 and the remarks that the problem of controllability in Section 2 is solvable for $T > 4$ if and only if a certain condition on the final position of the beam (see (3.6)) is satisfied.

In [1] Ullrich's Theorem is used for the solution of a problem of boundary controllability of a hybrid system of two flexible beams connected by a point mass.

2. The Problem of Controllability of a Rotating Timoshenko Beam

In [2] we have investigated the problem of controllability of a slowly rotating Timoshenko beam in a horizontal plane from a position of rest into a position of rest when the movement is controlled by the angular acceleration of the disk of a driving motor into which the beam is clamped. Here we consider the problem of steering the beam from a position of rest into an arbitrary position at some given time.

The equations of motion in a dimension-free formulation are given by

$$\begin{aligned}\ddot{w}(x,t) - w''(x,t) - \xi'(x,t) &= -\ddot{\theta}(t)(r+x),\\ \ddot{\xi}(x,t) - \gamma^2\xi''(x,t) + \xi(x,t) + w'(x,t) &= \ddot{\theta}(t)\end{aligned} \tag{2.1}$$

for $x \in (0,1)$ and $t > 0$, where $w(x,t)$ denotes the deflection of the center line of the beam and $\xi(x,t)$ the rotation angle of the cross section area at the location $x \in [0,1]$ and time $t \geq 0$, respectively, $\dot{w} = w_t$, $\dot{\xi} = \xi_t$ and $w' = w_x$, $\xi' = \xi_x$, θ is the rotation angle of the motor disk, $\dot{\theta} = \dfrac{d\theta}{dt}$, and r is the radius of the disk. In addition, we have boundary conditions of the form

$$\begin{aligned}w(0,t) = \xi(0,t) &= 0,\\ w'(1,t) + \xi(1,t) = \xi'(1,t) &= 0\end{aligned} \tag{2.2}$$

for $t \geq 0$.

We assume the beam to be in the position of rest at $t = 0$ which leads to the initial conditions

$$w(x,0) = \dot{w}(x,0) = \xi(x,0) = \dot{\xi}(x,0) = 0, \quad x \in [0,1], \tag{2.3}$$

$$\theta(0) = \dot{\theta}(0) = 0.$$

In [2] we have considered the special case $\gamma = 1$ which will also be done in the sequel. In this paper we investigate the following

Problem of controllability: Given a time $T > 0$ and a position $(w_T, \xi_T, \dot{w}_T, \dot{\xi}_T, \theta_T, \dot{\theta}_T)$ of the beam where w_T, ξ_T, $\dot{w}_T$, $\dot{\xi}_T$ are chosen in suitable function spaces and θ_T, $\dot{\theta}_T$ are given real numbers, find

$$\theta \in H_0^2(0,T) = \left\{\theta \in H^2(0,T) \mid \theta(0) = \dot{\theta}(0) = 0\right\}$$

such that

$$\theta(T) = \theta_T, \quad \dot{\theta}(T) = \dot{\theta}_T \tag{2.4}$$

and the weak solution (w,ξ) of (2.1), (2.2), (2.3) satisfies the end conditions

$$w(\cdot,T) = w_T, \quad \xi(\cdot,T) = \xi_T, \quad \dot{w}(\cdot,T) = \dot{w}_T, \quad \dot{\xi}(\cdot,T) = \dot{\xi}_T. \tag{2.5}$$

Let $H = L^2((0,1),\mathbb{R}^2)$. Then we define a linear operator $A : D(A) \to H$ by

$$A\begin{pmatrix} y \\ z \end{pmatrix} = \begin{pmatrix} -y'' - z' \\ y' - z'' + z \end{pmatrix}$$

for $\begin{pmatrix} y \\ z \end{pmatrix} \in D(A)$ where

$$D(A) = \left\{ \begin{pmatrix} y \\ z \end{pmatrix} \in H^2((0,1),\mathbb{R}^2) \,\middle|\, \begin{matrix} y(0) = z(0) = 0 \\ y'(1) + z(1) = z'(1) = 0 \end{matrix} \right\}.$$

With this operator (2.1) can be rewritten in the form

$$\begin{pmatrix} \ddot{w}(\cdot,t) \\ \ddot{\xi}(\cdot,t) \end{pmatrix} + A \begin{pmatrix} w(\cdot,t) \\ \xi(\cdot,t) \end{pmatrix} = \begin{pmatrix} f_1(\cdot,t) \\ f_2(\cdot,t) \end{pmatrix}, \tag{2.6}$$

for $t > 0$ where

$$f_1(x,t) = -\ddot{\theta}(t)(r+x), \quad f_2(x,t) = \ddot{\theta}(t)$$

for $x \in (0,1)$, $t > 0$.

One can show that $A : D(A) \to H$ is positive, self-adjoint and has an orthonormal sequence of eigenelements $\begin{pmatrix} y_j \\ z_j \end{pmatrix} \in D(A)$ for $j \in \mathbb{N}$ and a corresponding sequence of eigenvalues $\lambda_j \in \mathbb{R}$ such that

$$1 < \lambda_1 < \lambda_2 < \cdots < \lambda_j \to \infty \quad \text{as} \quad j \to \infty.$$

In [2] it is further shown that for large n the eigenvalues of A are of the form

$$\lambda_n = \begin{cases} \frac{1}{4}[(2k-1)\pi - \varepsilon_{2k-1}]^2 & \text{for} \quad n = 2k-1, \\ \frac{1}{4}[(2k-1)\pi + \varepsilon_{2k}]^2 & \text{for} \quad n = 2k, \end{cases} \tag{2.7}$$

where $\varepsilon_{2k-1},\ \varepsilon_{2k} > 0$, and $\lim\limits_{n\to\infty} \varepsilon_n = 0$.

The unique weak solution of (2.6) corresponding to the initial conditions (2.3) is then given by

$$\begin{pmatrix} w(x,t) \\ \xi(x,t) \end{pmatrix} = \sum_{j=1}^{\infty} \frac{1}{\sqrt{\lambda_j}} \int_0^t \sin\sqrt{\lambda_j}(t-s) \left\langle \begin{pmatrix} f_1(\cdot,s) \\ f_2(\cdot,s) \end{pmatrix}, \begin{pmatrix} y_j \\ z_j \end{pmatrix} \right\rangle_H ds \begin{pmatrix} y_j \\ z_j \end{pmatrix} \tag{2.8}$$

for $x \in [0,1]$ and $t \geq 0$ and its time derivative reads

$$\begin{pmatrix} \dot{w}(x,t) \\ \dot{\xi}(x,t) \end{pmatrix} = \sum_{j=1}^{\infty} \int_0^t \cos\sqrt{\lambda_j}(t-s) \left\langle \begin{pmatrix} f_1(\cdot,s) \\ f_2(\cdot,s) \end{pmatrix}, \begin{pmatrix} y_j \\ z_j \end{pmatrix} \right\rangle_H ds \begin{pmatrix} y_j \\ z_j \end{pmatrix}. \tag{2.9}$$

From these representations we infer that the end conditions (2.5) are equivalent to

$$\begin{aligned} a_j \int_0^T \sin\sqrt{\lambda_j}(T-t)\ddot{\theta}(t)dt &= \sqrt{\lambda_j} \left\langle \begin{pmatrix} w_T \\ \xi_T \end{pmatrix}, \begin{pmatrix} y_j \\ z_j \end{pmatrix} \right\rangle_H, \\ a_j \int_0^T \cos\sqrt{\lambda_j}(T-t)\ddot{\theta}(t)dt &= \left\langle \begin{pmatrix} \dot{w}_T \\ \dot{\xi}_T \end{pmatrix}, \begin{pmatrix} y_j \\ z_j \end{pmatrix} \right\rangle_H, \end{aligned} \tag{2.10}$$

for $j \in \mathbb{N}$ where

$$a_j = -\int_0^1 (r+x)y_j(x)dx + \int_0^1 z_j(x)dx, \quad j \in \mathbb{N}. \tag{2.11}$$

Let us assume that

$$a_j \neq 0 \quad \text{for all} \quad j \in \mathbb{N}. \tag{2.12}$$

If we define

$$c_j^1 = \frac{\sqrt{\lambda_j}}{a_j}\left\langle \begin{pmatrix} w_T \\ \xi_T \end{pmatrix}, \begin{pmatrix} y_j \\ z_j \end{pmatrix} \right\rangle_H \quad \text{and} \quad c_j^2 = \frac{1}{a_j}\left\langle \begin{pmatrix} \dot{w}_T \\ \dot{\xi}_T \end{pmatrix}, \begin{pmatrix} y_j \\ z_j \end{pmatrix} \right\rangle_H \tag{2.13}$$

for $j \in \mathbb{N}$ and put

$$\widehat{u}(t) = \ddot{\theta}(T-t) \quad \text{for} \quad t \in [0,T],$$

then (2.10) can be rewritten in the form

$$\begin{aligned} \int_0^T \sin\sqrt{\lambda_j}\, t\, \widehat{u}(t)dt &= c_j^1, \\ \int_0^T \cos\sqrt{\lambda_j}\, t\, \widehat{u}(t)dt &= c_j^2, \quad j \in \mathbb{N}. \end{aligned} \tag{2.14}$$

The end conditions (2.4) turn out to be equivalent to

$$\begin{aligned} \int_0^T t\widehat{u}(t)dt &= \theta_T, \\ \int_0^T \widehat{u}(t)dt &= \dot{\theta}_T. \end{aligned} \tag{2.15}$$

Then the problem of controllability is equivalent to finding some $\widehat{u} \in L^2(0,T)$ which satisfies (2.14), (2.15). If such $\widehat{u}$ has been found, then

$$\theta(t) = \int_0^t (t-s)\widehat{u}(T-s)ds, \quad t \in [0,T]$$

is a solution of the problem of controllability.

3. On the Solution of the Problem of Controllability

We start our investigation by considering the system (2.14) of moment equations for $T = 4$. On defining

$$c_k = c_k^2 + ic_k^1 \quad \text{for} \quad k \in \mathbb{N} \tag{3.1}$$

we at first rewrite (2.14) for $T = 4$ in the form

$$\begin{aligned} \int_0^4 e^{i\sqrt{\lambda_k}t}\widehat{u}(t)dt &= c_k, \\ \int_0^4 e^{-i\sqrt{\lambda_k}t}\widehat{u}(t)dt &= \overline{c}_k, \quad \text{for} \quad k \in \mathbb{N} \end{aligned} \tag{3.2}$$

when $\overline{c}_k$ denotes the conjugate complex of c_k.

Let us define

$$w_{k,0} = \frac{1}{\pi}\left(\sqrt{\lambda_{2k-1}} + \frac{\pi}{2}\right) \Longleftrightarrow \sqrt{\lambda_{2k-1}} = \pi w_{k,0} - \frac{\pi}{2},$$

$$w_{k,1} = \frac{1}{\pi}\left(\sqrt{\lambda_{2k}} + \frac{\pi}{2}\right) \Longleftrightarrow \sqrt{\lambda_{2k}} = \pi w_{k,1} - \frac{\pi}{2},$$

for $k \in \mathbb{N}$ and

$$w_{-k+1,j} = -w_{k,j} + 1 \quad \text{for} \quad k \in \mathbb{N} \quad \text{and} \quad j = 0, 1.$$

Then it follows that for every $k \in \mathbb{Z}$, $w_{k,0}$ and $w_{k,1}$ are distinct real numbers and from (2.7) we infer that

$$\lim_{k \to \pm\infty} |w_{k,j} - k| = 0 \quad \text{for} \quad j = 0, 1.$$

Further we define

$$c_{k,0} = \pi c_{2k-1}, \quad c_{k,1} = \pi c_{2k} \quad \text{for} \quad k \in \mathbb{N}$$

and

$$c_{-k+1,j} = \overline{c}_{k,j}, \quad \text{for} \quad k \in \mathbb{N} \quad \text{and} \quad j = 0, 1.$$

According to Theorem 1.1 and Remark 1.2 there exists a unique solution of

$$\int_0^{4\pi} e^{iw_{k,j}t} v(t)dt = c_{k,j} \tag{3.3}$$

for $k \in \mathbb{Z}$ and $j = 0, 1$, if and only if

$$\sum_{k=-\infty}^{\infty} \left(|c_{k,0}|^2 + \left| \frac{c_{k,0} - c_{k,1}}{w_{k,0} - w_{k,1}} \right|^2 \right) < \infty. \tag{3.4}$$

Let us define

$$\widehat{u}(\tau) = v(\pi\tau)e^{i\pi\tau/2} \quad \text{for} \quad \tau \in [0, 4]. \tag{3.5}$$

Then it follows that $\widehat{u} \in L^2(0, 4)$ and

$$\int_0^4 e^{i\sqrt{\lambda_{2k-1}}\tau}\widehat{u}(\tau)d\tau = \int_0^4 e^{i(w_{k,0}\pi - \pi/2)\tau} v(\pi\tau)e^{i\pi\tau/2}d\tau$$

$$= \frac{1}{\pi} \int_0^{4\pi} e^{iw_{k,0}t} v(t)dt = \frac{c_{k,0}}{\pi} = c_{2k-1} \quad \text{for} \quad k \in \mathbb{N}.$$

Similarly it follows that

$$\int_0^4 e^{i\sqrt{\lambda_{2k}}\tau} \widehat{u}(\tau)d\tau = c_{2k} \quad \text{for} \quad k \in \mathbb{N}.$$

Further it follows that

$$\int_0^4 e^{-i\sqrt{\lambda_{2k-1}}\tau} \widehat{u}(\tau)d\tau = \int_0^4 e^{-i(\sqrt{\lambda_{2k-1}}+\frac{1}{2})\tau} \widehat{u}(\tau) e^{\frac{i}{2}\tau} d\tau =$$

$$= \int_0^4 e^{-\frac{i}{\pi}(\sqrt{\lambda_{2k-1}}+\frac{1}{2})\pi\tau} v(\pi\tau) e^{\frac{i}{2}\pi\tau} e^{\frac{i}{2}\tau} d\tau =$$

$$= \frac{1}{\pi} \int_0^{4\pi} e^{-\frac{i}{\pi}(\sqrt{\lambda_{2k-1}}+\frac{1}{2})t} v(t) e^{\frac{i}{2}t} e^{\frac{i}{2}\frac{t}{\pi}} dt =$$

$$= \frac{1}{\pi} \int_0^{4\pi} e^{-\frac{i}{\pi}(\sqrt{\lambda_{2k-1}}-\frac{\pi}{2})t} v(t)dt =$$

$$= \frac{1}{\pi} \int_0^{4\pi} e^{-i(w_{k,0}-1)t} v(t)dt = \frac{1}{\pi} \int_0^{4\pi} e^{iw_{-k+1,0}t} v(t)dt =$$

$$= \frac{1}{\pi} c_{-k+1,0} = \frac{1}{\pi} \overline{c}_{k,0} = \overline{c}_{2k-1} \quad \text{for} \quad k \in \mathbb{N}.$$

In a similar way one shows that

$$\int_0^4 e^{-i\sqrt{\lambda_{2k}}\tau} \widehat{u}(\tau)d\tau = \overline{c}_{2k} \quad \text{for} \quad k \in \mathbb{N}.$$

Thus $\widehat{u} \in L^2(0,4)$ defined by (3.5) satisfies (3.2) if and only if (3.4) holds true which is equivalent to

$$\sum_{k=1}^{\infty} \left(|c_{2k-1}|^2 + \left| \frac{c_{2k-1} - c_{2k}}{\sqrt{\lambda_{2k-1}} - \sqrt{\lambda_{2k}}} \right|^2 \right) < \infty. \tag{3.6}$$

The function $\widehat{u} \in L^2(0,4)$ defined by (3.5) is also the only solution of (3.2), if (3.6) is satisfied because

$$v(t) = \widehat{u}\left(\frac{t}{\pi}\right) e^{-\frac{i}{2}t} \quad \text{for} \quad t \in [0, 4\pi]$$

is the unique solution of (3.3) for $k \in \mathbb{Z}$ and $j = 0, 1$, if and only if (3.4) is satisfied which is equivalent to (3.6). Finally let us show that the only solution $\widehat{u}$ of (3.2) is a real function. In fact, if we put $\widehat{u} = \mathrm{Re}\widehat{u}(t) + i\mathrm{Im}\widehat{u}(t)$ then it follows from (3.2) that

$$\int_0^4 e^{\pm i\sqrt{\lambda_k}t}\mathrm{Im}\widehat{u}(t)dt = 0 \quad \text{for} \quad k \in \mathrm{N}.$$

Hence from (3.3) we have

$$\int_0^{4\pi} e^{iw_{k,j}t}v_0(t)dt = 0 \quad \text{for} \quad k \in \mathbb{Z} \quad \text{and} \quad j = 0, 1,$$

where $v_0(t) = \mathrm{Im}\widehat{u}\left(\frac{t}{\pi}\right)e^{-\frac{i}{2}t}$ for $t \in (0, 4\pi)$. Since due to Theorem 1.1 and Remark 1.2 the solution of (3.3) is unique, we conclude that $\mathrm{Im}\widehat{u}(t) \equiv 0$, $t \in (0, 4)$ and, therefore, $\widehat{u}$ is a real function. Thus, the solution of (3.2) is also a solution of (2.14) fot $T = 4$ and the condition (3.6) is necessary and sufficient for solvability (unique) of (2.14) on the interval $(0, 4)$.

Next we assume that $T > 4$. Let the moment problem (2.14) have a solution $\widehat{u} \in L^2(0, T)$. Then by analogy with the case $T = 4$ one observes that the problem

$$\int_0^{\pi T} e^{iw_{k,j}t}v(t)dt = c_{k,j}$$

for $k \in \mathbb{Z}$ and $j = 0, 1$ has a solution $v(t) = \widehat{u}\left(\frac{t}{\pi}\right)e^{-\frac{i}{2}t}$ for $t \in [0, \pi T]$. Due to Remark 1.3 that yields that the condition (3.4) and, as a consequence, (3.6) holds. Thus the condition (3.6) is necessary for solvability of (2.14) if $T > 4$ as well.

Let now (3.6) be satisfied. If we define

$$\widetilde{u}(t) = \begin{cases} \widehat{u}(t) & \text{for} \quad 0 \leq t \leq 4, \\ 0 & \text{for} \quad 4 < t \leq T, \end{cases}$$

where $\widehat{u} \in L^2(0, 4)$ is the unique solution of (3.2), then $\widetilde{u} \in L^2(0, T)$ and

$$\begin{aligned} \int_0^T e^{i\sqrt{\lambda_k}t}\widetilde{u}(t)dt &= c_k, \\ \int_0^T e^{-i\sqrt{\lambda_k}t}\widetilde{u}(t)dt &= \overline{c}_k, \quad \text{for} \quad k \in \mathrm{N} \end{aligned} \tag{3.7}$$

which is equivalent to

$$\begin{aligned} \int_0^T \sin\sqrt{\lambda_k} t \widetilde{u}(t)dt &= c_k^1, \\ \int_0^T \cos\sqrt{\lambda_k} t \widetilde{u}(t)dt &= c_k^2, \quad \text{for} \quad k \in \mathbb{N}. \end{aligned} \tag{3.8}$$

In [2] we have shown that the system

$$\{t, 1, \sin\sqrt{\lambda_k}t, \cos\sqrt{\lambda_k}t, \mid t \in [0,T],\ k \in \mathbb{N}\}$$

is minimal in $L^2(0,T)$ if $T > 4$.

This implies in particular the existence of two functions $\widehat{u}_1$, $\widehat{u}_2 \in L^2(0,T)$ such that

$$\int_0^T t\widehat{u}_1(t)dt = 1, \quad \int_0^T \widehat{u}_1(t)dt = 0,$$
$$\int_0^T t\widehat{u}_2(t)dt = 0, \quad \int_0^T \widehat{u}_2(t)dt = 1,$$

and

$$\int_0^T \widehat{u}_i(t)\sin\sqrt{\lambda_k}tdt = \int_0^T \widehat{u}_i(t)\cos\sqrt{\lambda_k}tdt = 0$$

for $k \in \mathbb{N}$ and $i = 1, 2$.

Now let us define

$$\widetilde{\theta}_T = \int_0^T t\widetilde{u}(t)dt \quad \text{and} \quad \dot{\widetilde{\theta}}_T = \int_0^T \widetilde{u}(t)dt.$$

With these definitions we put

$$u^*(t) = \widetilde{u}(t) + (\theta_T - \widetilde{\theta}_T)\widehat{u}_1(t) + (\dot{\theta}_T - \dot{\widetilde{\theta}}_T)\widehat{u}_2(t)$$

for $t \in [0,T]$ and conclude $u^* \in L^2(0,T)$ as well as

$$\int_0^T tu^*(t)dt = \theta_T,$$

$$\int_0^T u^*(t)dt = \dot{\theta}_T,$$

$$\int_0^T \sin\sqrt{\lambda_k}tu^*(t)dt = c_k^1,$$

$$\int_0^T \cos\sqrt{\lambda_k}\,t u^*(t)dt = c_k^2, \quad \text{for} \quad k \in \mathbb{N}.$$

Hence $u^* \in L^2(0,T)$ is a solution of (2.14), (2.15).

Summarizing we obtain the

Theorem 3.1. *Let $T > 4$. Then the problem of controllability has a solution if and only if (3.6) is satisfied. For $T = 4$ the function $\widetilde{u} = \widehat{u}$ is uniquely defined as solution of (3.8) and therefore the problem of controllability has a (unique) solution, if and only if (3.6) is satisfied and $\widetilde{\theta}_T = \theta_T$ and $\dot{\widetilde{\theta}}_T = \dot{\theta}_T$.*

References

[1] C. Castro and E. Zŭazŭa, *Boundary Controllability of a Hybrid System Consisting in Two Fuxible Beams Connected by a Point Mass.* SIAM J. Control Optim. 36(1998), 1576–1595.

[2] W. Krabs and G.M. Sklyar, *On the Controllability of a Slowly Rotating Timoshenko Beam.* Jornal for Analysis and its Applications 18 (1999), 437–448.

[3] R.E.A.C. Paley and N. Wiener, *Fourier Transforms in the Complex Domain.* Amer. Math. Soc., Providence, R.J., 1934; 3rd printing 1978.

[4] D Ullrich, *Divided Differences and Systems of Norharmonic Fourier Series.* Proc. Amer. Soc. 80(1980), 47–57.

Inst. Math. Univ., ul. Wielkopolska 15
70-451 Szczecin, Poland;
Dept. Diff. Equat. and Control, Kharkov National University
Svoboda sqr.4, Kharkov, 61077, Ukraine
E-mail address: `korobow@sus.univ.szczecin.pl`
`vkorobov@univer.kharkov.ua`

Department of Mathematics
Technical University of Darmstadt
Schlossgartenstrasse 7
64289 Darmstadt, Germany
E-mail address: `krabs@mathematik.tu-darmstadt.de`

Inst. Math. Univ., ul. Wielkopolska 15
70-451 Szczecin, Poland;
Dept. Math. Anal., Kharkov National University
Svoboda sqr.4, Kharkov, 61077, Ukraine
E-mail address: `sklar@sus.univ.szczecin.pl;`
`sklyar@univer.kharkov.ua`

International Series of Numerical Mathematics, Vol. 139, 157–169

Domain Decomposition in Optimal Boundary Control of Hyperbolic Equations with Boundary Damping

J. E. Lagnese

Abstract. This paper is concerned with optimal boundary final-value control of scalar second order linear hyperbolic equations in which the control variable acts through an impedance boundary condition. The dissipative nature of the boundary condition improves the regularity of solutions and allows penalization of final state in the natural energy norm of the system. An iterative domain decomposition algorithm is proposed for the associated optimality system and its convergence is established provided the leading coefficients of the hyperbolic equation are continuously differentiable on the subdomains employed in the decomposition.

1. Introduction

1.1. Setting the problem

This work is concerned with domain decomposition in optimal boundary control of second order linear hyperbolic differential equations with boundary damping and with penalization of the final state in the cost functional. More precisely, we consider the problem

$$\begin{gathered}\frac{\partial^2 w}{\partial t^2} - \frac{\partial}{\partial x_j}(a_{jk}\frac{\partial w}{\partial x_k}) + cw = 0 \text{ in } Q := \Omega \times (0,T) \\ w = 0 \text{ on } \Sigma^D := \Gamma^D \times (0,T) \\ \frac{\partial w}{\partial \nu_A} + \alpha\frac{\partial w}{\partial t} = f \text{ on } \Sigma^N := \Gamma^N \times (0,T) \\ w(0) = w_0, \quad \frac{\partial w}{\partial t}(0) = v_0 \text{ in } \Omega,\end{gathered} \tag{1.1}$$

where $T > 0$, Ω is a bounded domain in $\mathbb{R}^n$ with piecewise smooth, Lipschitz boundary $\Gamma := \Gamma^D \cup \Gamma^N$, $\Gamma^N \neq \emptyset$, $\overline{\Gamma}^D \cap \overline{\Gamma}^N = \emptyset$, ν is the unit exterior normal vector to Γ and $\partial w/\partial \nu_A = \nu \cdot (A\nabla w)$, $A = (a_{ij})$. We assume that the matrix A is symmetric and uniformly positive definite in Ω with $L^\infty(\Omega)$ entries that

Research supported by the National Science Foundation through grant DMS-9972034.

are continuous in a neighborhood of Γ^N. Later on, to establish convergence of the domain decomposition algorithm, it will be necessary to assume that these coefficients are of class C^1 on the subdomains. Thus we deal with a *transmission problem.* We further assume that $c \in L^\infty(\Omega)$ with $c(x) > 0$ a.e., and that $\alpha \in L^\infty(\Gamma^N)$, $\alpha(x) \geq \alpha_0 > 0$ a.e. If

$$(w_0, v_0) \in \mathcal{H} := H^1_D(\Omega) \times L^2(\Omega), \quad f \in L^2(\Sigma^N),$$

then (1.1) has a unique solution with regularity

$$(w, \frac{\partial w}{\partial t}) \in C(\mathcal{H}), \quad \frac{\partial w}{\partial t}|_{\Sigma^N} \in L^2(\Sigma^N).$$

(Here $H^1_D(\Omega)$ is the closure in $H^1(\Omega)$ of $C^\infty(\overline{\Omega})$ functions that vanish on Γ_D. Also, we write $C(\mathcal{H})$ in place of $C([0,T],\mathcal{H})$ and, in general, we suppress reference to the time interval $[0,T]$ when it is clear from context.) Therefore, for any $(z_0, z_1) \in \mathcal{H}$ we may formulate the optimal control problem

$$\inf_{f \in L^2(\Sigma^N)} J(f) \text{ subject to (1.1)},$$

where

$$J(f) = \frac{1}{2}\int_{\Sigma^N} |f|^2 d\Sigma + \frac{k}{2}\|(w(T), \frac{\partial w}{\partial t}(T)) - (z_0, z_1)\|^2_{\mathcal{H}}.$$

The existence of a unique optimal control f_{opt} is standard. It may be shown that f_{opt} is given by

$$f_{\text{opt}} = -p|_{\Sigma^N} \tag{1.2}$$

where p is the solution of the problem

$$\begin{aligned} \frac{\partial^2 p}{\partial t^2} - \frac{\partial}{\partial x_j}(a_{jk}\frac{\partial p}{\partial x_k}) + cp = 0 \text{ in } Q \\ p = 0 \text{ on } \Sigma^D \\ \frac{\partial p}{\partial \nu_A} - \alpha\dot{p} = 0 \text{ on } \Sigma^N \\ p(T) = k(\frac{\partial w}{\partial t}(T) - z_1) \in L^2(\Omega), \\ \frac{\partial p}{\partial t}(T) = -k\mathcal{A}(w(T) - z_0) \in (H^1_D(\Omega))', \end{aligned} \tag{1.3}$$

and where $\mathcal{A}$ is the canonical isomorphism of $H^1_D(\Omega)$ onto its dual space $(H^1_D(\Omega))'$. It may be proved (by transposition, for example) that (1.3) has a unique solution with regularity

$$(\frac{\partial p}{\partial t}, p) \in C(\mathcal{H}'), \quad p|_{\Sigma^N} \in L^2(\Sigma^N), \tag{1.4}$$

where $\mathcal{H}' := (H^1_D(\Omega))' \times L^2(\Omega)$ (that is, we identify $L^2(\Omega)$ with its dual space, so that $\mathcal{H}'$ is the dual space of $\mathcal{H}$ with respect to the pivot space $L^2(\Omega) \times L^2(\Omega)$). In (1.3), the term $\dot{p}$ is the element of $(H^1(L^2(\Gamma^N)))'$ defined by

$$\langle \dot{p}, \phi \rangle = -\int_{\Sigma^N} p\frac{\partial \phi}{\partial t}\, d\Sigma, \quad \forall \phi \in H^1(L^2(\Gamma^N)). \tag{1.5}$$

Property (1.4) remains true if the boundary condition on Σ^N is replaced by $\dfrac{\partial p}{\partial \nu_A} - a\dot{p} = \dot{\rho}$, where $\rho \in L^2(\Sigma^N)$ is given, a fact that we shall use below. The object of this paper is to describe a domain decomposition method (DDM) for approximating the solution of the *optimality system* (1.1)–(1.3)

1.2. Discussion

One effect of introducing the dissipative term $a\partial w/\partial t$ into the boundary condition in (1.1) is to improve the regularity of the solution, which allows penalization of the final state in the energy norm in the cost functional. Without this term, when the control $f \in L^2(\Sigma^N)$ in general the solution of (1.1) will not evolve in the energy space $\mathcal{H}$ regardless of how regular the data and the boundary Γ may be. Therefore, in this situation penalization of the final state must be done in a weaker topology, for instance in $\mathcal{H}'$. This case was considered in Lagnese and Leugering [4] (with $a_{jk} = a\delta_{jk}$ with a a piecewise constant function, and $c \equiv 0$), where a convergent DDM was presented even in the limiting case $k = \infty$ (i.e., when one has exact controllability to (z_0, z_1)). Technically, the present case turns out to be more challenging, and the iterative transmission conditions introduced below are different than those considered in [4]. The latter were based on a generalization of the nonoverlapping Schwarz alternating algorithm introduced by P. L. Lions [5] and which were first introduced into optimal control problems by Benamou; see [1] and references therein. In fact, in [4] it was necessary to use an under relaxation of the "standard" algorithm in order to establish convergence of the iterations. From this point of view, the present results are stronger since, by utilizing a unique continuation argument, convergence without relaxation is established. (Relaxation may, however, improve the rate of convergence.) The DDM presented here is analogous to one introduced in [3] in the context of optimal boundary control of the Maxwell system. Perhaps surprisingly, the convergence analysis for the DDM for (1.1)–(1.3) turns out to be more complicated than that for the Maxwell optimality system.

Another rationale for the introduction of dissipation into the boundary condition on Σ^N stems from the case where Ω is the exterior of a reflecting body Γ^D. In this situation, it is common to introduce an artificial boundary Γ^N, together with the "transparent" boundary condition on Γ^N. While this boundary condition is nonlocal, its first order approximation (in an appropriate sense) is of the form in (1.1).

2. The Domain Decomposition

Let $\{\Omega_i\}_{i=1}^m$ be bounded domains in $\mathbb{R}^n$ with piecewise smooth, Lipschitz boudaries such that

$$\Omega_i \cap \Omega_j = \emptyset,\ i \neq j, \quad \Omega_i \subset \Omega,\ i = 1, \ldots, m, \quad \overline{\Omega} = \bigcup_{i=1}^m \overline{\Omega}_i.$$

We set

$$\Gamma_{ij} = \partial\Omega_i \cap \partial\Omega_j = \Gamma_{ji},\ i \neq j, \quad \Gamma_i = \bigcup_{j:\Gamma_{ij}\neq\emptyset} \Gamma_{ij},$$

$$\Gamma_i^D = \partial\Omega_i \cap \Gamma^D, \quad \Gamma_i^N = \partial\Omega_i \cap \Gamma^N.$$

Then $\partial\Omega_i = \Gamma_i \cup \Gamma_i^D \cup \Gamma_i^N$. It is assumed that each Γ_i^D, Γ_i^N and Γ_{ij} is either empty or has a nonempty interior. We further set

$$\Sigma_{ij} = \Gamma_{ij} \times (0,T), \quad \Sigma_i = \Gamma_i \times (0,T),$$

$$Q_i = \Omega_i \times (0,T) \quad \Sigma_i^D = \Gamma_i^D \times (0,T), \quad \Sigma_i^N = \Gamma_i^N \times (0,T).$$

Let $A_i = A|_{\Omega_i}$. $c_i = c|_{\Omega_i}$, $\alpha_i = \alpha|_{\Gamma_i^N}$. We assume that

$$A_i \in C^1(\Omega_i) \text{ and has a } C^1 \text{ extension to } \overline{\Omega}_i.$$

Let ν_i denote the unit exterior normal vector to $\partial\Omega_i$ and $\partial/\partial_{\nu_{A_i}} = \nu_i \cdot (A_i\nabla)$. Set

$$H^1_{D_i}(\Omega_i) = \{\phi \in H^1(\Omega_i) : \phi|_{\Gamma_i^D} = 0\}.$$

It is assumed that $H^1_D(\Omega)$ and $H^1_{D_i}(\Omega_i)$ are normed in such a way that

$$\|\phi\|^2_{H^1_D(\Omega)} = \sum_{i=1}^m \|\phi_i\|^2_{H^1_{D_i}(\Omega_i)}, \quad \forall\phi \in H^1_D(\Omega),$$

where $\phi_i = \phi|_{\Omega_i}$. Then if $\mathcal{A}_i$ is the canonical isomorphism of $H^1_{D_i}(\Omega_i)$ onto $(H^1_{D_i}(\Omega_i))'$ we have

$$\mathcal{A}\phi = (\mathcal{A}_1\phi_1, \dots, \mathcal{A}_m\phi_m)$$

in the sense that

$$\langle \mathcal{A}\phi, \psi\rangle = \sum_{i=1}^m \langle \mathcal{A}_i\phi_i, \psi_i\rangle, \quad \forall\phi, \psi \in H^1_D(\Omega),$$

where the angle brackets are taken in the appropriate dualities. For definiteness, and to simplify the analysis, we take

$$\|\phi\|^2_{H^1_{D_i}(\Omega_i)} = \int_{\Omega_i} (A_i\nabla\phi \cdot \nabla\phi + c_i|\phi|^2)dx, \ \phi \in H^1_{D_i}(\Omega_i),$$

which we may do by virtue of the assumption that $c(x) > 0$ a.e. in Ω. We use the standard norms in $L^2(\Omega)$ and $L^2(\Omega_i)$, and set $\mathcal{H}_i := H^1_{D_i}(\Omega_i) \times L^2(\Omega_i)$, $\mathcal{H}_i' = (H_{D_i}(\Omega_i))' \times L^2(\Omega_i)$, the dual space of $\mathcal{H}_i$ with respect to $L^2(\Omega_i) \times L^2(\Omega_i)$.

Let (w,p) be the solution of the global optimality system (1.1)–(1.3). This system is formally equivalent to the coupled local systems

$$\begin{cases} \dfrac{\partial^2 w_i}{\partial t^2} - \nabla\cdot(A_i\nabla w_i) + c_i w_i = 0 \\ \dfrac{\partial^2 p_i}{\partial t^2} - \nabla\cdot(A_i\nabla p_i) + c_i p_i = 0 \text{ in } Q_i \end{cases} \tag{2.1}$$

$$w_i = p_i = 0 \text{ on } \Sigma_i^D \tag{2.2}$$

$$\begin{cases} \dfrac{\partial w_i}{\partial \nu_{A_i}} + \alpha_i \dfrac{\partial w_i}{\partial t} + p_i = 0 \\ \dfrac{\partial p_i}{\partial \nu_{A_i}} - \alpha_i \dot{p}_i = 0 \text{ on } \Sigma_i^N \end{cases} \tag{2.3}$$

$$\begin{cases} w_i(0) = w_{0i}, \quad \dfrac{\partial w_i}{\partial t}(0) = v_{0i} \\ p_i(T) = k(\dfrac{\partial w_i}{\partial t}(T) - z_{1i}) \\ \dfrac{\partial p_i}{\partial t}(T) = -k\mathcal{A}_i(w_i(T) - z_{0i}) \text{ in } \Omega_i \end{cases} \tag{2.4}$$

together with the *transmission conditions* on the interfaces Σ_{ij}:

$$\begin{cases} w_i = w_j, \quad \dfrac{\partial w_i}{\partial \nu_{A_i}} = -\dfrac{\partial w_j}{\partial \nu_{A_j}} \\ p_i = p_j, \quad \dfrac{\partial p_i}{\partial \nu_{A_i}} = -\dfrac{\partial p_j}{\partial \nu_{A_j}}. \end{cases} \tag{2.5}$$

We rewrite (2.5) as

$$\begin{aligned} \frac{\partial w_i}{\partial \nu_{A_i}} + \beta\frac{\partial w_i}{\partial t} + \gamma p_i &= -\frac{\partial w_j}{\partial \nu_{A_j}} + \beta\frac{\partial w_j}{\partial t} + \gamma p_j \\ \frac{\partial p_i}{\partial \nu_{A_i}} - \beta\dot{p}_i + \gamma\frac{\partial \dot{w}_i}{\partial t} &= -\frac{\partial p_j}{\partial \nu_{A_j}} - \beta\dot{p}_j + \gamma\frac{\partial \dot{w}_j}{\partial t} \end{aligned} \tag{2.6}$$

where β and γ are nonzero constants and where $\dot{p}_i$, $\partial\dot{w}_i/\partial t \in (H^1(L^2(\Gamma_i)))'$ are defined by

$$\langle \dot{p}_i, \phi\rangle = -\int_{\Sigma_i} p_i \frac{\partial \phi}{\partial t}\, d\Sigma, \quad \langle \frac{\partial \dot{w}_i}{\partial t}, \phi\rangle = -\int_{\Sigma_i} \frac{\partial w_i}{\partial t}\frac{\partial \phi}{\partial t}\, d\Sigma, , \ \forall \phi \in H^1(L^2(\Gamma_i)),$$

assuming that p_i, $\partial w_i/\partial t \in L^2(\Sigma_i)$. Clearly (2.5) implies (2.6). Conversely, by interchanging i and j in (2.6) and adding the result to (2.6) we obtain

$$\frac{\partial w_i}{\partial \nu_{A_i}} = -\frac{\partial w_j}{\partial \nu_{A_j}}, \quad \frac{\partial p_i}{\partial \nu_{A_i}} = -\frac{\partial p_j}{\partial \nu_{A_j}} \text{ on } \Sigma_{ij}$$

and then

$$\beta\frac{\partial w_i}{\partial t} + \gamma p_i = \beta\frac{\partial w_j}{\partial t} + \gamma p_j, \quad \gamma\frac{\partial \dot{w}_i}{\partial t} - \beta\dot{p}_i = \gamma\frac{\partial \dot{w}_j}{\partial t} - \beta\dot{p}_j$$

on Σ_{ij}. Upon differentiating the first equation in t (in the $(H^1(L^2(\Gamma_i)))'$ sense) one finds that

$$\frac{\partial \dot{w}_i}{\partial t} = \frac{\partial \dot{w}_j}{\partial t}, \quad \dot{p}_i = \dot{p}_j \text{ on } \Sigma_{ij}$$

that is,

$$\int_{\Sigma_{ij}} p_i \frac{\partial \phi}{\partial t}\, d\Sigma = \int_{\Sigma_{ij}} p_j \frac{\partial \phi}{\partial t}\, d\Sigma, \ \forall \phi \in H^1(L^2(\Gamma_{ij}))$$

and, similarly, for $\partial \dot{w}_i/\partial t$, $\partial \dot{w}_j/\partial t$. Thus

$$p_i = p_j, \quad \frac{\partial w_i}{\partial t} = \frac{\partial w_j}{\partial t} \text{ on } \Sigma_{ij}$$

in the $L^2(\Sigma_{ij})$ sense, and then $w_i = w_j$ on Σ_{ij} since $w_{0i} = w_{0j}$ there.

Set

$$\lambda_{ij} = -\frac{\partial w_j}{\partial \nu_{A_j}} + \beta \frac{\partial w_j}{\partial t} + \gamma p_j \Big|_{\Sigma_{ij}}$$

$$\mu_{ij} = -\frac{\partial p_j}{\partial \nu_{A_j}} - \beta \dot{p}_j + \gamma \frac{\partial \dot{w}_j}{\partial t} \Big|_{\Sigma_{ij}}$$

Lemma 2.1. *Assume that $\beta > 0$, $\gamma > 0$, that $\lambda_{ij}, \rho_{ij} \in L^2(\Sigma_{ij})$, and that $\mu_{ij} = \dot{\rho}_{ij}$, $\forall j : \Gamma_{ij} \neq \emptyset$. Then (2.1)–(2.4), (2.6) is well posed. The solution has regularity*

$$(w_i, \frac{\partial w_i}{\partial t}) \in C(\mathcal{H}_i), \quad \frac{\partial w_i}{\partial t}|_{\Sigma_i^N \cup \Sigma_i} \in L^2(\Sigma_i^N \cup \Sigma_i),$$
$$(\frac{\partial p_i}{\partial t}, p_i) \in C(\mathcal{H}_i'), \quad p_i|_{\Sigma_i^N \cup \Sigma_i} \in L^2(\Sigma_i^N \cup \Sigma_i).$$

Proof. The lemma follows from the fact that (2.1)–(2.4), (2.6) is the optimality system for the LQR problem

$$\inf_{f_i, g_{ij}} \{ \int_{\Sigma_i^N} |f_i|^2 d\Sigma + \frac{1}{\gamma} \sum_{j:\Gamma_{ij} \neq \emptyset} \int_{\Sigma_{ij}} (|g_{ij}|^2 + |\gamma \frac{\partial w_i}{\partial t} + \rho_{ij}|^2) d\Sigma$$
$$+ k \|(w_i(T), \frac{\partial w_i}{\partial t}(T)) - (z_{0i}, z_{1i})\|^2_{\mathcal{H}_i} \}$$

subject to

$$\frac{\partial^2 w_i}{\partial t^2} - \nabla \cdot (A_i \nabla w_i) + c_i w_i = 0 \text{ in } Q_i$$
$$w_i = 0 \text{ on } \Sigma_i^D$$
$$\frac{\partial w_i}{\partial \nu_{A_i}} + \alpha_i \frac{\partial w_i}{\partial t} = f_i \text{ on } \Sigma_i^N$$
$$\frac{\partial w_i}{\partial \nu_{A_i}} + \beta \frac{\partial w_i}{\partial t} = \lambda_{ij} + g_{ij} \text{ on } \Sigma_{ij}, \forall j : \Gamma_{ij} \neq \emptyset$$
$$w_i(0) = w_{0i}, \quad \frac{\partial w_i}{\partial t}(0) = v_{0i} \text{ in } \Omega_i,$$

where $f_i \in L^2(\Sigma_i^N)$, $g_{ij} \in L^2(\Sigma_{ij})$. This problem has the same structure as the global optimal control problem. So for $f_i \in L^2(\Sigma_i^N)$ and $g_{ij} \in L^2(\Sigma_{ij})$, the system has a unique solution w_i with the indicated regularity and the optimal controls are given by

$$f_i = -p_i|_{\Sigma_i^N}, \quad g_{ij} = -\gamma p_i|_{\Sigma_{ij}},$$

where p_i is the solution of

$$\begin{aligned}
\frac{\partial^2 p_i}{\partial t^2} - \nabla \cdot (A_i \nabla p_i) + c_i p_i = 0 \text{ in } Q_i \\
p_i = 0 \text{ on } \Sigma_i^D \\
\frac{\partial p_i}{\partial \nu_{A_i}} - \alpha_i \dot{p}_i = 0 \text{ on } \Sigma_i^N \\
\frac{\partial p_i}{\partial \nu_{A_i}} - \beta \dot{p} = -\gamma \frac{\partial \dot{w}_i}{\partial t} + \dot{\rho}_{ij} \text{ on } \Sigma_{ij} \\
p_i(T) = k(\frac{\partial w_i}{\partial t}(T) - z_{1i}) \in L^2(\Omega_i) \\
\frac{\partial p_i}{\partial t}(T) = -k\mathcal{A}_i(w_i(T) - z_{0i}) \in (H^1_{D_i}(\Omega_i))'
\end{aligned}$$

Since $-\gamma \frac{\partial \dot{w}_i}{\partial t} + \dot{\rho}_{ij} \in (H^1(L^2(\Gamma_{ij})))'$, the last problem has a unique solution p_i with the indicated regularity.

The transmission conditions (2.6) suggest the following domain decomposition iteration:

$$\begin{cases}
\dfrac{\partial^2 w_i^{n+1}}{\partial t^2} - \nabla \cdot (A_i \nabla w_i^{n+1}) + c_i w_i^{n+1} = 0 \\
\dfrac{\partial^2 p_i^{n+1}}{\partial t^2} - \nabla \cdot (A_i \nabla p_i^{n+1}) + c_i p_i^{n+1} = 0 \text{ in } Q_i
\end{cases} \tag{2.7}$$

$$w_i^{n+1} = p_i^{n+1} = 0 \text{ on } \Sigma_i^D \tag{2.8}$$

$$\begin{cases}
\dfrac{\partial w_i^{n+1}}{\partial \nu_{A_i}} + \alpha_i \dfrac{\partial w_i^{n+1}}{\partial t} + p_i^{n+1} = 0 \\
\dfrac{\partial p_i^{n+1}}{\partial \nu_{A_i}} - \alpha_i \dot{p}_i^{n+1} = 0 \text{ on } \Sigma_i^N
\end{cases} \tag{2.9}$$

$$\begin{cases}
\dfrac{\partial w_i^{n+1}}{\partial \nu_{A_i}} + \beta \dfrac{\partial w_i^{n+1}}{\partial t} + \gamma p_i^{n+1} = \lambda_{ij}^n \\
\dfrac{\partial p_i^{n+1}}{\partial \nu_{A_i}} - \beta \dot{p}_i^{n+1} + \gamma \dfrac{\partial \dot{w}_i^{n+1}}{\partial t} = \mu_{ij}^n \text{ on } \Sigma_{ij}
\end{cases} \tag{2.10}$$

$$\begin{cases}
w_i^{n+1}(0) = w_{0i}, \quad \dfrac{\partial w_i^{n+1}}{\partial t}(0) = v_{0i} \\
p_i^{n+1}(T) = k(\dfrac{\partial w_i^{n+1}}{\partial t}(T) - z_{1i}) \\
\dfrac{\partial p_i^{n+1}}{\partial t}(T) = -k\mathcal{A}_i(w_i^{n+1}(T) - z_{0i}) \text{ in } \Omega_i
\end{cases} \tag{2.11}$$

where

$$\lambda_{ij}^n = -\frac{\partial w_j^n}{\partial \nu_{A_j}} + \beta \frac{\partial w_j^n}{\partial t} + \gamma p_j^n \Big|_{\Sigma_{ij}}$$

$$\mu_{ij}^n = -\frac{\partial p_j^n}{\partial \nu_{A_j}} - \beta \dot{p}_j^n + \gamma \frac{\partial \dot{w}_j^n}{\partial t} \Big|_{\Sigma_{ij}} \tag{2.12}$$

Lemma 2.2. *Assume that $\beta > 0$, $\gamma > 0$, that $\lambda_{ij}^0 \in L^2(\Sigma_{ij})$ and $\mu_{ij}^0 = \dot{\rho}_{ij}^0$ where $\rho_{ij}^0 \in L^2(\Sigma_{ij})$. Then for $n = 0, 1, \ldots$, $\lambda_{ij}^n \in L^2(\Sigma_{ij})$ and $\mu_{ij}^n = \dot{\rho}_{ij}^n$ where $\rho_{ij}^n \in L^2(\Sigma_{ij})$. As a consequence, the iteration (2.7)–(2.12) is well posed and*

$$(w_i^{n+1}, \frac{\partial w_i^{n+1}}{\partial t}) \in C(\mathcal{H}_i), \quad \frac{\partial w_i^{n+1}}{\partial t}|_{\Sigma_i^N \cup \Sigma_i} \in L^2(\Sigma_i^N \cup \Sigma_i),$$

$$(\frac{\partial p_i^{n+1}}{\partial t}, p_i^{n+1}) \in C(\mathcal{H}_i'), \quad p_i^{n+1}|_{\Sigma_i^N \cup \Sigma_i} \in L^2(\Sigma_i^N \cup \Sigma_i).$$

Proof. From Lemma 2.1, for $n = 0$ the system has a unique solution with the indicated regularity. It follows immediately that $\lambda_{ij}^1 \in L^2(\Sigma_{ij})$. Further $\mu_{ij}^1 = \dot{\rho}_{ij}^1 \in (H^1(L^2(\Gamma_{ij})))'$, where $\rho_{ij}^1 \in L^2(\Sigma_{ij})$ (see (3.3) below). The conclusion follows by induction.

3. Convergence of the Iterations

To establish convergence of the solutions $\{(w_i^{n+1}, p_i^{n+1})\}_{i=1}^m$ of the local optimality systems to the solution of the global optimality system, we consider the differences $w_i^{n+1} - w_i$, $p_i^{n+1} - p_i$, which we again denote by w_i^{n+1}, p_i^{n+1}. These satisfy (2.7)–(2.12) with

$$w_{0i} = v_{0i} = z_{0i} = z_{1i} = 0. \tag{3.1}$$

We introduce a new variable q_i^{n+1}, $i = 1, \ldots, m$, as the solution of the problem

$$\begin{aligned} \frac{\partial^2 q_i^{n+1}}{\partial t^2} - \nabla \cdot (A_i \nabla q_i^{n+1}) + c_i q_i^{n+1} &= 0 \text{ in } Q_i \\ q_i^{n+1} &= 0 \text{ on } \Sigma_i^D \\ \frac{\partial q_i^{n+1}}{\partial \nu_{A_i}} - \alpha_i \frac{\partial q_i^{n+1}}{\partial t} &= 0 \text{ on } \Sigma_i^N \\ \frac{\partial q_i^{n+1}}{\partial \nu_{A_i}} - \beta \frac{\partial q_i^{n+1}}{\partial t} + \gamma \frac{\partial w_i^{n+1}}{\partial t} &= \rho_{ij}^n \text{ on } \Sigma_{ij} \\ q_i^{n+1}(T) = k w_i^{n+1}(T), \quad \frac{\partial q_i^{n+1}}{\partial t}(T) &= k \frac{\partial w_i^{n+1}}{\partial t}(T) \text{ in } \Omega, \end{aligned} \tag{3.2}$$

where $\rho_{ij}^0 \in L^2(\Sigma_{ij})$ is arbitrary and, for $n \geq 1$,

$$\rho_{ij}^n = -\frac{\partial q_j^n}{\partial \nu_{A_j}} - \beta \frac{\partial q_j^n}{\partial t} + \gamma \frac{\partial w_j^n}{\partial t}. \tag{3.3}$$

The above problem has a unique solution with regularity

$$(q_i^{n+1}, \frac{\partial q_i^{n+1}}{\partial t}) \in C(\mathcal{H}_i), \quad \left.\frac{\partial q_i^{n+1}}{\partial t}\right|_{\Sigma_i^N \cup \Sigma_i} \in L^2(\Sigma_i^N \cup \Sigma_i)$$

and, what is more important to observe, the adjoint variable p_i^{n+1} in the iteration (2.7)–(2.12) is given by

$$p_i^{n+1} = \frac{\partial q_i^{n+1}}{\partial t} \tag{3.4}$$

so, as a result, $\mu_{ij}^{n+1} = \dot{\rho}_{ij}^{n+1}$. The following recursion formula is fundamental to proving that $w_i^{n+1} \to 0$ and $p_i^{n+1} \to 0$ in appropriate norms.

Lemma 3.1. *With the same assumptions as in Lemma 2.2, the following recursion relation holds:*

$$\mathcal{E}^{n+1} + \mathcal{F}^{n+1} = \mathcal{E}^n - \mathcal{F}^n, \tag{3.5}$$

where

$$\mathcal{E}^n = \frac{1}{2}\sum_{i=1}^m \int_{\Sigma_i} \left\{ \frac{\beta^2+\gamma^2}{\gamma}\left(|p_i^n|^2 + \left|\frac{\partial w_i^n}{\partial t}\right|^2\right) + \frac{1}{\gamma}\left(\left|\frac{\partial w_i^n}{\partial \nu_{A_i}}\right|^2 + \left|\frac{\partial q_i^n}{\partial \nu_{A_i}}\right|^2\right)\right\} d\Sigma, \tag{3.6}$$

$$\begin{aligned}\mathcal{F}^n = \sum_{i=1}^m \Big\{ [k + \frac{\beta}{2\gamma}(1-k^2)] \|(w_i^n(T), \frac{\partial w_i^n}{\partial t}(T))\|^2_{\mathcal{H}_i} + \frac{\beta}{2\gamma}\|(q_i^n(0), \frac{\partial q_i^n}{\partial t}(0))\|^2_{\mathcal{H}_i}\Big\} \\ + \sum_{i=1}^m \int_{\Sigma_i^N} \left\{ \left(1 + \frac{\alpha_i\beta}{\gamma}\right)|p_i^n|^2 + \frac{\beta}{\gamma} p_i^n \frac{\partial w_i^n}{\partial t} + \frac{\alpha_i \beta}{\gamma}\left|\frac{\partial w_i^n}{\partial t}\right|^2\right\} d\Sigma. \end{aligned}\tag{3.7}$$

Relation (3.5) is typical of recursion formulae arising in domain decomposition algorithms based on Robin type iterations at the interfaces. Its proof is elementary, but lengthy, and space does not permit its inclusion in these proceedings.

We now show how (3.5) implies convergence of the sequence of solutions of the local optimality systems to the solution of the global optimality system.

Theorem 3.1. *In addition to the assumptions of Lemma 3.1, suppose that*

$$\alpha_0 > \frac{1}{2}\left(-\frac{\gamma}{\beta} + \sqrt{\frac{\gamma^2}{\beta^2}+1}\right), \quad \frac{\beta}{\gamma} < 2k/(k^2-1) \text{ if } k > 1, \tag{3.8}$$

Then for $i = 1, \ldots, m$,

$$\begin{aligned}
&(w_i^n, \frac{\partial w_i^n}{\partial t}) \to 0, \quad (q_i^n, \frac{\partial q_i^n}{\partial t}) \to 0 \text{ weakly* in } L^\infty(\mathcal{H}_i)\\
&(w_i^n(T), \frac{\partial w_i^n}{\partial t}(T)) \to 0, \quad (q_i^n(0), \frac{\partial q_i^n}{\partial t}(0)) \to 0 \text{ strongly in } \mathcal{H}_i\\
&p_i^n\big|_{\Sigma_i^N} = \frac{\partial q_i^n}{\partial t}\bigg|_{\Sigma_i^N} \to 0, \quad \frac{\partial w_i^n}{\partial t}\bigg|_{\Sigma_i^N} \to 0 \text{ strongly in } L^2(\Sigma_i^N)\\
&\frac{\partial w_i^n}{\partial t}\bigg|_{\Sigma_i} \to 0, \quad \frac{\partial w_i^n}{\partial \nu_{A_i}}\bigg|_{\Sigma_i} \to 0 \text{ weakly in } L^2(\Sigma_i)\\
&\frac{\partial q_i^n}{\partial t}\bigg|_{\Sigma_i} \to 0, \quad \frac{\partial q_i^n}{\partial \nu_{A_i}}\bigg|_{\Sigma_i} \to 0 \text{ weakly in } L^2(\Sigma_i).
\end{aligned}$$

Remark 3.1. One immediately deduces from Theorem 3.1 the following convergence properties of the adjoint variables $p_i^n = \dfrac{\partial q_i^n}{\partial t}$:

$$\begin{aligned}
&p_i^n \to 0 \text{ weakly* in } L^\infty(L^2(\Omega_i))\\
&(p_i^n(T), \frac{\partial p_i^n}{\partial t}(T)) \to 0 \text{ strongly in } L^2(\Omega_i) \times (H^1_{D_i}(\Omega_i))'\\
&p_i^n\big|_{\Sigma_i^N} \to 0 \text{ strongly in } L^2(\Sigma_i^N), \quad p_i^n\big|_{\Sigma_i} \to 0 \text{ weakly in } L^2(\Sigma_i).
\end{aligned}$$

Remark 3.2. Note that for a given $k > 1$ and $\alpha_0 > 0$, (3.8) can always be satisfied by choosing γ/β sufficiently large.

Proof. It follows from (3.5) that

$$\mathcal{E}^{n+1} = \mathcal{E}^1 - 2\sum_{p=1}^{n+1}{}' \mathcal{F}^p,$$

where $\sum_{p=1}^{n+1}{}' c_p = (c_1 + c_{n+1})/2 + \sum_{p=2}^{n} c_p$. Since $\alpha_i \geq \alpha_0$, under the stated condition on α_0 the quadratic form

$$\left(1 + \frac{\alpha_i \beta}{\gamma}\right) |p_i^n|^2 + \frac{\beta}{\gamma} p_i^n \frac{\partial w_i^n}{\partial t} + \frac{\alpha_i \beta}{\gamma} \left|\frac{\partial w_i^n}{\partial t}\right|$$

is positive definite. (Note that any $\alpha_0 > 1/2$ will satisfy the hypothesis.) If $k > 1$ and if we further restrict β/γ according to (3.8), then

$$\sum_{p=1}^{\infty} \mathcal{F}^p \text{ converges and } \{\mathcal{E}^n\}_{n=1}^{\infty} \text{ is a bounded sequence.}$$

The convergence of $\sum \mathcal{F}^p$ then implies that

$$(w_i^n(T), \frac{\partial w_i^n}{\partial t}(T)) \to 0, \ (q_i^n(0), \frac{\partial q_i^n}{\partial t}(0)) \to 0 \text{ strongly in } \mathcal{H}_i$$
$$p_i^n|_{\Sigma_i^N} = \frac{\partial q_i^n}{\partial t}\Big|_{\Sigma_i^N} \to 0, \ \frac{\partial w_i^n}{\partial t}\Big|_{\Sigma_i^N} \to 0 \text{ strongly in } L^2(\Sigma_i^N).$$

From (2.9) and (3.2) we then have

$$\frac{\partial w_i^n}{\partial \nu_{A_i}}\Big|_{\Sigma_i^N} \to 0, \ \frac{\partial q_i^n}{\partial \nu_{A_i}}\Big|_{\Sigma_i^N} \to 0 \text{ strongly in } L^2(\Sigma_i^N)$$
$$(q_i^n(T), \frac{\partial q_i^n}{\partial t}(T)) \to 0 \text{ strongly in } \mathcal{H}_i.$$

The boundedness of $\mathcal{E}^n$ implies that λ_{ij}^n, ρ_{ij}^n are bounded in $L^2(\Sigma_{ij})$. Therefore

$$(w_i^n, \frac{\partial w_i^n}{\partial t}), \quad (q_i^n, \frac{\partial q_i^n}{\partial t}) \text{ are bounded in } L^\infty(\mathcal{H}_i)$$

which implies, in particular, that $\{w_i^n\}$, $\{q_i^n\}$ are uniformly bounded and equicontinuous on $[0,T]$ in the $L^2(\Omega_i)$ topology. It follows that on a subsequence $n = n_k$ of the positive integers,

$$\begin{aligned}
&w_i^n \to \tilde{w}_i, \quad q_i^n \to \tilde{q}_i \text{ weakly* in } L^\infty(H^1_{D_i}(\Omega_i)) \text{ and strongly in } C(L^2(\Omega_i))\\
&\frac{\partial w_i^n}{\partial t} \to \frac{\partial \tilde{w}_i}{\partial t}, \ \frac{\partial q_i^n}{\partial t} \to \frac{\partial \tilde{q}_i}{\partial t} \text{ weakly* in } L^\infty(L^2(\Omega_i))\\
&\frac{\partial w_i^n}{\partial t}\Big|_{\Sigma_i} \to f_i, \ \frac{\partial w_i^n}{\partial \nu_{A_i}}\Big|_{\Sigma_i} \to g_i \text{ weakly in } L^2(\Sigma_i)\\
&\frac{\partial q_i^n}{\partial t}\Big|_{\Sigma_i} \to F_i, \ \frac{\partial q_i^n}{\partial \nu_{A_i}}\Big|_{\Sigma_i} \to G_i \text{ weakly in } L^2(\Sigma_i)
\end{aligned} \tag{3.9}$$

for some f_i, g_i, F_i, G_i in $L^2(\Sigma_i)$, where both $\tilde{w}_i$ and $\tilde{q}_i$ satisfy

$$\begin{aligned}
&\frac{\partial^2 u_i}{\partial t^2} - \nabla \cdot (A_i \nabla u_i) + c_i u_i = 0 \text{ in } Q_i\\
&u_i|_{\Sigma_i^D} = 0\\
&\frac{\partial u_i}{\partial \nu_{A_i}}\Big|_{\Sigma_i^N} = \frac{\partial u_i}{\partial t}\Big|_{\Sigma_i^N} = 0\\
&u_i(0) = \frac{\partial u_i}{\partial t}(0) = u_i(T) = \frac{\partial u_i}{\partial t}(T) = 0 \text{ in } \Omega_i,
\end{aligned} \tag{3.10}$$

and

$$\frac{\partial \tilde{w}_i}{\partial t}\Big|_{\Sigma_i} = f_i, \ \frac{\partial \tilde{w}_i}{\partial \nu_{A_i}}\Big|_{\Sigma_i} = g_i, \ \frac{\partial \tilde{q}_i}{\partial t}\Big|_{\Sigma_i} = F_i, \ \frac{\partial \tilde{q}_i}{\partial \nu_{A_i}}\Big|_{\Sigma_i} = G_i \tag{3.11}$$

Since $\tilde{w}_i$ and $\tilde{q}_i$ vanish at $t = 0$ we also have $\tilde{w}_i|_{\Sigma_i^N} = \tilde{q}_i|_{\Sigma_i^N} = 0$. Thus $\tilde{w}_i$ and $\tilde{q}_i$ have zero Cauchy on Σ_i^N.

We now use a unique continuation argument to conclude that $\tilde{w}_i = \tilde{q}_i = 0$ in Q_i for $i = 1, \ldots, m$. First of all, suppose that Ω_i is a region for which $\Gamma_i^N \neq \emptyset$. Since $\tilde{w}_i$ and $\tilde{q}_i$ have zero Cauchy at $t = 0$ and $t = T$, they may be continued by zero to $\Omega \times \{t < 0\}$ and to $\Omega \times \{t > T\}$ as solutions of the hyperbolic equation (3.10) that have zero Cauchy data on $\Gamma_i^N \times (-\infty, +\infty)$. It then follows by unique continuation (see, for example, Hormander [2]) that $\tilde{w}_i = \tilde{q}_i = 0$ in $\Omega \times (-\infty, +\infty)$. In particular, $f_i = g_i = F_i = G_i = 0$ and therefore the convergence in (3.9) is through the entire sequence of positive integers.

Now suppose Ω_j is a region adjacent to a region Ω_i such that $\Gamma_i^N \neq \emptyset$. Then $\Sigma_{ij} \neq \emptyset$ and from (2.10), (2.12), (3.2)–(3.4) we have on Σ_{ij}

$$\frac{\partial w_i^{n+2}}{\partial \nu_{A_i}} + \beta \frac{\partial w_i^{n+2}}{\partial t} + \gamma \frac{\partial q_i^{n+2}}{\partial t} = -\frac{\partial w_j^{n+1}}{\partial \nu_{A_j}} + \beta \frac{\partial w_j^{n+1}}{\partial t} + \gamma \frac{\partial q_j^{n+1}}{\partial t} \tag{3.12}$$

$$\frac{\partial q_i^{n+2}}{\partial \nu_{A_i}} - \beta \frac{\partial q_i^{n+2}}{\partial t} + \gamma \frac{\partial w_i^{n+2}}{\partial t} = -\frac{\partial q_j^{n+1}}{\partial \nu_{A_j}} - \beta \frac{\partial q_j^{n+1}}{\partial t} + \gamma \frac{\partial w_j^{n+1}}{\partial t} \tag{3.13}$$

$$\frac{\partial w_j^{n+1}}{\partial \nu_{A_j}} + \beta \frac{\partial w_j^{n+1}}{\partial t} + \gamma \frac{\partial q_j^{n+1}}{\partial t} = -\frac{\partial w_i^{n}}{\partial \nu_{A_i}} + \beta \frac{\partial w_i^{n}}{\partial t} + \gamma \frac{\partial q_i^{n}}{\partial t} \tag{3.14}$$

$$\frac{\partial q_j^{n+1}}{\partial \nu_{A_j}} - \beta \frac{\partial q_j^{n+1}}{\partial t} + \gamma \frac{\partial w_j^{n+1}}{\partial t} = -\frac{\partial q_i^{n}}{\partial \nu_{A_i}} - \beta \frac{\partial q_i^{n}}{\partial t} + \gamma \frac{\partial w_i^{n}}{\partial t}. \tag{3.15}$$

Since, for the index i, convergence is through the entire sequence of positive integers, if we pass to the weak $L^2(\Sigma_{ij})$ limit in (3.12) - (3.15) through the subsequence $n = n_k$ we obtain

$$-g_j + \beta f_j + \gamma F_j = 0, \quad -G_j - \beta F_j + \gamma f_j = 0$$
$$g_j + \beta f_j + \gamma F_j = 0, \quad G_j - \beta F_j + \gamma f_j = 0$$

on Σ_{ij}. Therefore $f_j = g_j = F_j = G_j = 0$, that is

$$\frac{\partial \tilde{w}_j}{\partial t} = \frac{\partial \tilde{w}_j}{\partial \nu_{A_j}} = \frac{\partial \tilde{q}_j}{\partial t} = \frac{\partial \tilde{q}_j}{\partial \nu_{A_j}} = 0 \text{ on } \Sigma_{ij}.$$

The same unique continuation argument as above gives $\tilde{w}_j = \tilde{q}_j = 0$ in Q_j. One may now proceed step-by-step into the remaining interior regions Ω_j and conclude that $\tilde{w}_j = \tilde{q}_j = 0$ in Q_j and $f_j = g_j = F_j = G_j = 0$ for $j = 1, \ldots, m$. □

References

[1] J.-D. Benamou, "Domain decomposition, optimal control of systems governed by partial differential equations and synthesis of feedback laws," *J. Opt. Theory Appl.*, **102** (1999), 15–36.

[2] L. Hormander, "A uniqueness theorem for second order hyperbolic differential equations," Commun. PDEs, **17** (1992), 699–714.

[3] J. E. Lagnese, "A nonoverlapping domain decomposition for optimal boundary control of the dynamic Maxwell system," in *Control of Nonlinear Distributed Parameter Systems*, G. Chen, I. Lasiecka and J. Zhou, Eds., Marcel Dekker, to appear.

[4] J. Lagnese and G. Leugering, "Dynamic domain decomposition in approximate and exact boundary control in problems of transmission for wave equations," *SIAM J. Control Opt.*, **38** (2000), 503–537.

[5] P.-L. Lions, "On the Schwarz alternating method 3", in *The Third International Symposium on Domain Decomposition Methods for Partial Differential Equations*, T. Chan and R. Glowinski, Eds.), Society for Industrial and Applied Mathematics, Philadelphia, PA, 1990, 202–223.

Department of Mathematics
Georgetown University
Washington, DC 20057
USA

International Series of Numerical Mathematics, Vol. 139, 171–182

Optimal Regularity of Elastic and Thermoelastic Kirchhoff Plates with Clamped Boundary Control

Irena Lasiecka and Roberto Triggiani

Abstract. We consider mixed problems for the Kirchhoff elastic and thermoelastic systems, subject to boundary control in the clamped Boundary Conditions B.C. ("clamped control"). If w denotes elastic displacement and θ temperature, we establish optimal regularity of $\{w, w_t, w_{tt}\}$ in the elastic case, and of $\{w, w_t, w_{tt}, \theta\}$ in the thermoelastic case. Our results complement those in [L-L.1], where sharp (optimal) trace regularity results are obtained for the corresponding boundary homogeneous cases. The passage from the boundary homogeneous cases to the corresponding mixed problems involves a duality argument. However, in the present case of clamped B.C., the duality argument in question is both delicate and technical. Indeed, it produces new phenomena which are accounted for by introducing new, untraditional factor (quotient) spaces. These are critical in describing both interior regularity and exact controllability of mixed Kirchhoff problems with clamped controls.

1. Introduction

1.1. Introduction, preliminaries

Elastic Kirchhoff equation. Let Ω be an open bounded domain in $\mathbb{R}^n$ with smooth boundary Γ. Consider the following Kirchhoff elastic mixed problem with clamped boundary control in the unknown $w(t,x)$:

$$\begin{cases} w_{tt} - \gamma\Delta w_{tt} + \Delta^2 w = 0 & \text{in } (0,T] \times \Omega \equiv Q; & (1.1.1a) \\ w(0,\cdot) = w_0,\ w_t(0,\cdot) = w_1 & \text{in } \Omega; & (1.1.1b) \\ w|_\Sigma \equiv 0,\ \left.\dfrac{\partial w}{\partial \nu}\right|_\Sigma \equiv u & \text{in } (0,T] \times \Gamma \equiv \Sigma. & (1.1.1c) \end{cases}$$

In (1.1.1a), γ is a positive constant kept fixed throughout this paper: $\gamma > 0$. When $n = 2$, problem (1.1.1) describes the evolution of the displacement w of the elastic Kirchhoff plate model, which accounts for rotational inertia. In it, γ is proportional to the square of the thickness of the plate [Lag.1], [L-L.1].

Research partially supported by the National Science Foundation, under Grant DMS-9804056.

Thermoelastic Kirchhoff equations. With Ω, Γ and $\gamma > 0$ as above, consider now the corresponding thermoelastic mixed problem with clamped boundary control in the unknown $\{w(t,x), \theta(t,x)\}$:

$$\begin{cases} w_{tt} - \gamma\Delta w_{tt} + \Delta^2 w + \Delta\theta = 0 \quad \text{in } (0,T] \times \Omega \equiv Q; & (1.1.2a) \\ \theta_t - \Delta\theta - \Delta w_t = 0 \quad \text{in } Q; & (1.1.2b) \\ w(0,\cdot) = w_0,\ w_t(0,\cdot) = w_1,\ \theta(0,\cdot) = \theta_0 \quad \text{in } \Omega; & (1.1.2c) \\ w|_\Sigma \equiv 0;\ \left.\frac{\partial w}{\partial \nu}\right|_\Sigma \equiv u;\ \theta|_\Sigma \equiv 0 \quad \text{in } (0,T] \times \Gamma \equiv \Sigma. & (1.1.2d) \end{cases}$$

When $n = 2$, problem (1.1.2) describes the evolution of the displacement w and temperature θ (with respect to the stress-free temperature) of the thermoelastic Kirchhoff plate model, which accounts for rotational inertia [Lag.1], [L-L.1].

Function spaces. To state our results on optimal regularity of both mixed problems (1.1.1) and (1.1.2), we need to introduce some (non-classical, new) spaces. To begin with, define the following positive, self-adjoint operators

$$Af = \Delta^2 f,\ \mathcal{D}(A) = H^4(\Omega) \cap H_0^2(\Omega); \tag{1.1.3}$$

$$\mathcal{A}f = -\Delta f,\ \mathcal{A}_\gamma = I + \gamma\mathcal{A};\ \mathcal{D}(\mathcal{A}_\gamma) = \mathcal{D}(\mathcal{A}) = H^2(\Omega) \cap H_0^1(\Omega); \tag{1.1.4}$$

so that the following identifications hold true (with equivalent norms):

$$\mathcal{D}(A^{\frac{1}{2}}) \equiv H_0^2(\Omega);\ \mathcal{D}(A^{\frac{1}{4}}) \equiv \mathcal{D}(\mathcal{A}^{\frac{1}{2}}) \equiv \mathcal{D}(\mathcal{A}_\gamma^{\frac{1}{2}}) \equiv H_0^1(\Omega); \tag{1.1.5a}$$

$$\mathcal{D}(A^{\frac{3}{4}}) \equiv \left\{ f \in H^3(\Omega) : f|_\Gamma = 0,\ \left.\frac{\partial f}{\partial \nu}\right|_\Gamma = 0 \right\} \equiv H^3(\Omega) \cap H_0^2(\Omega). \tag{1.1.5b}$$

The space $\mathcal{D}(\mathcal{A}_\gamma^{\frac{1}{2}})$ will *always* be endowed with the following inner product, unless specifically noted otherwise:

$$(f_1, f_2)_{\mathcal{D}(\mathcal{A}_\gamma^{\frac{1}{2}})} = \left(\mathcal{A}_\gamma^{\frac{1}{2}} f_1, \mathcal{A}_\gamma^{\frac{1}{2}} f_2\right)_{L_2(\Omega)} = (\mathcal{A}_\gamma f_1, f_2)_{L_2(\Omega)},\ \forall\ f_1, f_2 \in H_0^1(\Omega), \tag{1.1.5c}$$

where, at this stage, we denote with the same symbol the $L_2(\Omega)$-inner product and the duality pairing $(\cdot,\cdot)_{V' \times V}$, $V \equiv H_0^1(\Omega)$, $V' = H^{-1}(\Omega)$ with $L_2(\Omega)$ as a pivot space [A.1, Thm. 1.5, p. 51], for the last term in (1.1.5c).

We then introduce the following (non-classical, untraditional or new) Hilbert spaces:

(i) we first define the Hilbert space $\tilde{L}_2(\Omega)$ (depending on γ) by:

$$\tilde{L}_2(\Omega) \equiv \text{dual of the space } \mathcal{D}(A^{\frac{1}{2}}) \text{ with respect to the space } \mathcal{D}(\mathcal{A}_\gamma^{\frac{1}{2}}) \text{ as a pivot space, endowed with the norm of (1.1.5c).} \tag{1.1.6}$$

This means the following: let $f \in \mathcal{D}(A^{\frac{1}{2}}) \equiv H_0^2(\Omega) \subset \mathcal{D}(\mathcal{A}_\gamma)$, or $\phi = A^{\frac{1}{2}} f \in L_2(\Omega)$. Then:

$$g \in \tilde{L}_2(\Omega) \iff (f,g)_{\mathcal{D}(\mathcal{A}_\gamma^{\frac{1}{2}})} = (\mathcal{A}_\gamma f, g)_{L_2(\Omega)} = \text{finite}, \ \forall\, f \in H_0^2(\Omega) \tag{1.1.7a}$$

$$= (f, \mathcal{A}_\gamma g)_{L_2(\Omega)} = (A^{-\frac{1}{2}}\phi, \mathcal{A}_\gamma g)_{L_2(\Omega)}$$

$$= \left(\phi, A^{-\frac{1}{2}}\mathcal{A}_\gamma g\right)_{L_2(\Omega)} = \text{finite}, \ \forall\, \phi \in L_2(\Omega) \tag{1.1.7b}$$

$$\iff A^{-\frac{1}{2}}\mathcal{A}_\gamma g \in L_2(\Omega), \tag{1.1.7c}$$

where we write in the same way inner products and corresponding duality pairings.

(ii) We next define the Hilbert space $\tilde{H}_{-1}(\Omega)$ (depending on γ) by

$$\tilde{H}_{-1}(\Omega) \equiv \text{dual of the space } \mathcal{D}(A^{\frac{3}{4}}) \text{ with respect to the space } \mathcal{D}(\mathcal{A}_\gamma^{\frac{1}{2}}) \text{ as a pivot space, endowed with the norm of (1.1.5c).} \tag{1.1.8}$$

This means the following: let $f \in \mathcal{D}(A^{\frac{3}{4}}) \equiv H^3(\Omega) \cap H_0^2(\Omega) \subset \mathcal{D}(\mathcal{A}_\gamma) \equiv H^2(\Omega) \cap H_0^1(\Omega)$, or $\phi = A^{\frac{3}{4}} f \in L_2(\Omega)$. Then:

$$g \in \tilde{H}_{-1}(\Omega) \iff (f,g)_{\mathcal{D}(\mathcal{A}_\gamma^{\frac{1}{2}})} = (\mathcal{A}_\gamma f, g)_{L_2(\Omega)} = \text{finite}, \ \forall f \in \mathcal{D}(A^{\frac{3}{4}}) \tag{1.1.9a}$$

$$= (f, \mathcal{A}_\gamma g)_{L_2(\Omega)} = (A^{-\frac{3}{4}}\phi, \mathcal{A}_\gamma g)_{L_2(\Omega)}$$

$$= \left(\phi, A^{-\frac{3}{4}}\mathcal{A}_\gamma g\right)_{L_2(\Omega)} = \text{finite}, \ \forall\, \phi \in L_2(\Omega) \tag{1.1.9b}$$

$$\iff A^{-\frac{3}{4}}\mathcal{A}_\gamma g \in L_2(\Omega), \tag{1.1.9c}$$

where we write the same way inner products and corresponding duality pairings.

Further description of the spaces $\tilde{L}_2(\Omega)$ and $\tilde{H}_{-1}(\Omega)$ is given in [L-T.4, Sects. 2 and 3, respectively]. Here we provide two key characterizations.

Proposition 1.1.1. *The space $\tilde{L}_2(\Omega)$ as defined in (1.1.6) is isometrically isomorphic [congruent, in the terminology of [T-L.1, p. 53]] to the factor, or quotient, space $L_2(\Omega)/\mathcal{H}$: in symbols*

$$\tilde{L}_2(\Omega) \cong L_2(\Omega)/\mathcal{H} \cong \mathcal{H}^\perp, \tag{1.1.10}$$

where $\mathcal{H}$ is the null space of the operator $(1-\gamma\Delta): \ L_2(\Omega) \to H^{-2}(\Omega) = [\mathcal{D}(A^{\frac{1}{2}})]'$:

$$\mathcal{H} \equiv \{h \in L_2(\Omega): \ (1-\gamma\Delta)h = 0 \ \textit{in} \ [\mathcal{D}(A^{\frac{1}{2}})]' \equiv H^{-2}(\Omega)\}, \tag{1.1.11a}$$

$$L_2(\Omega) = \mathcal{H} + \mathcal{H}^\perp; \ \mathcal{H}^\perp = \Pi L_2(\Omega), \ \Pi = \Pi^* \ \textit{orthogonal projection of } L_2(\Omega) \textit{ onto } \mathcal{H}^\perp. \tag{1.1.11b}$$

(hence, $\mathcal{H}$ is a space of generalized harmonic functions in $L_2(\Omega)$). If J denotes the isometric isomorphism between $\tilde{L}_2(\Omega)$ and $L_2(\Omega)/\mathcal{H}$, we then have for $g \in \tilde{L}_2(\Omega)$:

$$\|g\|_{\tilde{L}_2(\Omega)} = \|[Jg]\|_{L_2(\Omega)/\mathcal{H}} = \inf_{h\in\mathcal{H}} \|Jg - h\|_{L_2(\Omega)} = \|g_1\|_{L_2(\Omega)}, \tag{1.1.12}$$

for the unique element $g_1 = \Pi g \in \mathcal{H}^\perp$, $g_1 \in [Jg]$ (the latter being the coset or equivalence class of $L_2(\Omega)/\mathcal{H}$ containing the element Jg).

Proposition 1.1.2. *(i) The space $\tilde{H}_{-1}(\Omega)$ as defined in (1.1.8) admits the following alternative characterization*

$$\tilde{H}_{-1}(\Omega) \equiv [H^1(\Omega) \cap \mathcal{H}^{\perp}]', \tag{1.1.13}$$

with duality with respect to $L_2(\Omega)$ as a pivot space.

(ii) The space $\tilde{H}_{-1}(\Omega)$ in (1.1.8) is isometrically isomorphic [congruent, in the terminology of [T-L.1, p. 53]] to the factor, or quotient, space $[H^1(\Omega)]'/\mathbb{H}$: in symbols

$$\tilde{H}_{-1}(\Omega) \equiv [H^1(\Omega) \cap \mathcal{H}^{\perp}]' \cong [H^1(\Omega)]'/\mathbb{H} \cong \mathbb{H}^{\perp}, \tag{1.1.14}$$

where $\mathbb{H}$ is the null space of the operator $(1-\gamma\Delta) : \; [H^1(\Omega)]' \to [\mathcal{D}(A^{\frac{3}{4}})]'$,

$$\mathbb{H} \equiv \{h \in [H^1(\Omega)]' : \; (1-\gamma\Delta)h = 0 \textit{ in } [\mathcal{D}(A^{\frac{3}{4}})]'\}; \tag{1.1.15a}$$

$$\begin{aligned} [H^1(\Omega)]' &= \mathbb{H} + \mathbb{H}^{\perp}, \; \pi[H^1(\Omega)]' \equiv \mathbb{H}^{\perp}, \pi = \pi^* \\ &\qquad \textit{orthogonal projection of } [H^1(\Omega)]' \textit{ onto } \mathbb{H}^{\perp}. \end{aligned} \tag{1.1.15b}$$

Thus, if J now denotes the isometric isomorphism between $[H^1(\Omega) \cap \mathcal{H}^{\perp}]'$ and $[H^1(\Omega)]'/\mathbb{H}$, we then have for $g \in \tilde{H}_{-1}(\Omega) = [H^1(\Omega) \cap \mathcal{H}^{\perp}]'$:

$$\|g\|_{\tilde{H}_{-1}(\Omega)} = \|[Jg]\|_{[H^1(\Omega)]'/\mathbb{H}} = \inf_{h \in \mathbb{H}} \|Jg - h\|_{[H^1(\Omega)]'} = \|g_1\|_{[H^1(\Omega)]'} \tag{1.1.16}$$

for the unique element $g_1 = \pi g \in \mathbb{H}^{\perp}$, $g_1 \in [Jg]$ (the coset or equivalence class containing the element Jg).

1.2. Statement of main results: Optimal interior regularity

The following results provide optimal regularity properties for the mixed problems (1.1.1) and (1.1.2), at least for the mechanical variables. They justify the introduction of the spaces $\tilde{L}_2(\Omega)$ and $\tilde{H}_{-1}(\Omega)$.

Theorem 1.2.1. *Consider the Kirchhoff elastic problem (1.1.1) with $\{w_0, w_1\} = 0$ subject to the hypothesis that*

$$u \in L_2(0,T; L_2(\Gamma)) \equiv L_2(\Sigma). \tag{1.2.1}$$

Then, continuously,

$$\begin{cases} w \in C([0,T]; H_0^1(\Omega) \equiv \mathcal{D}(A^{\frac{1}{4}})); & (1.2.2) \\ w_t \in C([0,T]; \tilde{L}_2(\Omega)); & (1.2.3) \\ w_{tt} \in L_2(0,T; \tilde{H}_{-1}(\Omega)). & (1.2.4) \end{cases}$$

Theorem 1.2.2. *Consider the Kirchhoff thermoelastic problem (1.1.2) with* $\{w_0, w_1, \theta_0\} = 0$, *subject to the hypothesis (1.2.1) on u. Then, continuously*

$$\begin{cases} w \in C([0,T]; H_0^1(\Omega) \equiv \mathcal{D}(A^{\frac{1}{4}})); & (1.2.5) \\ w_t \in C([0,T]; \tilde{L}_2(\Omega)); & (1.2.6) \\ \left[w_{tt} - \frac{1}{\gamma}\theta\right] \in L_2(0,T; \tilde{H}_{-1}(\Omega)); & (1.2.7) \\ \theta \in L_p(0,T; H^{-1}(\Omega)) \cap C([0,T]; H^{-1-\epsilon}(\Omega), \ 1 < p < \infty; \ \epsilon > 0; & (1.2.8) \\ [\theta + w_t] \in C([0,T]; [H_{00}^{\frac{1}{2}}(\Omega)]'), H_{00}^{\frac{1}{2}}(\Omega) = \mathcal{D}(\mathcal{A}^{\frac{1}{4}}). & (1.2.9) \end{cases}$$

However, in addition, we have that

$$\begin{cases} \theta \in C([0,T]; L_2(\Omega)) \text{ and } w_{tt} \in L_2(0,T; \tilde{H}_{-1}(\Omega)), \text{ however, not continu-} \\ \text{ously in } u \in L_2(0,T; L_2(\Gamma)). \end{cases} \quad (1.2.10)$$

The full proof of Theorems 1.2.1 and 1.2.2 is given in [L-T.4], of which the present paper is a selected condensed version.

1.3. Literature

Kirchhoff elastic problem (1.1.1). With reference, at first, to the homogeneous Kirchhoff system ($\gamma > 0$):

$$\begin{cases} \phi_{tt} - \gamma\Delta\phi_{tt} + \Delta^2\phi = 0 & \text{in } Q \equiv (0,T] \times \Omega; & (1.3.1a) \\ \phi(T, \cdot) = \phi_0, \ \phi_t(T, \cdot) = \phi_1 & \text{in } \Omega; & (1.3.1b) \\ \phi|_\Sigma \equiv 0; \ \left.\frac{\partial\phi}{\partial\nu}\right|_\Sigma \equiv 0 & \text{in } \Sigma \equiv (0,T] \times \Gamma. & (1.3.1c) \end{cases}$$

$$\{\phi_0, \phi_1\} \in H_0^2(\Omega) \times H_0^1(\Omega). \quad (1.3.2)$$

Sharp trace estimates were obtained in [L-L.1]. More precisely, [L-L.1] establishes, by multiplier techniques, both of the following results:

(i) the trace regularity inequality for any $T > 0$,

$$\int_0^T \int_\Gamma |\Delta\phi|^2 d\Sigma \le c_T \|\{\phi_0, \phi_1\}\|^2_{H_0^2(\Omega)\times H_0^1(\Omega)}, \quad (1.3.3)$$

see [L-L.1, Eqn. (2.2.4), p. 123], for some constnat $c_T > 0$, as well as

(ii) the continuous observability inequality for all $T >$ some $T_0 > 0$:

$$c(T - T_0)\|\{\phi_0, \phi_1\}\|^2_{H_0^2(\Omega)\times H_0^1(\Omega)} \le \int_0^T \int_{\Gamma(x_0)} |\Delta\phi|^2 d\Sigma, \ c > 0, \quad (1.3.4)$$

[L-L.1, Eqn. (2.2.4), p. 123]. Here T_0 is a suitable positive constant depending on γ and the domain, and $\Gamma(x_0) = \{x \in \Gamma : (x - x_0) \cdot \nu(x) \ge 0\}$, $\nu(x) =$ unit outward normal at $x \in \Gamma$.

As is well known, it is a common duality or transposition argument that converts, as usual, inequalities such as (1.3.3) and (1.3.4), into, respectively:

(i) an interior regularity result $u \to \{w, w_t\}$ of the w-problem (1.1.1) (see [L-T.2]);

(ii) an exact controllability result (surjectivity or ontoness of the map

$$u \in L_2(0,T; L_2(\Gamma(x_0)) \to \{w(T), w_t(T)\}$$

onto a suitable state space (see [L-T.1]).

However, in the present case, the duality or transposition argument is non-standard, due to the special function spaces involved related to the B.C. The details, taken from [Las.1], [Tr.1], [E-L-T.1], [L-T.4], are given in the subsequent Section 2 in both a systematic functional analytic treatment and a PDE-version of the transposition argument.

The conclusion (1.2.3): $w_t \in C([0,T]; \tilde{L}_2(\Omega))$ for problem (1.1.1) was already noted in [Las.1], via, however, a functional analytic (rather than PDE's) approach, but the space $\tilde{L}_2(\Omega)$ was not clarified there beyond its definition (1.1.6). In particular, the characterization (1.1.10) of Proposition 1.1.1 is a new result of the present paper. [Las.1] was motivated by [L-L.1], where the space $\tilde{L}_2(\Omega)$ for w_t does not appear. The point that we wish to make is that it is the space $\tilde{L}_2(\Omega)$ [not $L_2(\Omega)$] that describes the optimal regularity – as well as the controllability – of the velocity w_t of the mixed problem (1.1.1). Then conclusions (1.2.2), (1.2.3) of Theorem 1.2.1 can be complemented with the following (exact controllability) result, arising this time from the continuous observability inequality (1.3.4) by transposition or duality [L-T.1].

Theorem 1.3.1. *With reference to the mixed problem (1.1.1), we have: the map*

$$\{w_0, w_1\} = 0, \quad u \in L_2(0,T; L_2(\Gamma(x_0)) \to \{w(T), w_t(T)\}$$

is surjective (onto) $H_0^1(\Omega) \times \tilde{L}_2(\Omega)$, for all $T > T_0 > 0$, with T_0 defined in (1.3.4).

To further elaborate: any (target) state $\{v_1, v_2\}$ with $v_1 \in H_0^1(\Omega)$ and $v_2 \in \mathcal{H}$ (the null space of generalized harmonic functions defined in (1.1.11a)) *cannot* be reached from the origin over a time interval $[0,T]$, $T > T_0$, by using an $L_2(0,T; L_2(\Gamma(x_0))$-control. Similarly, by reversing time, an initial condition $\{w_0, w_1\}$ with $w_0 \in H_0^1(\Omega)$ and $w_1 \in \mathcal{H}$ cannot be steered to rest $(0,0)$ over the time interval $[0,T]$, by using an $L_2(0,T; L_2(\Gamma(x_0))$-control. This is so since, by Proposition 1.1.1, Eqn. (1.1.10), any element $h \in \mathcal{H}$ has zero norm in $\tilde{L}_2(\Omega)$: $\|h\|_{\tilde{L}_2(\Omega)} = 0$, as the null space $\mathcal{H}$ of the operator $(1 - \gamma\Delta)$ consisting of generalized harmonic functions defined in (1.1.11a) acts as the zero element in $\tilde{L}_2(\Omega)$. Thus, Theorem 1.2.1 for $\{w, w_t\}$, as well as Theorem 1.3.1 do add new critical insight over the literature [L-L.1], [Las.1]. In addition, the regularity (1.2.4) for w_{tt} is entirely new.

Kirchhoff thermoelastic problem (1.1.2). For brevity, we limit our comments to the following considerations. The regularity (1.2.5), (1.2.6) for $\{w, w_t\}$ is the same as that given (and proved) in [Tr.2]. However, regarding the regularity of w_{tt} and θ, the statements in (1.2.7)–(1.2.10) of Theorem 1.2.2 represent a clarification over the literature [L-L.1], [Tr.2], [E-L-T.1-2].

As the spaces $\tilde{L}_2(\Omega)$ and $\tilde{H}_{-1}(\Omega)$ [their definitions, their properties, such as (1.1.10) and (1.1.14), and surrounding considerations] are not present in [L-L.1], we are unable to justify the claims for w_t asserted to be in $L_\infty(0,T;L_2(\Omega))$, and for θ asserted to be in $L_\infty(0,T;L_2(\Omega))$ continuously in $u \in L_2(0,T;L_2(\Gamma))$, which are made in [L-L.1, p. 160, Eqns. (3.30), (3.31)]: this reference states that they simply follow by a duality or transposition argument (such as the one from (2.18) to (2.25) below, in the elastic case) over the trace inequality in [L-L.1, (3.11), p. 157], which is the counterpart of (1.3.3) in the thermoelastic case. [Even if w_t *were* in $L_\infty(0,T;L_2(\Omega))$, then $\int_0^t e^{-\mathcal{A}(t-\tau)}\Delta w_t(\tau)d\tau$ would *not* be in $C([0,T];L_2(\Omega))$, but *only* in $L_p(0,T;L_2(\Omega)) \cap C([0,T];H^{-\epsilon}(\Omega))$, for any $1 \leq p < \infty$, and any $\epsilon > 0$. There is no "maximal regularity" for the $L_\infty(0,T;\cdot)$-spaces. Moreover, our claim in this paper is that $w_t \in C([0,T];\tilde{L}_2(\Omega))$ instead, as in (1.2.6).] By contrast we find that the regularity of $\theta \in C([0,T];L_2(\Omega))$ is *not* continuous in $u \in L_2(0,T;L_2(\Gamma))$, see (1.2.10), and requires the analysis of w_{tt}, which involves the new space $\tilde{H}_{-1}(\Omega)$. To get continuity in $u \in L_2(0,T;L_2(\Gamma))$, lower topologies are involved in our analysis for θ, as in (1.2.7) or (1.2.9). The regularity in (1.2.9) requires a delicate trace analysis, which is sketched in Section 5.2 of [L-T.4].

2. Proof of Theorem 1.2.1. Duality

Functional analytic proof. The abstract model of the mixed problem (1.1.1) is given by [L-T.2],

$$(I+\gamma\mathcal{A})w_{tt} = -Aw + AG_2u \quad \text{in, say, } [\mathcal{D}(A)]', \tag{2.1}$$

where G_2 is the following Green map defined by [L-T.2],

$$v = G_2u \iff \left\{\Delta^2 v = 0 \text{ in } \Omega;\ v|_\Gamma = 0,\ \left.\frac{\partial v}{\partial\nu}\right|_\Gamma = u\right\}, \tag{2.2}$$

and by elliptic regularity [L-M.1] and [G.1], see [L-T.2],

$$G_2:\ \text{continuous } L_2(\Gamma) \to H^{\frac{3}{2}}(\Omega) \cap H_0^1(\Omega)$$

$$\subset H^{\frac{3}{2}-4\epsilon}(\Omega) \cap H_0^1(\Omega) = \mathcal{D}(A^{\frac{3}{8}-\epsilon}) \tag{2.3a}$$

$$A^{\frac{3}{8}-\epsilon}G_2:\ \text{continuous } L_2(\Gamma) \to L_2(\Omega). \tag{2.3b}$$

Next, the corresponding solution of problem (1.1.1), or (2.1), with zero I.C., is then written as

$$\begin{bmatrix} w(t) \\ w_t(t) \end{bmatrix} = (Lu)(t) = \int_0^t e^{\mathbb{A}_{0,\gamma}(t-\tau)}\mathcal{B}_c \begin{bmatrix} 0 \\ u(\tau) \end{bmatrix} d\tau, \tag{2.4}$$

where the operator $Y_\gamma \supset \mathcal{D}(\mathbb{A}_{0,\gamma}) \to Y_\gamma$ is given by

$$\mathbb{A}_{0,\gamma} = \begin{bmatrix} 0 & I \\ -\mathcal{A}_\gamma^{-1}A & 0 \end{bmatrix}; \ \mathbb{A}_{0,\gamma}^* = \begin{bmatrix} 0 & -I \\ \mathcal{A}_\gamma^{-1}A & 0 \end{bmatrix}; \tag{2.5}$$

$$\mathcal{D}(\mathbb{A}_{0,\gamma}) \equiv \mathcal{D}(\mathbb{A}_{0,\gamma}^*) \equiv \mathcal{D}(A^{\frac{3}{4}}) \times \mathcal{D}(A^{\frac{1}{2}}) \equiv [H^3(\Omega) \cap H_0^2(\Omega)] \times H_0^2(\Omega); \tag{2.6}$$

$$Y_\gamma \equiv \mathcal{D}(A^{\frac{1}{2}}) \times \mathcal{D}(\mathcal{A}_\gamma^{\frac{1}{2}}) \equiv H_0^2(\Omega) \times H_0^1(\Omega) \tag{2.7}$$

(norm equivalence), see (1.1.5a-b). The * denotes the Y_γ-adjoint. In identifying the domains in (2.6), we have recalled from (1.1.5a) that $\mathcal{A}_\gamma^{-\frac{1}{2}}A^{\frac{1}{4}}$ is an isomorphism (bounded, with bounded inverse) on $L_2(\Omega)$. Moreover, [Tr.1], [E-L-T.1, Eqns. (3.34), (3.37)],

$$\mathcal{B}_c \begin{bmatrix} 0 \\ u \end{bmatrix} = \begin{bmatrix} 0 \\ \mathcal{A}_\gamma^{-1}AG_2u \end{bmatrix}, \ G_2^*Af = -\Delta f|_\Gamma, \ f \in \mathcal{D}(A). \tag{2.8}$$

The next step is the *key point* of the subsequent duality argument. If [ ]' denotes duality with respect to Y_γ as a pivot space, we have from (2.6), (2.7),

$$[\mathcal{D}(\mathbb{A}_{0,\gamma})]' \equiv [\mathcal{D}(\mathbb{A}_{0,\gamma}^*)]' \equiv \mathcal{D}(A^{\frac{1}{4}}) \times \tilde{L}_2(\Omega); \tag{2.9}$$

The second component space, $\tilde{L}_2(\Omega)$, in (2.9) arises, precisely by invoking its definition (1.1.6), by duality of the second component space, $\mathcal{D}(A^{\frac{1}{2}})$, in (2.6) with respect to the second component space, $\mathcal{D}(\mathcal{A}_\gamma^{\frac{1}{2}})$, in (2.7). Next, if $L_Tu = (Lu)(T)$, and $x = [x_1, x_2] \in [\mathcal{D}(\mathbb{A}_{0,\gamma}^*)]'$, then [E-L-T.1, Eqn. (5.2.6)],

$$\left(L_T \begin{bmatrix} 0 \\ u \end{bmatrix}, \begin{bmatrix} x_1 \\ x_2 \end{bmatrix} \right)_{[\mathcal{D}(\mathbb{A}_{0,\gamma}^*)]'} = \left(\begin{bmatrix} 0 \\ u \end{bmatrix}, L_T^\# \begin{bmatrix} x_1 \\ x_2 \end{bmatrix} \right)_{L_2(\Sigma)\times L_2(\Sigma)}, \tag{2.10}$$

where [E-L-T.1, Eqn. (5.2.8)],

$$\left(L_T^\# \begin{bmatrix} x_1 \\ x_2 \end{bmatrix} \right)(t) = \begin{bmatrix} 0 \\ \Delta\phi(T-t; y_0)|_\Gamma \end{bmatrix}, \tag{2.11}$$

and where [E-L-T.1, Eqn. (3.13)],

$$\begin{bmatrix} \phi(T-t; y_0) \\ -\phi_t(T-t; y_0) \end{bmatrix} = e^{\mathbb{A}_{0,\gamma}^*(T-t)}y_0; \tag{2.12}$$

$$y_0 = \begin{bmatrix} \phi_0 \\ -\phi_1 \end{bmatrix} = \mathbb{A}_{0,\gamma}^{-1} \begin{bmatrix} x_1 \\ x_2 \end{bmatrix} \in Y_\gamma \equiv \mathcal{D}(A^{\frac{1}{2}}) \times \mathcal{D}(\mathcal{A}_\gamma^{\frac{1}{2}}) \equiv H_0^2(\Omega) \times H_0^1(\Omega) \tag{2.13}$$

is the solution of the homogeneous problem (1.3.1). Thus, the aforementioned trace regularity result (1.3.3)

$$\{\phi_0, \phi_1\} \in H_0^2(\Omega) \times H_0^1(\Omega) \to \Delta\phi|_\Gamma \in L_2(0,T; L_2(\Gamma)), \tag{2.14}$$

due to [L-L.1, p. 123] is equivalent, via (2.11), (2.13), to

$$L_T^\# : \text{ continuous } [\mathcal{D}(\mathbb{A}_{0,\gamma}^*)]' \to L_2(0,T; L_2(\Gamma)), \tag{2.15}$$

which then is, in turn, equivalent to [L-T.2]

$$L: \quad \text{continuous } L_2(0,T;L_2(\Gamma)) \to (Lu) = \{w, w_t\}$$
$$\in C([0,T]; H_0^1(\Omega) \times \tilde{L}_2(\Omega)). \tag{2.16}$$

Then, properties (1.2.2), (1.2.3) of Theorem 1.2.1 are established.

We now establish (1.2.4). We return to the abstract model (2.1), which we rewrite as

$$A^{-\frac{3}{4}}\mathcal{A}_\gamma w_{tt} = -A^{\frac{1}{4}}w + A^{-\frac{1}{8}+\epsilon}\left(A^{\frac{3}{8}-\epsilon}G_2 u\right) \in L_2(0,T;L_2(\Omega)), \tag{2.17}$$

where the regularity noted in (2.17) follows from $A^{\frac{1}{4}}w \in C([0,T];L_2(\Omega))$ by (1.2.2), as well as from $A^{\frac{3}{8}-\epsilon}G_2 u \in L_2(0,T;L_2(\Omega))$ by (2.3b), as well as from $A^{\frac{3}{8}-\epsilon}G_2 u \in L_2(0,T;L_2(\Omega))$ by (2.3b) on G_2 and (1.2.1) on u. Thus, as usual via the characterization (1.1.9c), we see that (2.17) says that $w_{tt} \in L_2(0,T;\tilde{H}_{-1}(\Omega))$, as claimed in (1.2.4). [The above argument shows that, in the present circumstances, the term Aw is the *critical* one, while AG_2u is subordinated to it, in model (2.1).]

PDE proof of (1.2.2), (1.2.3). Multiplying the non-homogeneous w-problem (1.1.1) with $\{w_0, w_1\} = 0$ and $u \in L_2(0,T;L_2(\Gamma))$ by the solution ϕ of problem (1.3.1), we obtain after integration by parts in t, and use of Green's second theorem, once the appropriate boundary conditions are invoked:

$$\begin{aligned} 0 &= \int_0^T ((1-\gamma\Delta)w_{tt},\phi)_\Omega dt + \int_0^T (\Delta^2 w,\phi)_\Omega dt \\ &= [((1-\gamma\Delta)w_t,\phi)_\Omega]_0^T - [((1-\gamma\Delta)w,\phi_t)_\Omega]_0^T \\ &\quad + \int_0^T ((1-\gamma\Delta)w,\phi_{tt})_\Omega dt + \int_0^T (w,\Delta^2\phi)_\Omega dt + \int_0^T \left(\frac{\partial w}{\partial\nu},\Delta\phi\right)_\Gamma dt, \end{aligned} \tag{2.18}$$

where (,) denotes $L_2(\Omega)$ or $L_2(\Gamma)$-norms. Finally, since in the first integral term on the right of (2.18), $(1-\gamma\Delta)$ may be moved from the left (as acting on w) to the right (as acting on ϕ_{tt}) by Green's theorem, with no boundary terms by (1.3.1c), use of Eqn. (1.3.1a) finally yields from (2.18):

$$((1-\gamma\Delta)w_t(T),\phi(T))_{L_2(\Omega)} - ((1-\gamma\Delta)w(T),\phi_t(T))_{L_2(\Omega)} + \int_0^T (u,\Delta\phi|_\Gamma)_\Gamma dt = 0. \tag{2.19}$$

The boundary integral term in (2.19) is well-defined by u in (1.2.1) and $\Delta\phi|_\Gamma$ in (1.3.3). Thus, we need to investigate the well-posedness of the terms involving the initial conditions:

$$((1-\gamma\Delta)w_t(T),\phi(T))_{L_2(\Omega)} \quad \text{and} \quad ((1-\gamma\Delta)w(T),\phi_t(T))_{L_2(\Omega)}. \tag{2.20}$$

As $\phi_t(T) \in H_0^1(\Omega)$ by (1.3.2), the well-posedness of the second term in (2.20) then requires

$$(1 - \gamma\Delta)w(T) = \mathcal{A}_\gamma w(T) \in H^{-1}(\Omega) \equiv [\mathcal{D}(\mathcal{A}_\gamma^{\frac{1}{2}})]', \tag{2.21}$$

invoking the operator $\mathcal{A}_\gamma$ in (1.1.4) below [since $w(T)$ satisfies zero Dirichlet B.C., as in (1.1.1c)]; or finally

$$w(T) \in \mathcal{D}(\mathcal{A}_\gamma^{\frac{1}{2}}) \equiv H_0^1(\Omega). \tag{2.22}$$

So far all is essentially standard. Not so for the first term of (2.20), however. Indeed, as $\phi(T) \in H_0^2(\Omega)$ by (1.3.2), the well-posedness of the first term in (2.20) then requires

$$(1 - \gamma\Delta)w_t(T) = \mathcal{A}_\gamma w_t(T) \in H^{-2}(\Omega) \equiv [\mathcal{D}(A^{\frac{1}{2}})]', \tag{2.23}$$

invoking the operator $\mathcal{A}_\gamma$ in (1.1.4) below [since $w_t(T)$ satisfies zero Dirichlet B.C., by (1.1.1c)], as well as the elastic operator A in (1.1.3). Thus, (2.23) characterizes $w_t(T)$ as satisfying the condition

$$A^{-\frac{1}{2}}\mathcal{A}_\gamma w_t(T) \in L_2(\Omega). \tag{2.24}$$

But (2.24), in turn, characterizes $w_t(T)$ as belonging to the space which we called $\tilde{L}_2(\Omega)$ in (1.1.6), Eqn. (1.1.7c). We conclude that

$$w_t(T) \in \tilde{L}_2(\Omega). \tag{2.25}$$

3. The Space $\tilde{L}_2(\Omega)$ is Isometric to the Factor Space $L_2(\Omega)/\mathcal{H}$: Proof of Proposition 1.1.1

To prove (1.1.10), we shall use a standard result [A.1, Thm. 1.6, p. 53], [T-L.1, Thm. 3.5, p. 135]. Using Aubin's notation, we set

$$P \equiv \mathcal{D}(A^{\frac{1}{2}}): \text{ a closed subspace of } V \equiv \mathcal{D}(\mathcal{A}_\gamma) \tag{3.1}$$

equipped with the inner product $(w, v)_V = (\mathcal{A}_\gamma w, \mathcal{A}_\gamma v)_{L_2(\Omega)}$. By the above references, we have that

$$P' \text{ is isometrically isomorphic (congruent) to the factor space } V'/P^\perp, \tag{3.2}$$

$$P^\perp \equiv \{f \in V' : f(v) = (f, v)_{V' \times V} = 0, \quad \forall\, v \in P \subset V\}; \tag{3.3}$$

$$P' \equiv \text{space of continuous linear functionals on } P; \tag{3.4}$$

$$V' \equiv \text{space of continuous linear functionals on } V, \tag{3.5}$$

and $(\ ,\ )_{V' \times V}$ denotes the duality pairing on $V' \times V$. We now take

$\mathcal{D}(\mathcal{A}_\gamma^{\frac{1}{2}})$ (equipped with the norm as in (1.1.5c)) as a common pivot space for P and V, (3.6)

and we note that P is dense in $\mathcal{D}(\mathcal{A}_\gamma^{\frac{1}{2}})$. Then:

(i) P' can be isometrically identified with the space

$$\begin{aligned} \tilde{L}_2(\Omega) &\equiv \text{dual of } [\mathcal{D}(A^{\frac{1}{2}}) \equiv P] \text{ with respect to } \mathcal{D}(\mathcal{A}_\gamma^{\frac{1}{2}}) \\ &= [\mathcal{D}(A^{\frac{1}{2}})]'_{\text{w.r.t } \mathcal{D}(\mathcal{A}_\gamma^{\frac{1}{2}})} \quad \text{(see (1.1.6))}. \end{aligned} \tag{3.7}$$

(ii) V' can be isometrically identified with the space

$$L_2(\Omega) \equiv [\mathcal{D}(\mathcal{A}_\gamma)]'_{\text{w.r.t } \mathcal{D}(\mathcal{A}_\gamma^{\frac{1}{2}})}. \tag{3.8}$$

Next, we find the corresponding isometric identification for $P^\perp$ (which is a closed subspace of V'). By the Riesz representation theorem, if I denotes the canonical isometry from V onto V', then $P^\perp$ in (3.3) can be isometrically identified with the following subspace of V:

$$\{I^{-1}f \in V : \text{ inner product } (I^{-1}f, v)_V = 0, \ \forall\, v \in P \subset V\} \tag{3.9}$$

$$\equiv \{w \in V \equiv \mathcal{D}(\mathcal{A}_\gamma) : (w, v)_V = (\mathcal{A}_\gamma w, \mathcal{A}_\gamma v)_{L_2(\Omega)} = 0, \ \forall\, v \in P\} \tag{3.10}$$

$$= \{h \in L_2(\Omega) : (h, \mathcal{A}_\gamma v)_{L_2(\Omega)} = 0, \ \forall\, v \in P = \mathcal{D}(A^{\frac{1}{2}}) \equiv H_0^2(\Omega) \subset \mathcal{D}(\mathcal{A}_\gamma)\} \tag{3.11}$$

$$= \{h \in L_2(\Omega) : (h, (1 - \gamma\Delta)v)_{L_2(\Omega)} = 0, \ \forall\, v \in P \equiv H_0^2(\Omega)\} \tag{3.12}$$

$$= \{h \in L_2(\Omega) : ((1 - \gamma\Delta)h, v)_{L_2(\Omega)} = 0, \ \forall\, v \in P \equiv H_0^2(\Omega)\} \tag{3.13}$$

$$= \{h \in L_2(\Omega) : (1 - \gamma\Delta)h = 0, \text{ in } H^{-2}(\Omega)\} \equiv \mathcal{H} = \mathcal{N}\{(1 - \gamma\Delta)\}, \tag{3.14}$$

the null space of the operator $(1-\gamma\Delta) : L_2(\Omega) \to H^{-2}(\Omega)$, introduced in (1.1.11a). In going through the steps above, we have used: the inner product below (3.1a) from (3.9) to (3.10); the important property [L-T.4, Lemma 2.1.2],

$$\begin{cases} \mathcal{A}_\gamma : \text{ continuous from } H_0^2(\Omega) \equiv \mathcal{D}(A^{\frac{1}{2}}) \text{ onto } \mathcal{H}^\perp; \\ \text{equivalently} \\ \mathcal{A}_\gamma A^{-\frac{1}{2}} : \text{ continuous from } L_2(\Omega) \text{ onto } \mathcal{H}^\perp; \end{cases}$$

the Green's identity from (3.12) to (3.13), since $v \in H_0^2(\Omega)$. In conclusion:

$$\begin{cases} \text{the space } P^\perp \text{ in (3.3), as a closed subspace of } V', \text{ can be isometri-} \\ \text{cally identified with the space } \mathcal{H}, \text{ as a closed subspace of } L_2(\Omega). \end{cases} \tag{3.15}$$

Thus, we return to (3.2), (3.6): invoking further (3.7) and (3.9), we conclude that $\tilde{L}_2(\Omega)$ can be isometrically identified with $L_2(\Omega)/\mathcal{H}$ and (3.15) proves (1.1.10).

References

[A.1] J. P. Aubin, *Approximation of Elliptic Boundary-Value Problems,* Wiley-Interscience, 1972.

[E-L-T.1] M. Eller, I. Lasiecka, and R. Triggiani, Simultaneous exact/approximate boundary controllability of thermo-elastic plates with variable transmission coefficients, *Lecture Notes in Pure and Applied Mathematics*, vol. , J. Cagnol, M. Polis, J. P. Zolesio, editors, to appear.

[G.1] P. Grisvard, Caracterization de quelques espaces d'interpolation, *Arch. Rational Mech. Anal.* 25 (1967), 40–63.

[Lag.1] J. Lagnese, Boundary Stabilization of Thin Plates, SIAM, Philadelphia, 1989.

[L-L.1] J. Lagnese and J. L. Lions, *Modelling, Analysis and Control of Thin Plates,* Masson 1988.

[Las.1] I. Lasiecka, Controllability of a viscoelastic Kirchhoff plate, *Int. Series Num. Math.* 91 (1989), 237–247, Birkhäuser Verlag, Basel.

[L-T.1] I. Lasiecka and R. Triggiani, Exact boundary controllability for the wave equation with Neumann boundary control, *Appl. Math. & Optimiz.* 19 (1989), 243–290.

[L-T.2] I. Lasiecka and R. Triggiani, Differential and Algebraic Riccati Equations with applications to boundary/point control problems, vol. 164, Springer-Verlag Lecture Notes in Control & Information Sciences, 1991, 160 pp.

[L-T.3] I. Lasiecka and R. Triggiani, Simultaneous exact/approximate boundary controllability of thermoelastic plates with variable transmission coefficient, Marcel Dekker Lecture Notes in Pure and Applied Mathematics, to appear (110 pp).

[L-T.4] I. Lasiecka and R. Triggiani, Factor spaces and implications on Kirchhoff equations with clamped boundary conditions, submitted. Preprint available.

[L-M.1] J. L. Lions and E. Magenes, *Nonhomogeneous Boundary Value Problems,* Springer-Verlag, Vol. 1, p. 197.

[T-L.1] A. E. Taylor and D. C. Lay, *Introduction to Functional Analysis,* second edition, John Wiley, 1980.

[Tr.1] R. Triggiani, Sharp regularity theory of thermoelastic mixed problems, *Applicable Analysis*

Department of Mathematics
University of Virginia
Kerchof Hall
Charlottesville, VA 22904
E-mail address: I.L.: `il2v@virginia.edu`; R.T.: `rt7u@virginia.edu`

International Series of Numerical Mathematics, Vol. 139, 183–190

The Interplay of Convection, Conduction and Phase Transition in a Shape-Memory-Alloy

Ingo Müller, Norbert Papenfuß -Janzen, and Stefan Seelecke

Abstract. A shape memory wire dragged through a heating chamber undergoes external heating by conduction and internal heating due to the latent heat of the phase transition that is triggered by the supply of heat. Depending on parameters the wire may undergo a rapid transition to a stationary state, or a limit cycle in phase fraction and temperature, or a damped oscillation to a vortex point.

1. Introduction

Shape memory alloys have proved to be amenable to modelization by a fairly simple set of ordinary differential equations, at least for homogeneous deformations under uniaxial load conditions. The authors and various co-workers have developed a structural model of such alloys by simulating the phase transition as an activated process, see [1]–[5]. The model is by now so easy to handle that we have installed it in our web page as an interactive program on which a visitor may perform his own (computer) experiments, see O. Heintze, S. Seelecke and M. Petzoldt at http://www.thermodynamik.tu-berlin.de.

In the present paper we turn to a more ambitious case: While previously we have restricted the attention to the interacting effects of heat conduction and phase transition we are now taking convection into account. Thus we study the comportment of phase fractions and temperature in a wire that is loaded and dragged through a heating chamber. Since the governing equations are highly non-linear, we expect interesting non-trivial solutions. And so it occurs: The interaction of convection, conduction and phase transition provide a wealth of complex solutions, e.g. limit cycles and relaxation to vortex points.

2. Scope

We consider a combined case of convection, heat conduction and phase transition as it may occur when a shape memory wire under a tensile load is carried through a heat bath at a constant rate, see Fig.1.

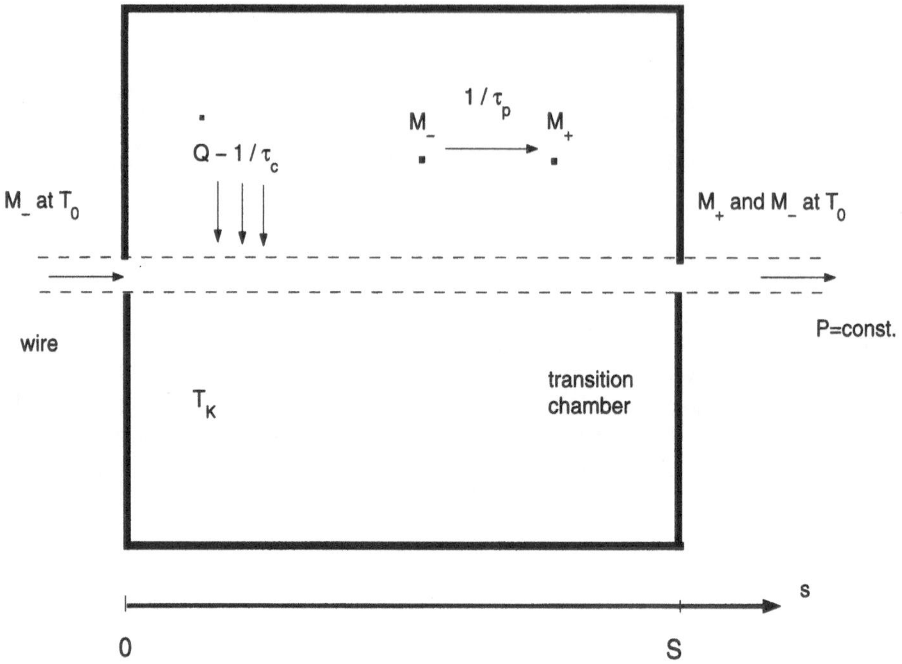

FIGURE 1. Wire pulled through a heat bath

The phase transition within the wire can be a martensitic twinning transition between the variants M_- and M_+. We assume that the wire is in the M_- phase as it enters the heating chamber and that the transition $M_- \to M_+$ may occur under the combined influence of the tensile load and the temperature increase. The latent heat liberated in the transition will further heat the wire and accelerate the transition as long as the M_- lasts or until the wire leaves the chamber. But fresh wire is continuously fed into the chamber so that the process may resume. Depending on the parameters of the wire and of the chamber and on the speeds of convection, conduction and transition we may create stationary conditions inside the chamber, or limit cycles, or damped oscillations tending to a vortex point.

3. Model for Wire

The constituent particles or layers of the phases M_- and M_+ may be imagined as occupying the two wells of a double-well-potential. Actually, at any finite temperature the particles will fluctuate about their respective minima and the mean kinetic energy is proportional to the temperature T, usually that energy is indicated by the height of the pools of particles that occupy the well, see Fig.2.

The double well is perfectly symmetric when the wire is not loaded. But upon application of a tensile load it is deformed as shown in Fig.2; the barrier to be surpassed for a particle in a $M_- \to M_+$ transition is now smaller than the barrier for the reverse transition. Therefore we assume that the latter does not

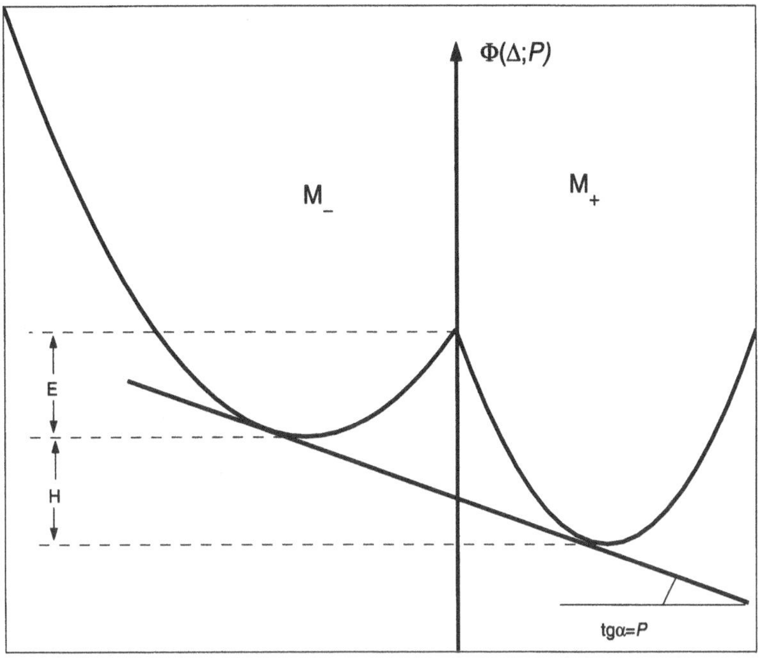

FIGURE 2. Double-well-potential for the phases $M_{\pm}$ under a load P

occur. For the transition probability $M_{-} \rightarrow M_{+}$ we assume

$$\overset{-+}{p} = \frac{1}{\tau_p} e^{-\frac{E}{kT}} , \tag{3.1}$$

where E is the height of the barrier and τ_p is a characteristic time for the phase transition. Obviously $\overset{-+}{p}$ depends strongly on T. The ansatz (3.1) is typically for thermally activated processes in chemistry and solid state physics; here we use if for the phase transition $M_{-} \rightarrow M_{+}$.

4. Balance of Mass of Phase M_{+}

We apply the balance of mass of the phase M_{+} to a control volume surrounding the wire inside the heating chamber shown in Fig.1. This represents an "open system", because the wire enters and leaves it, carrying the total mass flux $\dot{m}$ in and out of the system. In doing so, it brings the mass flux $\dot{m}_{+}^{in} = \dot{m}x_{+}(0)$ into the system on the left, i.e. at $s = 0$, and it carries the mass flux $\dot{m}_{+}^{out} = \dot{m}x_{+}(S)$ out of the system on the right, i.e. at $s = S$.

Also there is a production of M_{+} inside the system due to the phase transition. At every point s inside the chamber that production may reasonably be assumed to be proportional to $m'_{-}(s) = m'(1 - x_{+}(s))$, the mass density of M_{-}

per unit length, and the factor of proportionality is given by the transition probability (3.1). Therefore the equation of balance of the mass of the phase M_+ reads

$$\frac{d}{dt}\int_0^S \dot{m}' x_+(s)ds = \dot{m}\,(x_+(0) - x_+(S)) + \int_0^S (1 - x_+(s))\frac{1}{\tau_p}\, e^{-\frac{E}{kT(s)}}\, ds. \tag{4.1}$$

A drastic simplification results from the assumption of homogeneity inside the chamber. Thus $x_+(s)$ and $T(s)$ are considered independent of s and naturally, $x_+(S)$ is then also equal to the s-independent value x_+. We write $m'S = m$ and obtain for this homogeneous case

$$\frac{dx_+}{dt} = \frac{1}{\tau_K}\,(x_+(0) - x_+) + \frac{1}{\tau_p}\, e^{-\frac{E}{kT}}\,(1 - x_+), \tag{4.2}$$

where $\frac{1}{\tau_K} \equiv \frac{\dot{m}}{m}$ is a characteristic time associated with the wire being dragged through the system.

It is true that the assumption of homogeneity is a steep assumption. It reduces to an ordinary differential equation what would otherwise be a partial differential equation.

5. Balance of Energy

We neglect the kinetic energy of the wire and thus write the balance of energy for the open system under consideration

$$\begin{aligned}&\tfrac{d}{dt}\int_0^S m'(x_+(s)u_+(s) + x_-(s)u_-(s))ds = \\ &= \dot{m}\,[(x_+(0)u_+(0) + x_-(0)u_-(0)) - (x_+(S)u_+(S) + x_-(S)u_-(S))] + \dot{Q} + \dot{A}.\end{aligned} \tag{5.1}$$

$\dot{Q}$ is the heating and $\dot{A}$ the working applied to the surface of the system. The working is equal to zero as long as the load is considered constant and homogeneous along the wire and provided the phase transition does not affect the density of the wire. The heating is assumed proportional to the temperature difference $T_K - T(s)$ between wire and chamber, and the factor of proportionality is α, the heat transfer coefficient. $u_\pm(s)$ are the specific internal energies of the two phases and we may relate them to the temperature by

$$u_\pm(s) = cT(s) + u_\pm^R\,,$$

where c is the (common) specific heat of the phases and $u_\pm^R$ are their energy constants. Their difference $u_-^R - u_+^R$ is equal to the height difference H of the potential wells in Fig.2.

With all this and again assuming homogeneity of the fields $x_+(s)$ and $T(s)$ we obtain

$$c\frac{dT}{dt} = -\frac{1}{\tau_K}c(T - T(0)) + H\frac{1}{\tau_p}(1 - x_+)e^{-\frac{E}{kT}} - \frac{c}{\tau_C}(T - T_K)\,, \tag{5.2}$$

where $\frac{1}{\tau_C} = \frac{\alpha S}{mc}$ is a characteristic time associated with the wire being heated or cooled by heat conduction.

6. Solutions

We solve the equations (4.2) and (5.2) for $x_+(t)$ and $T(t)$ by assuming the initial conditions $x_+(0) = 0$ and $T(0) = T_0$. That is to say that we feed the wire into the chamber in the M_- −phase. We choose T_0 to make the temperature dimensionless and τ_K to make time dimensionless. As fixed parameters we select:

barrier height	$\frac{E}{kT_0} = 13.5$
latent heat	$\frac{H}{cT_0} = 6$
characteristic time for conduction	$\frac{\tau_c}{\tau_K} = 3.$

Fig. 3 shows the effect of different chamber temperature for a fixed characteristic time $\frac{\tau_p}{\tau_K} = 400$ of the phase transition. Inspection of the figure shows that even a slight change of T_K/T_0 provides a strong difference in the qualitative behaviour of the solutions. Below each line of graphs in Fig. 3 we have added comments on the salient points of convection, conduction and phase transition in each case and we suggest that the reader take a little time to digest our interpretation.

Here we stress only that three qualitatively different solutions may occur:

- a quick transition to a stationary case
- a limit cycle
- a spiral into a vortex point.

Fig. 4 focuses on the case $\frac{T_K}{T_0}$ – the middle one of Fig. 3 – and it shows that stationarity, limit cycle and vortex point solutions may be selected by different speeds of the phase transition. Again the salient points of the processes are listed next to the graphs and they are easy to appreciate.

7. Discussion

Since the phase transition is associated by a lenghtening of a wire, we have the possibility to produce a pulsating lenght change in the wire, – at least in the case of the limit cycle solution. It is conceivable that a method can be found by which the original state of the wire is restored and by which the wire is fed back into the heating chamber in a loop. Thus the wire could be put to work to drive a maschine.

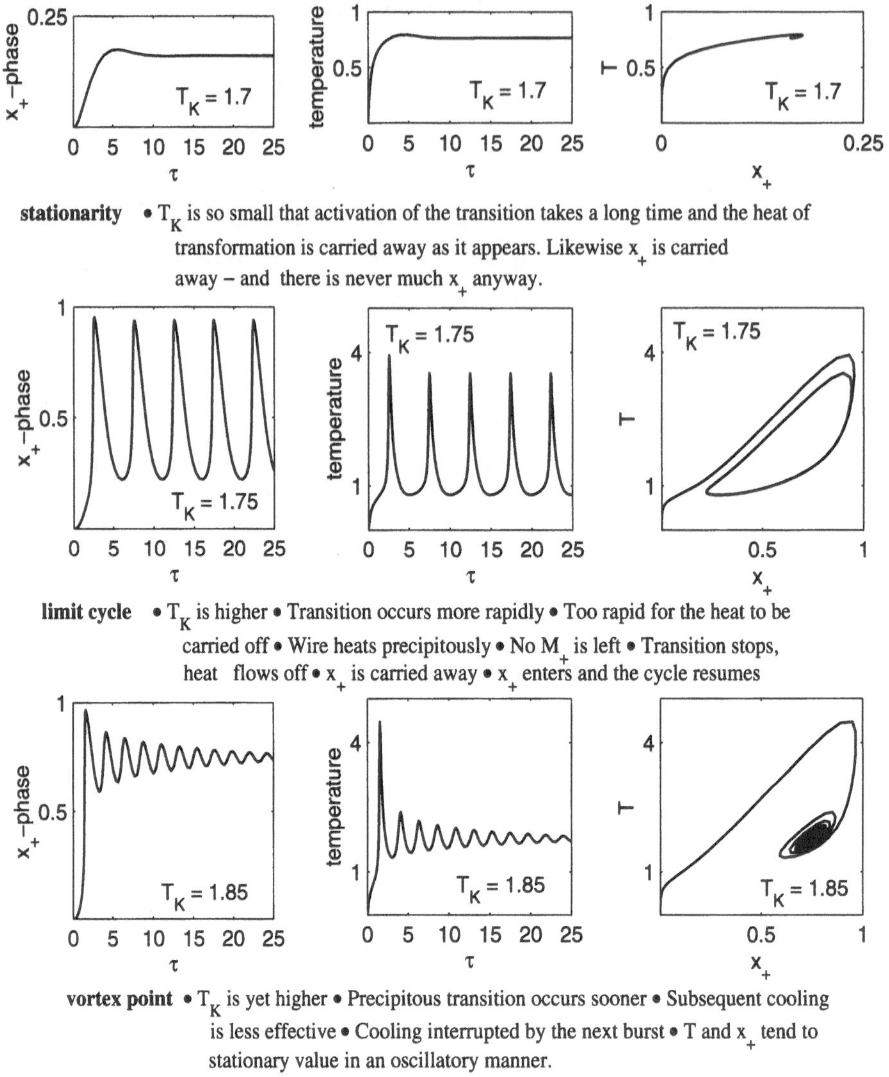
x₊-phase
temperature
T
τ
x₊
TK = 1.7
stationarity • TK is so small that activation of the transition takes a long time and the heat of transformation is carried away as it appears. Likewise x₊ is carried away – and there is never much x₊ anyway.
TK = 1.75
limit cycle • TK is higher • Transition occurs more rapidly • Too rapid for the heat to be carried off • Wire heats precipitously • No M₊ is left • Transition stops, heat flows off • x₊ is carried away • x₊ enters and the cycle resumes
TK = 1.85
vortex point • TK is yet higher • Precipitous transition occurs sooner • Subsequent cooling is less effective • Cooling interrupted by the next burst • T and x₊ tend to stationary value in an oscillatory manner.

FIGURE 3. The effect of the chamber temperature on the solution with a fixed characteristic time for the phase transition.

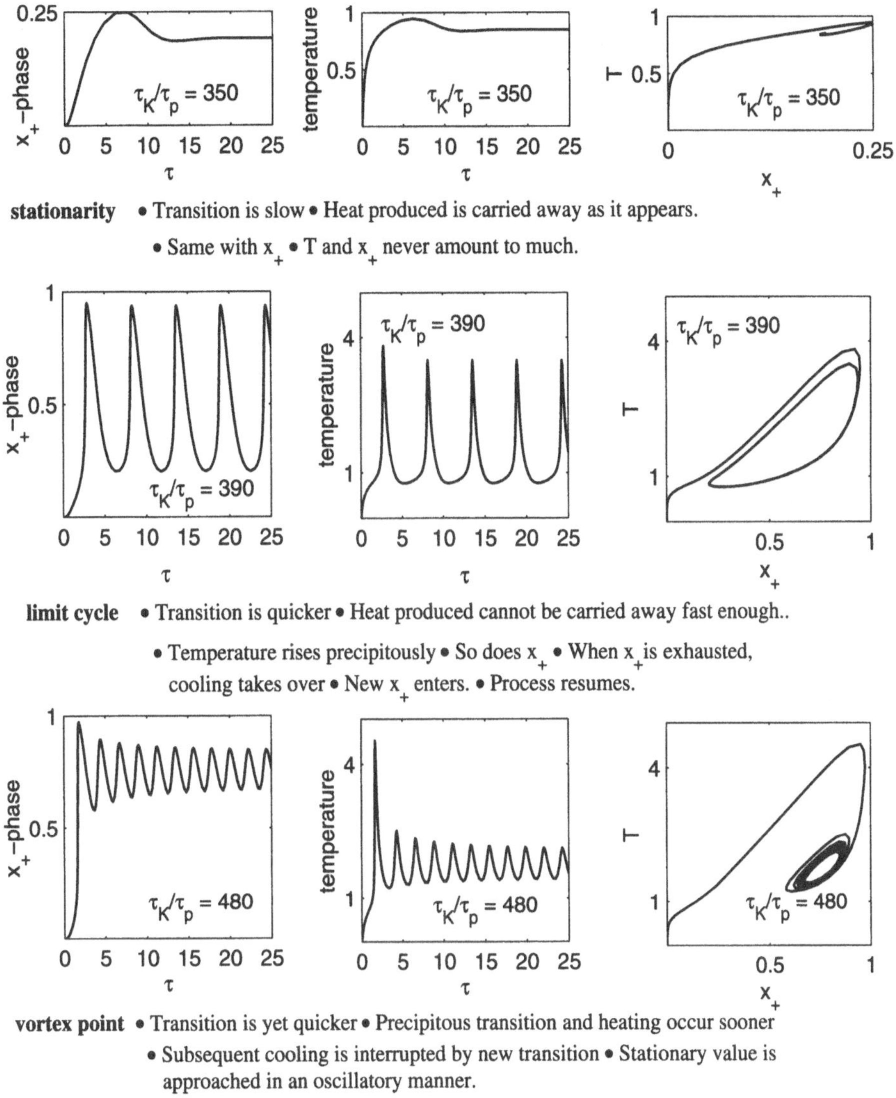
x+-phase
temperature
T
τ
x+
τK/τp = 350
τK/τp = 390
τK/τp = 480
stationarity • Transition is slow • Heat produced is carried away as it appears.
• Same with x+ • T and x+ never amount to much.
limit cycle • Transition is quicker • Heat produced cannot be carried away fast enough..
• Temperature rises precipitously • So does x+ • When x+ is exhausted, cooling takes over • New x+ enters. • Process resumes.
vortex point • Transition is yet quicker • Precipitous transition and heating occur sooner
• Subsequent cooling is interrupted by new transition • Stationary value is approached in an oscillatory manner.

FIGURE 4. The effect of the speed of the phase transition on the solutions with a fixed chamber temperature.

References

[1] I. Müller, *Pseudoelasticity in Shape Memory Alyos – An Extreme Case of Thermoelasticity,* Proc. Symp. Termoelasticità Finita. Mec. dei Lincei, Rome May/June 1985 (1985)

[2] M. Achenbach, *Ein Model zur Simulation des Last-Verformungs-Temperaturverhaltens von Legierungen mit Formerinnerungsvermögen,* Dissertation TU Berlin (1986)

[3] I. Müller and S. Seelecke, *Thermodynamic Aspects of Shape Memory Alloys,* Mathematical & Comp. Modeling 465 (in press).

[4] S. Seelecke and I. Müller, *Shape memory alloy actuators in smart structures – modelling and simulation,* Appl. Mech. Rev. (submitted).

[5] E. da Silva, *Zur Kalorimetrie von Gedächtnislegierungen und zu ihrer Anwendung als elektrisch aktivierte Aktuatoren,* Dissertation TU Berlin (2000).

(I. Müller) Faculty of Processing Sciences
Thermodynamics
TU-Berlin
Fasanenstr. 90
D-10623 Berlin Germany
E-mail address, I. Müller: `im@thermodynamik.tu-berlin.de`

(N. Papenfuß) Head of Modeling and Simulation Group
Materials Science Institute (WW)
Aachen University of Technology
52056 Aachen (Germany)
www.rwth-aachen.de/ww/

(S. Seelecke) Department of Mechanical and Aerospace Engineering
North Carolina State University
Campus Box 7910
Raleigh, NC 27695-7910
E-mail address, S. Seelecke: `stefan seelecke@ncsu.edu`

International Series of Numerical Mathematics, Vol. 139, 191–202

Semipermeable Curves and Level Sets of the Value Function in Differential Games with the Homicidal Chauffeur Dynamics

V.S. Patsko, V.L.Turova

Abstract. A classical and a modified (acoustic) variants of the differential game "homicidal chauffeur" are considered. An interesting peculiarity of the latter variant consists, in particular, in the presence of holes located strictly inside the victory domain of the pursuit-evasion game. In the paper, an explanation to this phenomenon is given. The explanation is based on an analysis of families of semipermeable curves that are determined from only the dynamics of the system. Results of the computation of level sets of the value function are presented.

1. Introduction

The homicidal chauffeur game [4], [6] is one of the most known model differential games of pursuit-evasion. In [1], [3], an acoustic capture variant of this game proposed by P.Bernhard was considered. The evader must reduce his speed when he comes close to the pursuer in order not to be heard. Mathematically, this can be expressed in taking the restriction on the velocity of the evader that depends on the distance between the evader and pursuer.

It was shown in [1],[3] that the solvability set (victory domain) of the acoustic problem can have holes located strictly inside this set. In such a case, it is impossible to compute the boundary of the solvability set using only barrier lines emitted from the usable part [4] of the terminal set.

This paper joins the paper [8] and is devoted to the description of families of semipermeable curves arising both in the classical homicidal chauffeur problem and its acoustic modification. The families of semipermeable curves are determined from only the dynamics of the system (including constraints on the controls) and do not depend on the form of the terminal set. The knowledge of the structure of these families can be very useful when studying different properties of solutions of time-optimal games. In particular, barrier lines which bound the solvability set are composed from arcs of smooth semipermeable curves.

It was found out that for some parameters of the problem, regions where semipermeable curves are absent can arise. It is shown that certain regions of this type can cause holes in solvability sets. The boundary of the solvability set can be completely described using semipermeable curves issued from the boundary of such regions and from the boundary of the terminal set.

In the final part of the paper, results of the computation of level sets of the value function for the acoustic problem are presented.

2. Games with the Homicidal Chauffeur Dynamics

The pursuer P has a fixed speed $w^{(1)}$ but his radius of turn is bounded by a given quantity R. The evader E is inertialess. He steers by choosing his velocity vector $v = (v_1, v_2)'$ from some set. The kinematic equations are:

$$P: \quad \dot{x}_p = w^{(1)} \sin \psi \qquad\qquad E: \quad \dot{x}_e = v_1$$
$$\dot{y}_p = w^{(1)} \cos \psi \qquad\qquad \dot{y}_e = v_2.$$
$$\dot{\psi} = w^{(1)} \varphi / R, \qquad |\varphi| \leq 1$$

The number of equations can be reduced to two (see [4]) if a coordinate system with the origin at P and the axis x_2 in the direction of P's velocity vector is used. The axis x_1 is orthogonal to the axis x_2.

The dynamics in the reduced coordinates is

$$\begin{aligned} \dot{x}_1 &= -w^{(1)} x_2 \varphi / R + v_1 \\ \dot{x}_2 &= \ \ w^{(1)} x_1 \varphi / R + v_2 - w^{(1)}, \qquad |\varphi| \leq 1. \end{aligned} \tag{1}$$

The state vector $(x_1,\ x_2)'$ gives the relative position of E with respect to P.

2.1. Classical homicidal chauffeur game

The control v is chosen from a circle of radius $w^{(2)} > 0$ with the center at the origin. The objective of the control φ of the pursuer is to minimize the time of attainment of a given terminal set M by the state vector of system (1). The objective of the control v of the evader is to maximize this time. Therefore the payoff of the game is the time of attaining the terminal set.

2.2. Acoustic game

The difference is that the constraint on the control of player E depends on x. It is given by the formula

$$\mathcal{Q}(x) = k(x) Q, \quad k(x) = \min\{|x|, s\}/s, \ s > 0.$$

Here s is a parameter. We have $\mathcal{Q}(x) = Q$ if $|x| \geq s$. The objective of the control φ is to minimize the time of attaining a terminal set M. The objective of the control v is to maximize this time.

For the unification of notation, let us agree that $\mathcal{Q}(x) = Q$ for the classical homicidal chauffeur game.

3. Semipermeable Curves in Differential Games with the Homicidal Chauffeur Dynamics

The families of smooth semipermeable curves are determined from only the dynamics of the system and the bounds on the controls of the players.

We explain now what semipermeable curves mean (see also [4]). Let

$$H(\ell,x) = \min_{|\varphi|\le 1}\max_{v\in\mathcal{Q}(x)} \ell' f(x,\varphi,v) = \max_{v\in\mathcal{Q}(x)}\min_{|\varphi|\le 1} \ell' f(x,\varphi,v), \quad x\in R^2,\ \ell\in R^2. \quad (2)$$

Here $f(x,\varphi,v) = p(x)\varphi + v + g$, $p(x) = (-x_2, x_1)'\cdot w^{(1)}/R$ and $g = (0, -w^{(1)})'$. Fix $x \in R^2$ and consider ℓ such that $H(\ell,x) = 0$. Letting $\varphi^* = \operatorname{argmin}\{\ell' p(x)\varphi\colon |\varphi|\le 1\}$ and $v^* = \operatorname{argmax}\{\ell' v\colon v\in\mathcal{Q}(x)\}$, it follows that $\ell' f(x,\varphi^*,v)\le 0$ holds for any $v\in\mathcal{Q}(x)$, and $\ell' f(x,\varphi,v^*)\ge 0$ holds for any $\varphi\in[-1,1]$. This means that the direction $f(x,\varphi^*,v^*)$, which is orthogonal to ℓ, separates the vectograms $U(v^*) = \{f(x,\varphi,v^*)\colon \varphi\in[-1,1]\}$ and $V(\varphi^*) = \{f(x,\varphi^*,v)\colon v\in\mathcal{Q}(x)\}$ of players P and E. Such a direction is called semipermeable. A smooth curve is called a semipermeable curve if the tangent vector at any point of this curve is a semipermeable direction.

The number of semipermeable directions depends on the form of the function $\ell\to H(\ell,x)$ at the point x. In the case considered, the function $H(\cdot,x)$ is composed of two convex functions:

$$H(\ell,x) = \begin{cases} \max\limits_{v\in\mathcal{Q}(x)} \ell' v + \ell' p(x) + \ell' g, & \text{if } \ell' p(x) < 0 \\ \max\limits_{v\in\mathcal{Q}(x)} \ell' v - \ell' p(x) + \ell' g, & \text{if } \ell' p(x) \ge 0. \end{cases}$$

The semipermeable directions are derived from the roots of the equation $H(\ell,x)=0$. We will distinguish the roots "$-$" to "$+$" and the roots "$+$" to "$-$". When classifying these roots, we suppose that $\ell\in\mathcal{E}$, where $\mathcal{E}$ is the boundary of a convex polygon containing the origin. We say that ℓ_* is a root $-$ to $+$ if $H(\ell_*,x) = 0$, and if $H(\ell,x) < 0$ ($H(\ell,x) > 0$) for $\ell < \ell_*$ ($\ell > \ell_*$) that are sufficiently close to ℓ_*, where the notation $\ell < \ell_*$ means that the direction of the vector ℓ can be obtained from the direction of the vector ℓ_* using a counterclockwise rotation through an angle not exceeding π. The roots $-$ to $+$ and the roots $+$ to $-$ are called roots of the first and second type, respectively.

We denote roots of the first type by $\ell^{(1),i}(x)$ and roots of the second type by $\ell^{(2),i}(x)$. The right index takes the value 1 or 2, and indicates the half-plane $\{\ell\in R^2 : \ell' p(x) < 0\}$ or $\{\ell\in R^2 : \ell' p(x)\ge 0\}$. Due to the above property of the piecewise convexity of the function $H(\cdot,x)$, the equation $H(\ell,x)=0$ can have at most two roots of each type for any given x.

We now describe how the families of smooth semipermeable curves can be constructed.

3.1. Constraint Q on the control of player E does not depend on x

Assume that the constraint Q does not depend on x that is $Q(x) = Q$. Denote

$$A_* = \{(x_1, x_2) : \; x_1 = \frac{v_2 R}{w^{(1)}} - R, \;\; x_2 = -\frac{v_1 R}{w^{(1)}}, \;\; (v_1, v_2)' \in Q\}, \tag{3}$$

$$B_* = \{(x_1, x_2) : \; x_1 = -\frac{v_2 R}{w^{(1)}} + R, \;\; x_2 = \frac{v_1 R}{w^{(1)}}, \;\; (v_1, v_2)' \in Q\}. \tag{4}$$

The set B_* is symmetric to the set A_* with respect to the origin. Let $C_* = A_* \bigcap B_*$.

3.1.1. ROOTS OF EQUATION $H(\ell, x) = 0$ Let us show for all $x \notin C_*$ that the equation $H(\ell, x) = 0$ has at least one root of the first type and one root of the second type. To prove this, it is sufficient to verify that, for any x, there exist vectors $\underline{\ell}$ and $\overline{\ell}$ such that $H(\underline{\ell}, x) < 0$ and $H(\overline{\ell}, x) > 0$.

Let $x \notin A_*$. Then there exists a vector $\widetilde{\ell}$ such that $\widetilde{\ell}'x > \widetilde{\ell}'z$ for any $z \in A_*$. That is

$$-\widetilde{\ell}'x + \max_{z \in A_*} \widetilde{\ell}'z < 0.$$

Denote by $\overline{x}$ the nearest to x point of A_*. The vector $x - \overline{x}$ can be considered as $\widetilde{\ell}$.

Assume $\underline{\ell} = \left(-\widetilde{\ell}_2 R/w^{(1)}, \; \widetilde{\ell}_1 R/w^{(1)}\right)'$. We have

$$H(\underline{\ell}, \; x) \leq \underline{\ell}' \left(\frac{w^{(1)} x_2}{R}, \; \frac{-w^{(1)} x_1}{R}\right)' + \underline{\ell}' g + \max_{v \in Q} \underline{\ell}' v =$$

$$-\widetilde{\ell}'x + \max_{v \in Q} \widetilde{\ell}' \left(\frac{v_2 R}{w^{(1)}} - R, \; \frac{-v_1 R}{w^{(1)}}\right)' = -\widetilde{\ell}'x + \max_{z \in A_*} \widetilde{\ell}'z < 0.$$

Similarly, one can show for $x \notin B_*$ that there exists a vector $\underline{\ell}$ such that $H(\underline{\ell}, x) < 0$. Hence, if $x \notin C_*$, then there exists a vector $\underline{\ell}$ such that $H(\underline{\ell}, x) < 0$.

Consider $\overline{\ell} \neq 0$ such that $\overline{\ell}'p(x) = 0$ and $\overline{\ell}'g \geq 0$. Since $0 \in \text{int}Q$, then $H(\overline{\ell}, x) = \max\{\overline{\ell}'v : \; v \in Q\} + \overline{\ell}'g > 0$. This completes the proof.

Let $x \in \text{int}C_*$. We show that $H(\ell, \; x) > 0$ for all $\ell \neq 0$. Take $\ell \neq 0$. Suppose that $\min\{\ell'p(x)\varphi : \varphi \in [-1, 1]\}$ occurs for $\varphi = -1$ ($\varphi = 1$). It follows from the definition of the set A_* (B_*) that for $x \in \text{int}A_*$ ($x \in \text{int}B_*$), there exists a vector $v_* \in \text{int}Q$ such that $f(x, \; -1, \; v_*) = 0$ ($f(x, \; 1, \; v_*) = 0$). Hence, $H(\ell, \; x) > 0$. Therefore, roots of the first and second type do not exist for $x \in \text{int}C_*$. Due to continuity of H, strict roots do not exist for $x \in \partial C_*$ too.

3.1.2. CASE $C_* = \emptyset$ We consider cones spanned onto the sets A_* and B_* with the apex at the origin. Denote these cones by coneA_* and coneB_*, respectively. The part of coneA_* after deleting the set

$$\{(x_1, x_2) : \; x_1 = \frac{v_2 R}{w^{(1)}\varphi} - R/\varphi, \; x_2 = -\frac{v_1 R}{w^{(1)}\varphi}, \;\; 1 < \varphi < \infty, \;\; (v_1, v_2)' \in Q\}$$

is denoted by A. Similarly, the set B as the part of coneB_* is introduced.

One can find the domains of the functions $\ell^{(j),i}(\cdot)$, $j = 1, 2$, $i = 1, 2$. Figure 1 presents the sets A and B and the domains of the functions $\ell^{(j),i}(\cdot)$, $j = 1, 2$, $i =$

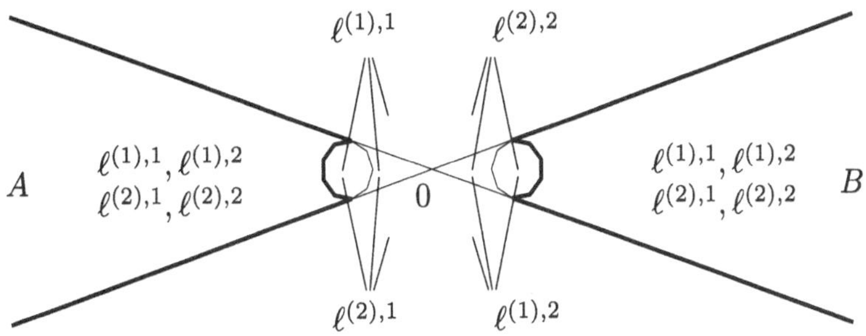

FIGURE 1. Domains of $\ell^{(j),i}$. Set Q does not depend on x; $C_* = \emptyset$.

1, 2, for the case where the set Q is a polygonal approximation of a circle of some radius $w^{(2)}$. The boundaries of A and B are drawn with the thick lines. There exist two roots of the first type and two roots of the second type at each internal point of the sets A and B. For any point in the exterior of A and B, there exist one root of the first type and one root of the second type.

The function $\ell^{(j),i}(\cdot)$ is Lipschitz continuous on any closed bounded subset of the interior of its domain. Consider the two-dimensional differential equation

$$dx/dt = \Pi \ell^{(j),i}(x), \tag{5}$$

where Π is the matrix of rotation through the angle $\pi/2$, the rotation being clockwise or counterclockwise if $j = 1$ or $j = 2$, respectively. Since the tangent vector at each point of the trajectory defined by this equation is a semipermeable direction, the trajectories are semipermeable curves. Therefore player P can keep the state vector x on one side of the curve (positive side), and player E can keep x on the other (negative) side. Equation (5) specifies a family $\Lambda^{(j),i}$ of smooth semipermeable curves. Pictures of the families $\Lambda^{(j),i}$ for the case $C_* = \emptyset$ are given in [7].

3.1.3. CASE $C_* \neq \emptyset$ There are no roots in the set C_*, there are four roots in the set $R^2 \setminus (A_* \bigcup B_*)$, and there are two roots (one root of the first type and one root of the second type) in the rest part of the plane. Figure 2 shows the domains of the functions $\ell^{(j),i}(\cdot)$ for this case. The set Q is a circle of some radius $w^{(2)} > w^{(1)}$. The digits 4, 2 and 0 state the number of roots. Using (5), one can produce the families $\Lambda^{(j),i}$ for the case where $C_* \neq \emptyset$.

The following important property holds true for any point $x \in C_* = A_* \bigcap B_*$: for any $\varphi \in [-1, 1]$ there exists $v \in Q$ such that $f(x, \varphi, v) = 0$. Therefore, in the region C_*, player E can counter any control of player P, so the state remains immovable all the time. Further, if a point x with the above property does not belong to the terminal set M, then M cannot be reached from x. We call regions of such points the superiority sets of player E.

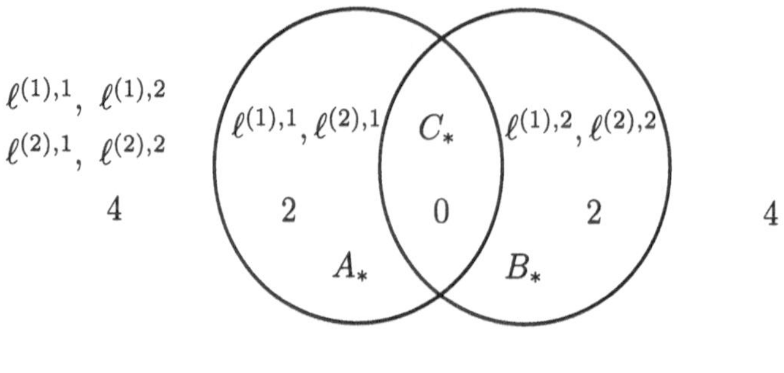

FIGURE 2. Domains of $\ell^{(j),i}(\cdot)$. Set $\mathcal{Q}$ does not depend on x; $C_* \neq \emptyset$.

3.2. Constraint $\mathcal{Q}$ on the control of player E depends on x

Using the form of the domains of $\ell^{(j),i}(\cdot)$ from section 3.1, one can construct the domains for the case $\mathcal{Q}(x) = k(x)Q$. Let us describe briefly how it can be done.

First note that $k(x) = \text{const}$ for the points x of any circumference of some fixed radius with the center at $(0,0)$. It holds $k(x) = 1$ outside the circle of radius s. Take a circumference $\Omega(r)$ of radius r with the center at $(0,0)$. Set $k(r) = \min\{r,\ s\}/s$ and $Q(r) = k(r)Q$. We have $\mathcal{Q}(x) = Q(|x|)$.

Form the sets $A_*(r)$ and $B_*(r)$ substituting the set $Q(r)$ instead of Q in formulae (3) and (4) for A_* and B_*. Let $C_*(r) = A_*(r) \bigcap B_*(r)$. Using $A_*(r)$ and $B_*(r)$, construct domains of $\ell^{(j),i}(\cdot)$, the cases $C_*(r) = \emptyset$ and $C_*(r) \neq \emptyset$ being distinguished. Put the circumference $\Omega(r)$ onto the constructed domains. As a result, a division of the circumference onto arcs is obtained. The number and the type of roots are the same for all points of each arc. This technique is applied for every r in $[0, s]$, and identically named division points are connected. Thus the circle of radius s is divided into parts according to the kinds of roots. Outside this circle, the dividing lines coincide with the lines constructed for the case when $\mathcal{Q}$ does not depend on x.

Since Q is a circle of radius $w^{(2)}$, then $Q(r)$ is a circle of radius $w^{(2)}(r) = \min\{r,\ s\}w^{(2)}/s$. The condition $C_*(r) = \emptyset$ means $w^{(2)}(r) < w^{(1)}$, and the condition $C_*(r) \neq \emptyset$ is equivalent to the relation $w^{(2)}(r) \geq w^{(1)}$. If $w^{(2)}(r) \leq w^{(1)}$, we put the points $x \in \Omega(r)$ onto the domains of Figure 1 constructed for $w^{(2)} = w^{(2)}(r)$. Otherwise, if $w^{(2)}(r) > w^{(1)}$, we put these points onto the domains of Figure 2.

Figures 3 and 4 were constructed in this way for the parameters $w^{(1)} = 1$, $R = 0.8$, $s = 0.75$ and $w^{(2)} = 1.8$ and 2. In Figure 3, two symmetric superiority sets of player E arise, the upper set being denoted by C_U and the lower set by C_L. If we increase $w^{(2)}$, the sets C_U and C_L expand and form a doubly connected region that is denoted by C_* in Figure 4. The number of roots of the equation $H(\ell, x) = 0$ is also given in Figures 3 and 4. A picture of the family $\Lambda^{(1),1}$ corresponding to the parameters of Figure 3 is given in [8].

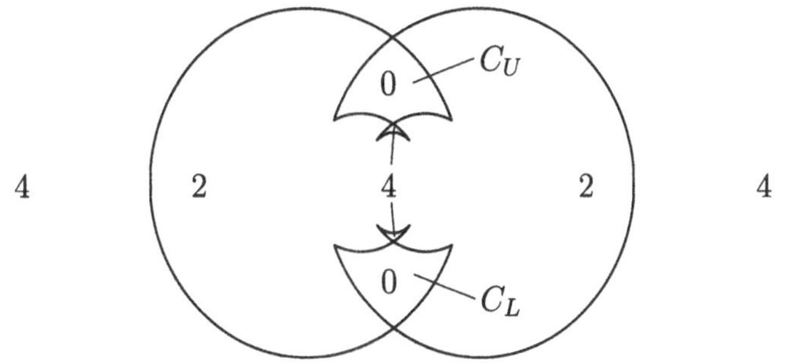

FIGURE 3. Superiority sets C_U and C_L of player E. Set $\mathcal{Q}$ depends on x; $w^{(2)} = 1.8$.

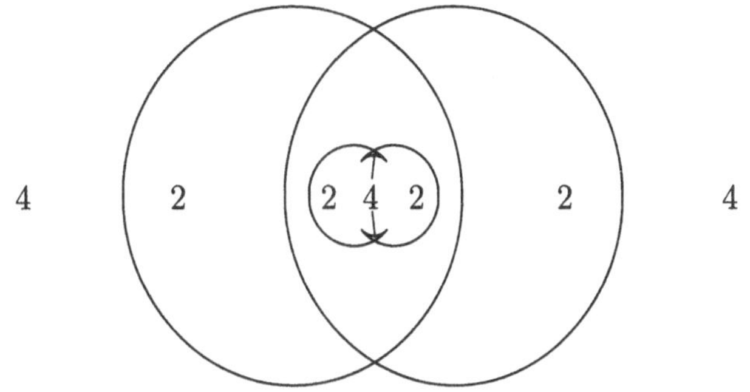

FIGURE 4. Superiority set C_* of player E. Set $\mathcal{Q}$ depends on x; $w^{(2)} = 2$.

4. Formation of Holes in Solvability Sets Due to Superiority Sets

The role of superiority sets in the appearance of holes within the solvability sets will be explained in this section. As noted above, in the case of problem 2.2, there can be one doubly connected superiority set C_* of player E, or two simply connected sets C_U and C_L, or the superiority set can be empty. In the case of problem 2.1, the superiority set of player E can be simply connected or empty.

4.1. Stable set $\hat{D}$

Let D be a closed set. Assume that the objective of player E is to bring the state of the system to the set D. Denote by $\hat{D}$ the solvability set (victory domain of player E) for this problem. It follows from the definition of $\hat{D}$ that E can bring the state of the system to D from any point $x \in \hat{D}$, but player P can prevent the state of the system from approaching the set D for any point $x \notin \hat{D}$. The boundary of $\hat{D}$ is composed of smooth semipermeable curves of the families $\Lambda^{(j),i}$. The sewing points possess the semipermeability property (see [2]). In some cases, a part of the boundary of $\hat{D}$ can coincide with a part of the boundary of D.

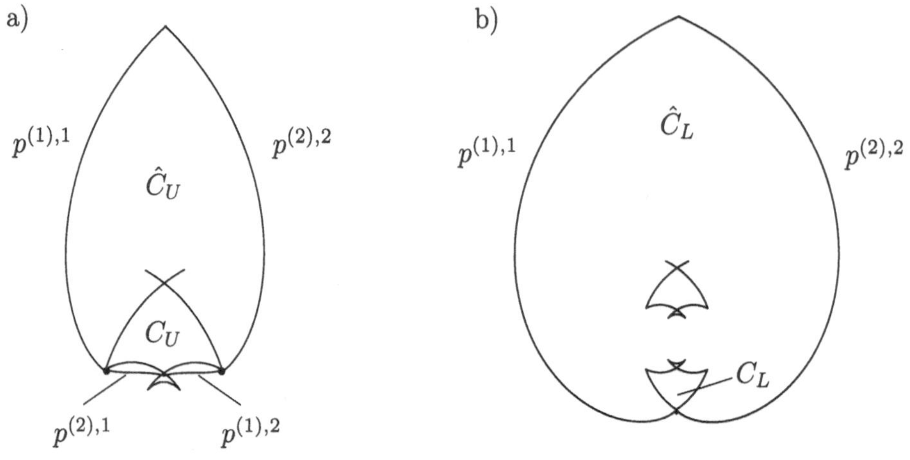

FIGURE 5. Construction of the sets $\hat{C}_U$ and $\hat{C}_L$ on the base of the superiority sets C_U and C_L.

Below, the set C_* or one of the sets C_U and C_L is used as the set D. Since in this case, D is a superiority set of E, it possesses the property of v-stability (see [5] for the definition) or, in other terms, the property of viability for E (see [1]), and the set $\hat{D}$ is v-stable too. This means [5] that player E can hold the trajectories of the system in $\hat{D}$ for infinite time. Hence, if $\hat{D} \bigcap M = \emptyset$, then the time for achieving the terminal set M in the main problem is infinite for any point x in $\hat{D}$. For this reason, level lines of the value function cannot "penetrate" into the set $\hat{D}$.

Due to the simple geometry of the sets D of the problems considered, the sets $\hat{D}$ can be obtained easily using the families of semipermeable curves. For example, Figure 5a presents the configuration of $\hat{C}_U$. The values of parameters correspond to Figure 3. The sewing point of the semipermeable curves $p^{(2),2}$ and $p^{(1),2}$ from the families $\Lambda^{(2),2}$ and $\Lambda^{(1),2}$, and symmetric to it sewing point of the curves $p^{(1),1}$ and $p^{(2),1}$ from the families $\Lambda^{(1),1}$ and $\Lambda^{(2),1}$, lie on the boundary of C_U. In Figure 5b, an example of the set $\hat{C}_L$ for the same values of parameters is given.

Since level lines of the value function cannot penetrate into the set $\hat{D}$ in the case $\hat{D} \bigcap M = \emptyset$, one can try to generate examples with holes in solvability sets using the knowledge of the geometry of the sets $\hat{D}$. We will show that the sets $\hat{C}_U$, but not the sets $\hat{C}_L$ or $\hat{C}_*$ can appear as holes.

4.2. Set $\hat{C}_L$ cannot be a hole

For the set C_L, a collection of expanding v-stable sets can be easily obtained. Figure 6a shows such a collection computed for the set C_L from Figures 3 and 5b. The first set of the collection is $\hat{C}_L$. The boundaries of the sets are formed by semipermeable curves $p^{(1),1}$ and $p^{(2),2}$.

Figure 6b shows the semipermeable curves that form the boundary of some set S from the above collection. The curve $p^{(1),1}$ corresponds to the control $\varphi = 1$,

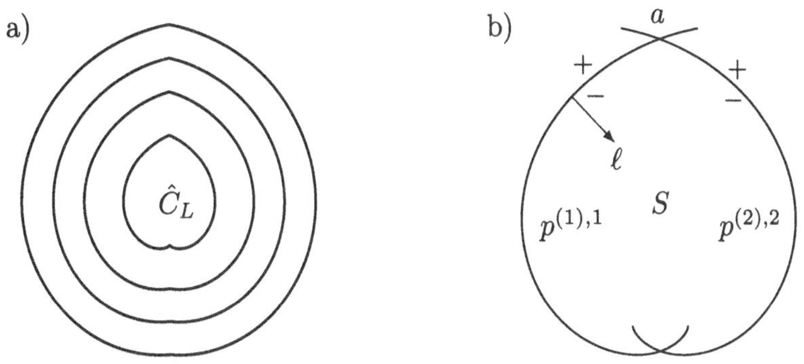

FIGURE 6. a: Collection of expanding v-stable sets for the set C_L.
b: Explanation of v-stability.

but the curve $p^{(2),2}$ corresponds to the control $\varphi = -1$. The sign "+" ("−") marks those sides of curves that player P (E) keeps. The curves $p^{(1),1}$ and $p^{(2),2}$ are faced with negative sides at the intersection point a. The property of v-stability means the following: for any $x \in \partial S$ and any $\varphi \in [-1,\ 1]$ there exists $v \in \mathcal{Q}(x)$ such that the vector $f(x,\ \varphi,\ v)$ is directed inside the set S or it is tangent to the boundary of S at x. For any point $x \in \partial S$ excluding the point a, a vector $v \in \mathcal{Q}(x)$ that gives the maximum in (2) would be appropriate. A normal vector to the curve in the negative side direction is considered as ℓ when computing the maximum in (2). For the point a, the choice of an appropriate v depends on φ.

Let us assume that there exists a hole $\hat{C}_L$ which is located strictly inside the solvability set. It follows from this assumption that: 1) $\hat{C}_L \bigcap M = \emptyset$, 2) for any boundary point x of $\hat{C}_L$, there exist points of the fronts that are arbitrarily close to x. Consider a v-stable set $\bar{S}$ from the expanding collection generated by the set C_L and such that $\bar{S}$ and M have common points on the boundaries of $\bar{S}$ and M only. Take a point x on a front strictly inside the set $\bar{S}$. Such a point exists because the set $\hat{C}_L$ belongs to the interior of the set $\bar{S}$ and the fronts come arbitrarily close to the set $\hat{C}_L$. Then, player E can keep the trajectories of the system within a set $\widetilde{S}$, which is a subset of $\bar{S}$ and contains the point x on its boundary, for infinite time. This contradicts to the fact that x lies on the front and, therefore, player P brings the system to M for a finite time.

Similar arguments are true for the sets $\hat{C}_*$ in the acoustic or classical game.

4.3. Set $\hat{C}_U$ can be a hole

We show that the set C_U cannot generate an expanding collection of v-stable sets.

Denote by $r^\flat = w^{(1)}s/w^{(2)}$ the minimal r for which $C_*(r) \neq \emptyset$. Consider the circle $F(\widetilde{r})$ of radius $\widetilde{r} = r^\flat/2$ with the center at the origin. We have

$$w^{(1)} - w^{(2)}(|x|) \geq w^{(1)} - w^{(2)}(\widetilde{r}) = w^{(1)}/2, \quad x \in F(\widetilde{r}). \tag{6}$$

Let $\xi(r) = -R + w^{(2)}(r)R/w^{(1)}, \quad r \geq 0.$

Since the set C_U is strictly above the axis x_1, then, for any $r \geq 0$, the set $C_*(r)$ does not contain the points of intersection of the circumference $\Omega(r)$ of radius r and the center at the origin with the axis x_1. Hence $r > \xi(r)$.

Let $x^{\#}(r,\ \alpha)$ and $x^{\diamondsuit}(r,\ \alpha)$ are right and left intersection points of the straight line $x_2 = \alpha$ with the circumference $\Omega(r)$, $0 \leq \alpha \leq \widetilde{r}$, $r \geq \widetilde{r}$. Using inequality $r > \xi(r)$, choose positive $\widetilde{\beta}$ and $\widetilde{\alpha}$, $\widetilde{\alpha} \leq \widetilde{r}$, so that $x_1^{\#}(r,\ \alpha) \geq \widetilde{\beta} + \xi(r)$, $r \geq \widetilde{r}$, $0 \leq \alpha \leq \widetilde{\alpha}$. We also obtain $x_1^{\diamondsuit}(r,\ \alpha) \leq -\widetilde{\beta} - \xi(r)$, $r \geq \widetilde{r}$, $0 \leq \alpha \leq \widetilde{\alpha}$.

Denote by $X(\alpha) = \{x :\ 0 < x_2 \leq \alpha\}$, $\alpha \leq \widetilde{\alpha}$, a horizontal strip of the width α over the axis x_1.

Using the inequality for $x_1^{\#}(r, \alpha)$, we obtain $x_1 \geq \widetilde{\beta} + \xi(|x|)$ for the points $x \in X(\widetilde{\alpha})$ on the right of the circle $F(\widetilde{r})$. Hence, it holds

$$\dot{x}_2|_{\varphi=-1} = -x_1 w^{(1)}/R + v_2 - w^{(1)} \leq -\widetilde{\beta} w^{(1)}/R + w^{(1)} -$$

$$w^{(2)}(|x|) + v_2 - w^{(1)} \leq -\widetilde{\beta} w^{(1)}/R \tag{7}$$

for any $v \in \mathcal{Q}(x)$ and $\varphi = -1$. Similarly, using the inequality for $x_1^{\diamondsuit}(r, \alpha)$, we get

$$\dot{x}_2|_{\varphi=1} = x_1 w^{(1)}/R + v_2 - w^{(1)} \leq -\widetilde{\beta} w^{(1)}/R \tag{8}$$

for $x \in X(\widetilde{\alpha})$ on the left of the circle $F(\widetilde{r})$, any $v \in \mathcal{Q}(x)$ and $\varphi = 1$.

If a point $x \in X(\widetilde{\alpha})$ belongs to the circle $F(\widetilde{r})$ and satisfies the inequality $x_1 \geq \xi(\widetilde{r})/2 = -R/4$, then we obtain

$$\dot{x}_2|_{\varphi=-1} = -x_1 w^{(1)}/R + v_2 - w^{(1)} \leq w^{(1)}/4 + v_2 - w^{(1)} \leq -w^{(1)}/4 \tag{9}$$

for any $v \in \mathcal{Q}(x)$ and $\varphi = -1$. It was taken into account here that, using (6), the relation $|v_2| \leq w^{(2)}(|x|) \leq w^{(2)}(\widetilde{r}) = w^{(1)}/2$ holds for $x \in F(\widetilde{r})$. Similarly, if a point $x \in X(\widetilde{\alpha})$ belongs to the circle $F(\widetilde{r})$ and satisfies the inequality $x_1 \leq R/4$, then for any $v \in \mathcal{Q}(x)$ and $\varphi = 1$, we get

$$\dot{x}_2|_{\varphi=1} = x_1 w^{(1)}/R + v_2 - w^{(1)} \leq -w^{(1)}/4. \tag{10}$$

Let $\widetilde{\gamma} = \min\{\widetilde{\beta} w^{(1)}/R,\ w^{(1)}/4\}$. Take positive $\overline{\alpha} \leq \min\{\widetilde{\alpha},\ \widetilde{r}/2\}$ such that

$$\overline{\alpha} w^{(2)}/\widetilde{\gamma} \leq \min\{R/4,\ \widetilde{r}/2\}. \tag{11}$$

Put $\varphi \equiv -1$ for the states $x_0 \in X(\overline{\alpha})$ with $x_{01} \geq 0$. Taking into account (11) and the estimate $\dot{x}_1 = w^{(1)} x_2/R + v_1 \geq v_1 \geq -w^{(2)}$ for $x_2 \geq 0$, we obtain that any trajectory emanated from the point x_0 remains on the right side from the vertical straight line $x_1 = \max\{-R/4,\ -\widetilde{r}/2\}$ within the time $\overline{\alpha}/\widetilde{\gamma}$. Using (7) and (9), we get from here that the trajectory arrives at the axis x_1 within this time. Similarly, setting $\varphi = 1$ and using (8), (10) and (11), one obtains that any trajectory emanated from the point $x_0 \in X(\overline{\alpha}), x_{01} \leq 0$, arrives at the axis x_1 within the time $\overline{\alpha}/\widetilde{\gamma}$ remaining on the left side from the straight line $x_1 = \min\{R/4,\ \widetilde{r}/2\}$.

Thus player P can bring trajectories to the axis x_1 from any initial point x that belongs to the strip $X(\overline{\alpha})$. It follows from this property that $\hat{C}_U \bigcap X(\overline{\alpha}) = \emptyset$. Using the latter, one obtains that there is not any collection of v-stable sets that monotonically expands from the set $\hat{C}_U$ and fills out the whole plane.

5. Holes and the Boundary of Solvability Set

The results of the previous section show that holes located strictly inside solvability sets (victory domains) cannot be formed due to the sets $\hat{C}_*$ and $\hat{C}_L$. On the contrary, the set $\hat{C}_U$ being generated in a way similar to that for $\hat{C}_*$ and $\hat{C}_L$ can be a hole in the victory domain.

In the papers [1], [3], the acoustic game is considered for a terminal set M in the form of a rectangle $\{(x_1, x_2) : -3.5 \leq x_1 \leq 3.5, \ -0.2 \leq x_2 \leq 0\}$. Figure 7 presents level sets of the value function of the acoustic game with the above terminal set and the following values of parameters: $w^{(1)} = 1$, $w^{(2)} = 1.5$ and $s = 0.8$. The computation was done using an algorithm [7], [8] developed by the authors. One can see that in fact the hole coincides with the set $\hat{C}_U$.

It is emphasized in [1], [3] that the victory domain in similar examples with holes cannot be obtained using semipermeable curves (barriers) emitted from the boundary of the terminal set only. Now this conclusion can be formulated more precisely: the boundary of the victory domain is composed not only of semipermeable curves issued from the boundary of the terminal set but also of semipermeable curves emitted from the boundary of the set C_U.

If one increases $w^{(2)}$, the hole is being inflated and becomes "open" (see, for example, Figure 22 in [8]). The boundary of the victory domain transforms into

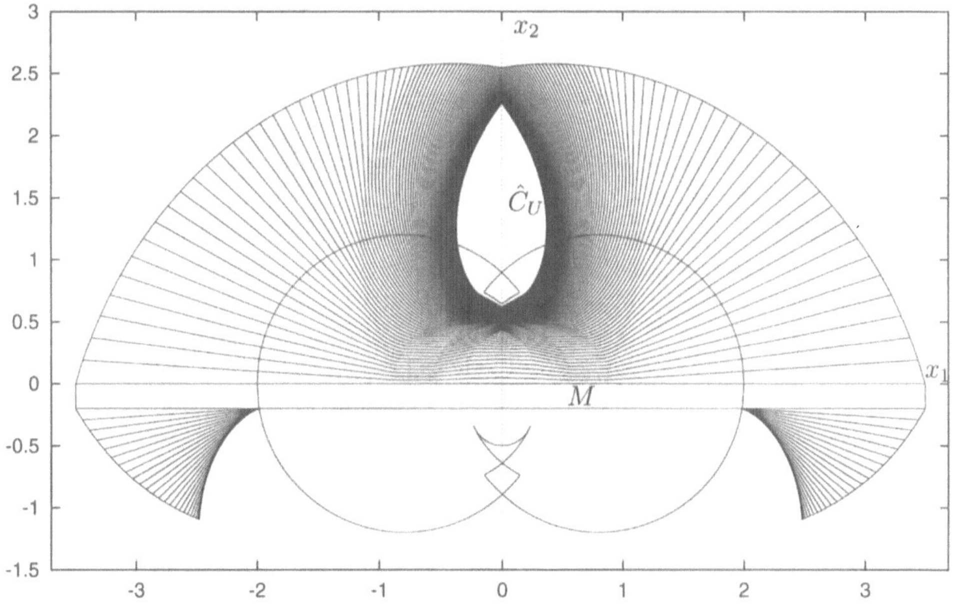

FIGURE 7. Level sets for $w^{(2)} = 1.5$; 746 upper fronts, 340 lower fronts, every 10th front is plotted.

a connected curve but even in this case, it is composed of semipermeable curves emitted both from the boundary of the terminal set and boundary of the set C_U.

The following question can be formulated. Does an example with the homicidal chauffeur dynamics exist where a hole, which is strictly inside the victory domain, does not coincide with the set $\hat{C}_U$? (In this paper, it is shown that such holes cannot coincide with the sets $\hat{C}_*$ and $\hat{C}_L$.)

Acknowledgement. This research was partially supported by the Russian Foundation for Basic Researches under Grant No. 00-01-00348.

References

[1] P. Cardaliaguet, M.Quincampoix and P.Saint-Pierre, *Numerical Methods for Optimal Control and Differential Games*, Ceremade CNRS URA 749, University of Paris-Dauphine (1995).

[2] P. Cardaliaguet, *Nonsmooth Semipermeable Barriers, Isaacs' Equation, and Application to a Differential Game with One Target and Two Players*, Applied Mathematics and Optimization, **36** (1997), 125–146.

[3] P. Cardaliaguet, M.Quincampoix and P.Saint-Pierre, *Set-Valued Numerical Analysis for Optimal Control and Differential Games*, in: M.Bardi, T.E.S.Raghavan, and T.Parthasarathy, Eds., Stochastic and Differential Games: Theory and Numerical Methods, Annals of the International Society of Dynamic Games (Birkhäuser, Boston) **4** (1999), 177–247.

[4] R. Isaacs, *Differential Games*, John Wiley, New York, **1965**.

[5] N.N. Krasovskii and A.I.Subbotin, *Game-Theoretical Control Problems*, Springer-Verlag, New York, **1988**.

[6] A.W. Merz, *The Homicidal Chauffeur – a Differential Game*, PhD Dissertation, Stanford University (1971).

[7] V.S. Patsko and V.L. Turova, *Homicidal chauffeur game. Computation of level sets of the value function*, in: Preprints of the Eighth International Symposium on Differential Games and Applications held in Maastricht, July 5–8, 1998, 466–473.

[8] V.S. Patsko and V.L. Turova, *Acoustic homicidal chauffeur game*, in: M.J.D. Powell and S.Scholtes, Eds., System Modelling and Optimization: methods, theory, and applications: 19th IFIP TC7 Conference on System Modelling and Optimization, July 12–16, 1999, Cambridge, UK (Kluwer Academic Publishers, Boston-Dordrecht-London) (2000), 227–249.

V.S.Patsko,
Institute of Mathematics and Mechanics,
S. Kovalevskaya str. 16,
620219 Ekaterinburg
Russia
E-mail address: patsko@imm.uran.ru

V.L.Turova
Research Center caesar,
Friedensplatz 16,
53129 Bonn,
Germany
turova@caesar.de

International Series of Numerical Mathematics, Vol. 139, 203–216

Sensitivity Calculations for 2D-Optimization of Turbomachine Blading

Peter Rentrop, Sven-Olaf Stoll, and Utz Wever

Abstract. In power plants the blading of turbomachinery is crucial for efficiency considerations. For the aerodynamic optimization of the blades, the profile of the turbine blade is described by Beziér polynomials. Their coefficients are used as design variables in a nonlinear optimization procedure. The steam flow is modelled by the 2D Euler equations. This paper deals with the question of sensitivity calculations, which is needed for the gradient based optimization algorithm. Several discretizations of the sensitivity equation are discussed and compared. Numerical results and simulations are based on CLAWPACK by LeVeque.

1. Introduction

The last blade row of large steam turbines shares considerably in the total power output, e.g. 7% of a 1000MW coal-fired power plant. Therefore, any improvements of the energy conversion from steam flow into rotation can be very profitable. In order to enhance the efficiency of the last blade row, the corresponding transonic and supersonic flow field must be considered in more detail. The steam flow through the blade row suffers from the occurence of shock-waves. These shock-waves produce high losses of efficiency. By changing the profile shock-waves can nearly be avoided or remarkably reduced in their strengths. Efficient optimization algorithms are based on gradient information. Thus a main task for setting up the optimization procedure is to provide the flow sensitivities, see [BB95, BB97].

This paper discusses discretizations for the sensitivity equation. Several approaches such as splitting methods and generalized advection methods are presented and compared. Numerical results are presented for the shock-tube problem and for a profile of turbomachine blading within the framework of [BWZ].

2. Modeling of the Profile Geometry

The profile of the turbine blades is described by two Beziér curves $X_1(s), X_2(t)$, which have a geometric C^2-connection at the nose:

$$X_1(s) = \begin{pmatrix} x_1(s) \\ y_1(s) \end{pmatrix} = \sum_{i=0}^{m} \begin{pmatrix} \alpha_i \\ \beta_i \end{pmatrix} B_i^m(s), \quad s \in [0,1], \tag{1}$$

and

$$X_2(t) = \begin{pmatrix} x_2(t) \\ y_2(t) \end{pmatrix} = \sum_{i=0}^{n} \begin{pmatrix} \alpha_{i+m} \\ \beta_{i+m} \end{pmatrix} B_i^n(t), \quad t \in [0,1]. \tag{2}$$

It holds $(\alpha_m, \beta_m)^T = X_1(1) = X_2(0)$. The $2(n+m+1)$ Bézier coefficients (α_i, β_i) are the design parameters of an optimization problem. Figure 1 shows the profile of a turbine blade and its corresponding Bézier polygon.

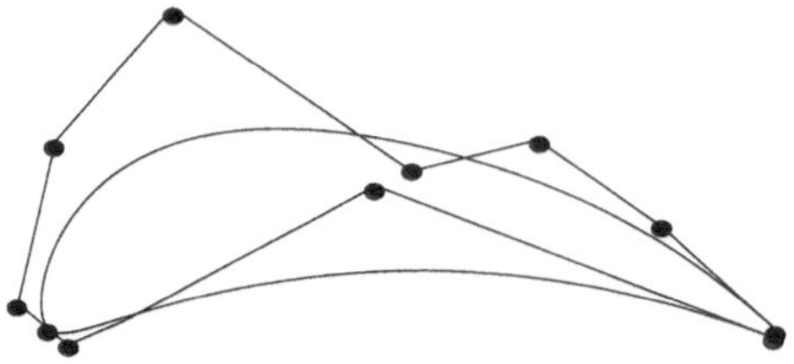

FIGURE 1. Blade profile and corresponding Bézier Polygon

Bézier curves are given in the parameter description $X(t) = \sum_{i=0}^{n} \mathbf{b}_i B_i^n(t)$ based on the Bernstein polynomials

$$B_i^n(t) = \binom{n}{i} (1-t)^{n-i} t^i.$$

The Bézier points $\mathbf{b}_i = (\alpha_i, \beta_i)^T$ give information about the geometric properties, their connection Bézier polygon. The curve lies in the convex hull of the associated polygon and it has a variation diminishing property. Bézier curves can be evaluated numerically stable by the de Casteljau algorithm, see [HL92]. Theoretically they are a special case of B-Splines.

3. Euler Gas Equation

The flow around the turbine blade is described by the Euler equations, a system of nonlinear hyperbolic conservation laws. The formulation depends on the choice of the variables. Density $\rho(x,y,t)$, velocities $u(x,y,t), v(x,y,t)$ and static pressure $p(x,y,t)$ are called the *primitive* variables. Usually they are used for the physical boundary conditions. With density, momentum in x-direction $m = \rho u$, momentum in y-direction $n = \rho v$ and total energy E, the so-called *conservative* variables, the gas equations can be written as a conservation law

$$F(\mathbf{q}) = \frac{\partial \mathbf{q}}{\partial t} + \frac{\partial}{\partial x} F(\mathbf{q}) + \frac{\partial}{\partial y} G(\mathbf{q}) = 0 \tag{3}$$

with

$$\mathbf{q} = \begin{pmatrix} \rho \\ m \\ n \\ E \end{pmatrix}, \quad F(\mathbf{q}) = \begin{pmatrix} m \\ m\,u + p \\ m\,v \\ (E+p)\,u \end{pmatrix}, \quad G(\mathbf{q}) = \begin{pmatrix} n \\ n\,u \\ n\,v + p \\ (E+p)\,v \end{pmatrix}. \tag{4}$$

Equations (3), (4) describe the physical conservation of mass, momentum and energy.

To complete these equations, a supplementary equation defining the thermodynamic property must be added. For an ideal gas it holds

$$E = \frac{p}{\gamma - 1} + \frac{1}{2}\rho(u^2 + v^2), \tag{5}$$

where $\gamma = c_p/c_v$ is the ratio of specific heat coefficients under constant pressure c_p and constant volume c_v.

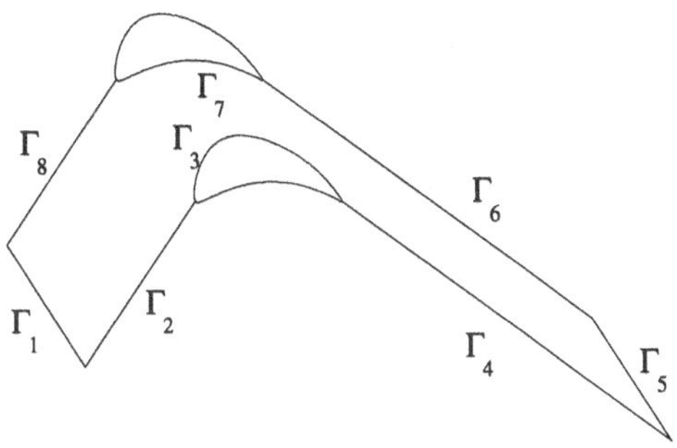

FIGURE 2. Simulation domain: Inlet Γ_1, Outlet Γ_5, periodic boundary conditions (Γ_2, Γ_8) and (Γ_4, Γ_6), wall condition Γ_3, Γ_7

Figure 2 describes the simulation domain between two blades. The fluid enters at the inlet Γ_1 and leaves at the outlet Γ_5. At (Γ_2, Γ_8) and (Γ_4, Γ_6) periodic boundary conditions to neighboured blades are prescribed. At the inlet Γ_1 the temperature T, the total pressure P_t and the angle of the flow $\angle(u, v)$ are given. At the outlet Γ_8 the static pressure p is given. At the profile sections Γ_3, Γ_7 the slip condition

$$(u, v)(-\dot{y}, \dot{x})^T = 0 \tag{6}$$

holds. The geometry of Γ_3, Γ_7 is determined by the shape of the profile, i. e. the Bézier coefficients. The total pressure P_t is discussed in the next section.

4. Optimization Problem

According to [Tra82] a good size to measure the efficiency of a turbine blade is the total pressure. It is a direct function of the fluid variables:

$$P_t = p\left(\frac{\gamma - 1}{2}M^2 + 1\right)^{\frac{\gamma}{\gamma-1}}, \tag{7}$$

where

$$M = \sqrt{\frac{(u^2 + v^2)\rho}{\gamma p}} \tag{8}$$

is the Mach number. The total pressure is prescribed at the inlet and it can be computed at the outlet. The difference between the total pressure at the inlet and the normalized total pressure at the outlet defines the total pressure loss, which forms the objective function of the optimization problem

$$\min_{\alpha} P_t(\mathbf{q}(\alpha)) := P_t|_{\Gamma_1} - \frac{\int_{\Gamma_5} \rho P_t \langle (u,v), \mathbf{n}_5 \rangle \, ds}{\int_{\Gamma_5} \rho \langle (u,v), \mathbf{n}_5 \rangle \, ds}, \tag{9}$$

where $\mathbf{n}_5$ is the outer normal direction at the outlet and α the vector of shape parameters $(\ldots, \alpha_i, \beta_i, \ldots)^T$. The total pressure loss is caused by the shocks in the transonic case.

Equation (9) hast to be minimized under some additional constraints like given range for the centre of gravity, given range of area of the blade, given range of momentum of inertia and some limitations on the curvature of the blade.

The Sequential Quadratic Programming (SQP) methods is applied as an efficient algorithm for nonlinear optimization problems under constraints, see [Spe93]. The following terms must be provided:

- The Jacobian of the constraints,
- The gradient of the objective function: $\nabla P_t(\mathbf{q}(\alpha))$.

The Jacobian of the constraints can be achieved by explicit differentiation, the evaluation of the gradient of the objective function $\nabla P_t(\mathbf{q}(\alpha))$ is discussed in the next section. The optimization problem follows the approach in [BWZ].

5. Sensitivity Equation

The total pressure is an explicit function of the state variables. Assuming the states to be continuous, its gradient with respect to α is given by

$$\nabla P_t(\mathbf{q}(\alpha)) = \frac{\partial P_t}{\partial \mathbf{q}} \frac{\partial \mathbf{q}}{\partial \alpha}. \tag{10}$$

$\frac{\partial P_t}{\partial \mathbf{q}}$ can easily be computed. The evaluation of $\frac{\partial \mathbf{q}}{\partial \alpha}$ by finite differences

$$\frac{\partial \mathbf{q}}{\partial \alpha} \approx \frac{\mathbf{q}(\alpha + \Delta\alpha) - \mathbf{q}(\alpha)}{\Delta\alpha}$$

is too imprecise and inefficient. It requires in addition a solution of the flow equations with a pertubed geometry Γ_3, Γ_6.

Explicit differentiation of the Euler gas equations yields

$$\frac{\partial}{\partial \alpha}\frac{\partial \mathbf{q}}{\partial t} + \frac{\partial}{\partial \alpha}\frac{\partial}{\partial x} F(\mathbf{q}) + \frac{\partial}{\partial \alpha}\frac{\partial}{\partial y} G(\mathbf{q}) = 0. \tag{11}$$

By interchanging the order of derivatives with respect to α and with respect to (x, y), it holds

$$\frac{\partial \mathbf{s}}{\partial t} + \frac{\partial}{\partial x}\left(\frac{dF(\mathbf{q})}{dq}\mathbf{s}\right) + \frac{\partial}{\partial y}\left(\frac{dG(\mathbf{q})}{dq}\mathbf{s}\right) = 0 \tag{12}$$

with

$$\mathbf{s} = \frac{\partial \mathbf{q}}{\partial \alpha} = \begin{pmatrix} \partial\rho/\partial\alpha \\ \partial m/\partial\alpha \\ \partial n/\partial\alpha \\ \partial E/\partial\alpha \end{pmatrix} = \begin{pmatrix} \rho_\alpha \\ m_\alpha \\ n_\alpha \\ E_\alpha \end{pmatrix} \tag{13}$$

The derivative of the Euler gas equations with respect to one of the input parameters is noted as *sensitivity equation.* This input parameter may be an inlet/outlet- or a geometry parameter. The sensitivities as a solution of the sensitivity equation describe the variation of the flow field due to a varying parameter. Especially there holds the relation between the sensitivities of the pressure p_α and the total energy E_α

$$p_\alpha = (\gamma - 1)\left(E_\alpha - \frac{1}{2}\rho_\alpha(u^2 + v^2) - \rho(uu_\alpha + vv_\alpha)\right)$$

with $u_\alpha = \frac{m_\alpha \rho - m\rho_\alpha}{\rho^2}$ and $v_\alpha = \frac{n_\alpha \rho - n\rho_\alpha}{\rho^2}$.

The initial and boundary conditions must also be differentiated with respect to α. The main boundary condition for the Euler gas equation in this context is the wall condition

$$\langle \mathbf{w}, \mathbf{n} \rangle = 0, \tag{14}$$

where $\mathbf{n} = (n_x, n_y)^T$ is the normal to the profile and $\mathbf{w} = (u, v)^T$ is the velocity. The derivative with respect to the design parameter α gives

$$\begin{aligned} D_\alpha\langle \mathbf{w}, \mathbf{n} \rangle &= \langle \mathbf{w}, \frac{\partial \mathbf{n}}{\partial \alpha} \rangle + \langle D_\alpha \mathbf{w}, \mathbf{n} \rangle \\ &= \langle \mathbf{w}, \frac{\partial \mathbf{n}}{\partial \alpha} \rangle + \langle \frac{\partial \mathbf{w}}{\partial x}\frac{\partial x}{\partial \alpha} + \frac{\partial \mathbf{w}}{\partial y}\frac{\partial y}{\partial \alpha} + \frac{\partial \mathbf{w}}{\partial \alpha}, \mathbf{n} \rangle \\ &= \langle \mathbf{w}, \frac{\partial \mathbf{n}}{\partial \alpha} \rangle + \langle \frac{\partial \mathbf{w}}{\partial x}\frac{\partial x}{\partial \alpha} + \frac{\partial \mathbf{w}}{\partial y}\frac{\partial y}{\partial \alpha}, \mathbf{n} \rangle + \langle \frac{\partial \mathbf{w}}{\partial \alpha}, \mathbf{n} \rangle \\ &= 0, \end{aligned}$$

where D_α means the total derivative. Rearranging the terms gives the inhomogeneous wall condition denoted in Table 1.

Again equation (12) is a conservation law, a linear hyperbolic pde, which depends on the solution of the Euler gas equation.

Applying the chain rule on equation (12) yields

$$\frac{\partial s}{\partial t} + \frac{dF(\mathbf{q})}{d\mathbf{q}}\frac{\partial s}{\partial x} + \frac{dG(\mathbf{q})}{d\mathbf{q}}\frac{\partial s}{\partial y} = -\left(\frac{d^2F(\mathbf{q})}{d\mathbf{q}^2}\frac{\partial \mathbf{q}}{\partial x}s + \frac{d^2G(\mathbf{q})}{d\mathbf{q}^2}\frac{\partial \mathbf{q}}{\partial y}s\right), \tag{15}$$

leading to an advection equation with source term.

The Euler equations have been differentiated and the order of differentiation has been interchanged, which only is possible if all derivatives

$$\frac{\partial F}{\partial x}, \frac{\partial F}{\partial \alpha}, \frac{\partial G}{\partial y}, \frac{\partial G}{\partial \alpha}$$

exist and are continuous. The flow field consists of smooth regions, which are seperated by jumps. At these jumps, the sensitivies contain δ-peaks, which of course generate numerical problems. For control problems of hyperbolic conservation laws a weak solution can be defined and it can be shown, that the numerical approximation converges to the exact solution, see [Ulb00].

	Euler gas equation	sensitivity equation
Γ_1	$T,\ P_t,\ \angle(u,v)$	$T_\alpha = 0,\ P_{t,\alpha} = 0,\ \angle(u,v)$
Γ_5	p	$p_\alpha = 0$
$\Gamma_3,\ \Gamma_7$	$\langle \mathbf{w}, \mathbf{n} \rangle = 0$	$\langle \mathbf{w}_\alpha, \mathbf{n} \rangle = -\langle \mathbf{w}_x x_\alpha + \mathbf{w}_y y_\alpha, \mathbf{n} \rangle - \langle \mathbf{w}, \mathbf{n}_\alpha \rangle$
$\Gamma_2,\ \Gamma_8$	$\mathbf{q}_{\Gamma_2} = \mathbf{q}_{\Gamma_8}$	$\mathbf{q}_{\alpha,\Gamma_2} = \mathbf{q}_{\alpha,\Gamma_8}$
$\Gamma_4,\ \Gamma_6$	$\mathbf{q}_{\Gamma_4} = \mathbf{q}_{\Gamma_6}$	$\mathbf{q}_{\alpha,\Gamma_4} = \mathbf{q}_{\alpha,\Gamma_6}$

TABLE 1. Boundary conditions

6. Numerical Methods for Conservation Laws

Theoretically the solution of hyperbolic pde's is characterized by discontinuities like shocks, contact discontinuities or rarefaction waves. A conservative formulation leads to weak solutions. Their uniqueness can be achieved via entropy conditions, which enforce the physically reasonable solution.

The simplest example for a nonlinear conservation law shows that consistency and stability is not sufficient: Burger's equation

$$u_t + \left(\frac{1}{2}u^2\right)_x = 0 \tag{16}$$

with initial values (Cauchy problem)

$$u_0(x) = u(x,0) = \begin{cases} u_l, & x < 0, \\ u_r, & x \geq 0. \end{cases} \tag{17}$$

A natural discretization gives

$$U_j^{n+1} = U_j^n - \frac{\Delta t}{\Delta x} U_j^n (U_j^n - U_{j-1}^n), \tag{18}$$

where U_j^n is the numerical approximation to the solution $u(x_j, t_n)$ at the discrete grid points $x_j = j\Delta x$ and time $t_n = n\Delta t$. The initial condition

$$U_j^0 = \begin{cases} 1, & j < 0, \\ 0, & j \geq 0 \end{cases} \tag{19}$$

enforces $U_j^n = U_j^0$ for all n, j, Δt and Δx. For all $\Delta t, \Delta x$ the scheme converges to $u(x,t) = u_0(x)$, which is no weak solution of Burger's equation.

In general a conservation law has the form $u_t + f(u)_x = 0$. The numerical scheme

$$\mathbf{U}_j^{n+1} = \mathbf{U}_j^n - \frac{\Delta t}{\Delta x}\left(\mathbf{F}(\mathbf{U}_{j-p}^n, \ldots, \mathbf{U}_{j+q}^n) - \mathbf{F}(\mathbf{U}_{j-p-1}^n, \ldots, \mathbf{U}_{j+q-1}^n)\right) \tag{20}$$

is in conservation form. $\mathbf{F}$ is called the numerical flux function which is consistent if

$$\mathbf{F}(\hat{\mathbf{u}}, \ldots, \hat{\mathbf{u}}) = f(\hat{\mathbf{u}}) \qquad \hat{\mathbf{u}} \in \mathbf{R}. \tag{21}$$

The following theorem ensures, that a scheme converges to a physically correct weak solution.

Theorem 1 (Lax and Wendroff). *If a sequence $\{\mathbf{U}_i(x,t)\}$ of solutions computed by a consistent and conservative scheme with $\Delta t, \Delta x \to 0$ for $i \to \infty$ converges to a function $\mathbf{u}(x,t)$, then $\mathbf{u}(x,t)$ is a weak solution of the conservation law.*

6.1. Diffusion and dispersion

Two classical schemes are the first-order Lax-Friedrichs-scheme with

$$\mathbf{F}(\mathbf{U}_{j+1}, \mathbf{U}_j) = \frac{h}{2k}\left(\mathbf{U}_j - \mathbf{U}_{j+1}\right) + \frac{1}{2}\left(f(\mathbf{U}_j) + f(\mathbf{U}_{j+1})\right), \tag{22}$$

where $k = \Delta t, h = \Delta x$, and the second-order Lax-Wendroff-scheme

$$\mathbf{F}(\mathbf{U}_{j+1}, \mathbf{U}_j) = \frac{h}{2k}\left(f(\mathbf{U}_{j+1}) - f(\mathbf{U}_j)\right) + \frac{k^2}{2h^2}\left(A_{j+1/2}(f(\mathbf{U}_{j+1}) - f(\mathbf{U}_j))\right), \tag{23}$$

where $A_{j+1/2}$ is the Jacobian evaluated at $\frac{1}{2}(\mathbf{U}_{j+1} + \mathbf{U}_j)$.
Applying (22) to the linear advection equation $\mathbf{u}_t + A\mathbf{u}_x = 0$ and replacing the numerical approximation by the exact values yields

$$\mathbf{u}_t + \mathbf{A}\mathbf{u}_x = \frac{h^2}{2k}\left(\mathbf{I} - \left(\frac{k}{h}\mathbf{A}\right)^2\right)\mathbf{u}_{xx} + \mathcal{O}(h^2) + \mathcal{O}(k^2). \tag{24}$$

The Lax-Friedrichs-scheme is of first order for the conservation law, but of second order for an equation with additional diffusion. This is called *numerical diffusion.*

In general methods with positive diffusion smear the discontinuities. Figure 3 shows the exact solution of a scalar advection law and the numerical approximations by the Lax-Friedrichs- and the Lax-Wendroff-scheme.

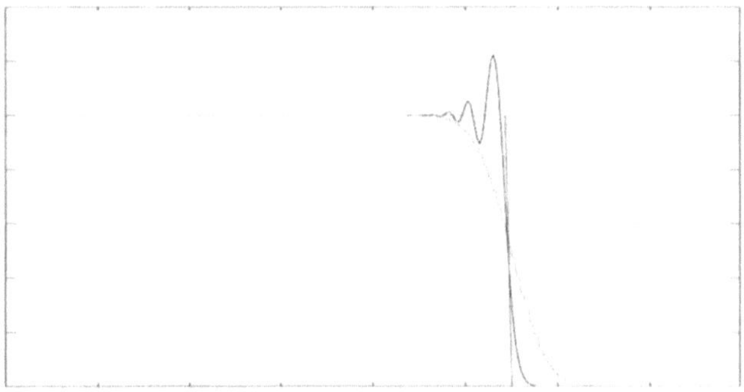

FIGURE 3. Diffusion and dispersion

The Lax-Wendroff-scheme shows another phenomenon. In smooth regions it is a second order approximation, but near discontinuities it produces strong oscillations. Actually it approximates of third order the equation

$$\mathbf{u}_t + \mathbf{A}\mathbf{u}_x = \frac{h^2}{6}\mathbf{A}\left(\frac{k^2}{h^2}\mathbf{A}^2 - \mathbf{I}\right)\mathbf{u}_{xxx}. \tag{25}$$

According to the dispersion term this effect is called *numerical dispersion.*

To handle these difficulties, usually limiters are used, see [LeV92]. In smooth regions the solution is approximated with higher order schemes, in regions of greater changes the numerical method switches to a first order scheme. The so-called ENO-schemes (essentially non oscillating) try to avoid, that the stencils extend over the discontinuities, see [Shu97]. Another approach to stabilize methods of higher order is the use of artificial viscosity terms.

We have used the two-step Richtmyer-Lax-Wendroff scheme

$$\mathbf{q}_{i+1/2}^{n+1/2} = \frac{1}{2}\left(\mathbf{q}_i^n + q_{i+1}^n\right) - \frac{\Delta t}{2\Delta x}\left(f(\mathbf{q}_{i+1}^n) - f(\mathbf{q}_{i-1}^n)\right) \tag{26}$$

$$\mathbf{q}_i^{n+1} = \mathbf{q}_i^n - \frac{\Delta t}{\Delta x}\left(f(\mathbf{q}_{i+1/2}^{n+1/2}) - f(\mathbf{q}_{i-1/2}^{n+1/2})\right) \tag{27}$$

and the method of Godunov

$$\mathbf{q}_{i+1/2}^{n+1/2} = \frac{1}{2}\left(\mathbf{q}_i^n + q_{i+1}^n\right) - \frac{\Delta t}{\Delta x}\left(f(\mathbf{q}_{i+1}^n) - f(\mathbf{q}_{i-1}^n)\right) \tag{28}$$

$$\mathbf{q}_i^{n+1} = \mathbf{q}_i^n - \frac{\Delta t}{\Delta x}\left(f(\mathbf{q}_{i+1/2}^{n+1/2}) - f(\mathbf{q}_{i-1/2}^{n+1/2})\right). \tag{29}$$

To stabilize these schemes an artificial viscosity term was added:

$$\mathbf{q}_i^{n+1} = \mathbf{q}_i^{n+1} + \frac{\nu\Delta t}{\Delta x}\Delta_1\left[|\Delta_1\mathbf{q}_{i+1}^{n+1}|\Delta_1\mathbf{q}_{i+1}^{n+1}\right], \tag{30}$$

where $\Delta_1\mathbf{q}_i^n = \mathbf{q}_i^n - \mathbf{q}_{i-1}^n$ and ν is a constant.

Another method is the Roe linearization, see [Roe81], where the modified linear conservation law

$$\hat{\mathbf{q}}_t + \hat{A}(\mathbf{q}_L, \mathbf{q}_R)\,\hat{\mathbf{q}}_x = 0 \tag{31}$$

is solved. For $\hat{A}$ the following conditions must hold:

- $\hat{A}(\mathbf{q}_L, \mathbf{q}_R)(\mathbf{q}_R - \mathbf{q}_L) = f(\mathbf{q}_R) - f(\mathbf{q}_L)$,
- $\hat{A}(\mathbf{q}_L, \mathbf{q}_R)$ is diagonalizable with real eigenvalues and
- $\hat{A}(\mathbf{q}_L, \mathbf{q}_R) \longrightarrow \frac{df}{d\mathbf{q}}(\bar{\mathbf{q}}), \quad \mathbf{q}_L, \mathbf{q}_R \longrightarrow \bar{\mathbf{q}}$.

The first condition ensures, that the method is conservative, the second that the linearized equation is hyperbolic and the third guarantees the consistence.

6.2. Special schemes for the sensitivity equation

The sensitivity equation (12) has the form of an advection equation:

$$\mathbf{q}_t + (A(x)\,\mathbf{q})_x = 0. \tag{32}$$

The problem of the numerical schemes (26,27) and (28,29) is that they are of second order and thus oscillations at the discontinuities may occur. A Roe scheme (31) cannot be used because there exists no Roe matrix for (32). Furthermore the structure of nonlinearity should be used. For scalar advection equations for incompressible flow exist a lot of algorithms. We generalized LeVeque's algorithm for scalar advection equations, see [LeV95], to systems. In LeVeque's `CLAWPACK` software the solution is updated by

$$\mathbf{q}_i^{n+1} = \mathbf{q}_i^n - \frac{\Delta t}{\Delta x}\left(\mathcal{A}^-\Delta\mathbf{q}_{i+1}^n + \mathcal{A}^+\Delta\mathbf{q}_i^n\right) \tag{33}$$

The numerical flux has the form $F(\mathbf{q}_i, \mathbf{q}_{i-1}) = f(\mathbf{q}_i) - \mathcal{A}^+\Delta\mathbf{q}_i$ and $F(\mathbf{q}_{i+1}, \mathbf{q}_i) = f(\mathbf{q}_i) + \mathcal{A}^-\Delta\mathbf{q}_{i+1}$. We provide the terms $\mathcal{A}^-\Delta\mathbf{q}_i$, $\mathcal{A}^+\Delta\mathbf{q}_i$ and λ_i^p as follows:

$$\begin{aligned}
\lambda_i^p &= \lambda^p\left(A(x_{i-1/2})\right), & (34)\\
\mathcal{A}^-\Delta q_i &= f_i^0 - A(x_{i-1})\,q_{i-1}, & (35)\\
\mathcal{A}^+\Delta q_i &= A(x_i)\,q_i - f_i^0, & (36)\\
f_i^0 &= A^+(x_{i-1/2})\,q_{i-1} + A^-(x_{i-1/2})\,q_i. & (37)
\end{aligned}$$

This algorithm is of first order consistent, conservative and stable for CFL numbers less than 1.

Another approach is applying a splitting method to equation (15):

$$\mathbf{s}_t + A(\mathbf{q})\mathbf{s}_x = 0 \tag{38}$$

and

$$\mathbf{s}_t = -\frac{\partial F(\mathbf{q},\mathbf{s})}{\partial \mathbf{q}}\mathbf{q}_x. \tag{39}$$

The source term is discretized by

$$\mathbf{s}_i^{n+1/2} = \mathbf{s}_i^n - \frac{\Delta t}{\Delta x}\frac{\partial F(\mathbf{q}_i,\mathbf{s}_i)}{\partial \mathbf{q}}(\mathbf{q}_i - \mathbf{q}_{i-1}). \tag{40}$$

The advantage is, that the decomposition of the Roe matrix can be recovered for the computation of the sensitivity.

7. Numerical Results

First a one dimensional Riemann shock-tube problem for the Euler equations is studied:

$$\mathbf{q}_t + F(\mathbf{q})_x = 0, \quad \mathbf{q}(x,0) = \begin{cases} \mathbf{q}_L & \text{if} \quad x < 0, \\ \mathbf{q}_R & \text{if} \quad x < 0. \end{cases} \tag{41}$$

with initial data

$$\begin{aligned} \rho_L &= 1.0, & \rho_R &= 0.125, \\ u_L &= 0.0, & u_R &= 0.0, \\ p_L &= 1.0 & p_R &= 0.1. \end{aligned}$$

The exact piecewise defined solution of this Riemann problem can be found in [LeV92]. By differentiating them with respect to the left initial pressure an exact sensitivity is derived, see [Sto98]. Figure 4 shows the exact solution of the Riemann problem and the exact sensitivity at time $t = 0.148$. Figure 5 shows approximated

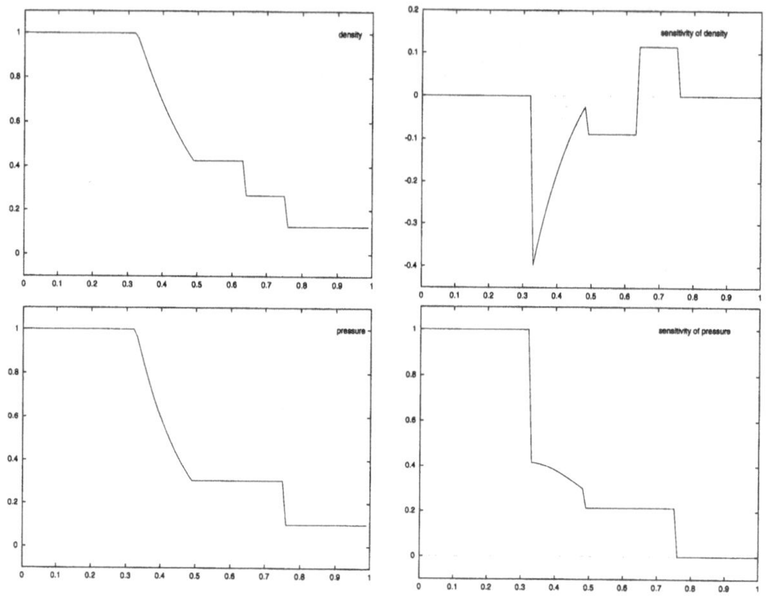

FIGURE 4. Exact density and pressure of the shock-tube problem (left column) and their sensitivity with respect to the pressure at the left side of the diaphragm.

sensitivities, which are computed by several numerical schemes. In the case of the splitting scheme we set the source term equal to zero, if the jump is greater than a given tolerance. This limiting is done in order to reduce the size of the spikes, see also [App97].

The presented techniques are also tested on a slightly changed profile from Traupel. In the initial configuration the normalized total pressure loss is about 6.65%. For the optimized profile it decreases to 1.83%. Figure 6 shows the pressure distribution of a turbine profile and the sensitivity of the pressure with respect to one of the Bézier coefficients at the suction side. As expected, the biggest influence is in the surrounding of the Bézier coefficient. The initial and the optimized profile are shown in Figure 7.

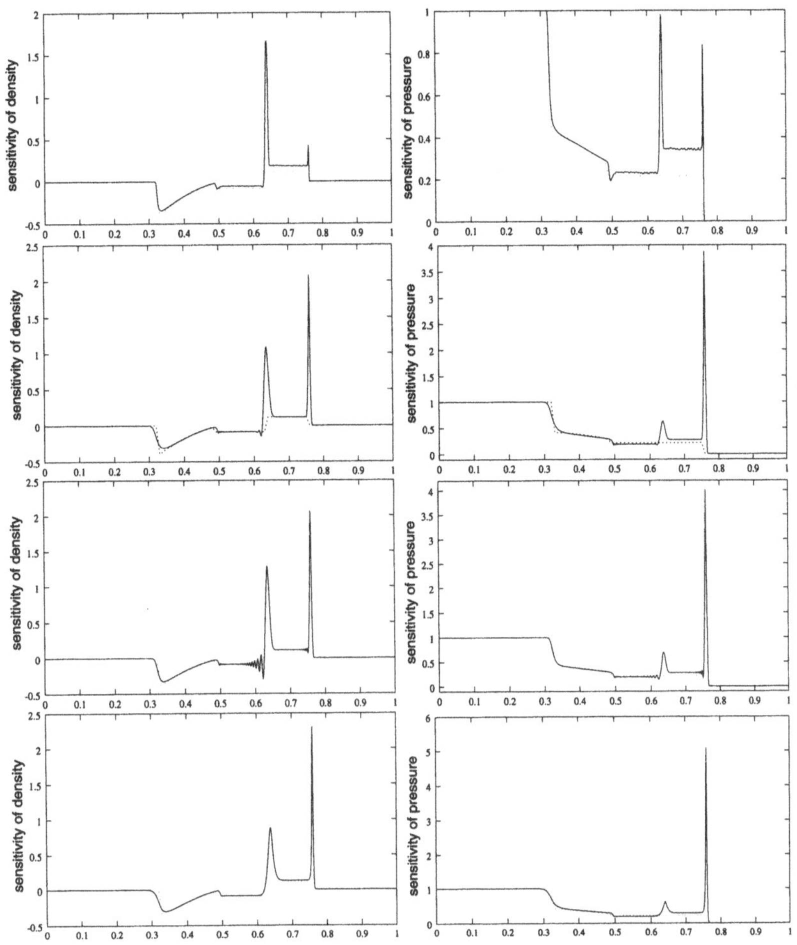

FIGURE 5. Sensitivities of the Shocktube problem: computed with splitting method (first row), Godunov scheme (second row), Lax-Wendroff scheme (third row) and advection algorithm (last row).

8. Conclusion

We tested the Richtmyer-Lax-Wendroff-, the Godunov-, the limited splitting scheme and our generalized advection algorithm. The results for the three first schemes are comparable to those in [AG96]. The limited splitting scheme gives the worst results. Even in smooth regions it is no good approximation. The other schemes give good approximations for the sensivity in smooth regions of the fluid flow. At the discontinuities the δ−function is approximated by large spikes. For the Godunov- and the Richtmyer-Lax-Wendroff-scheme we observe oscillations at the jumps. It costs a lot of effort to find a good parameter ν for (30). With the advection algorithm no oscillations are produced and the spikes need not to be

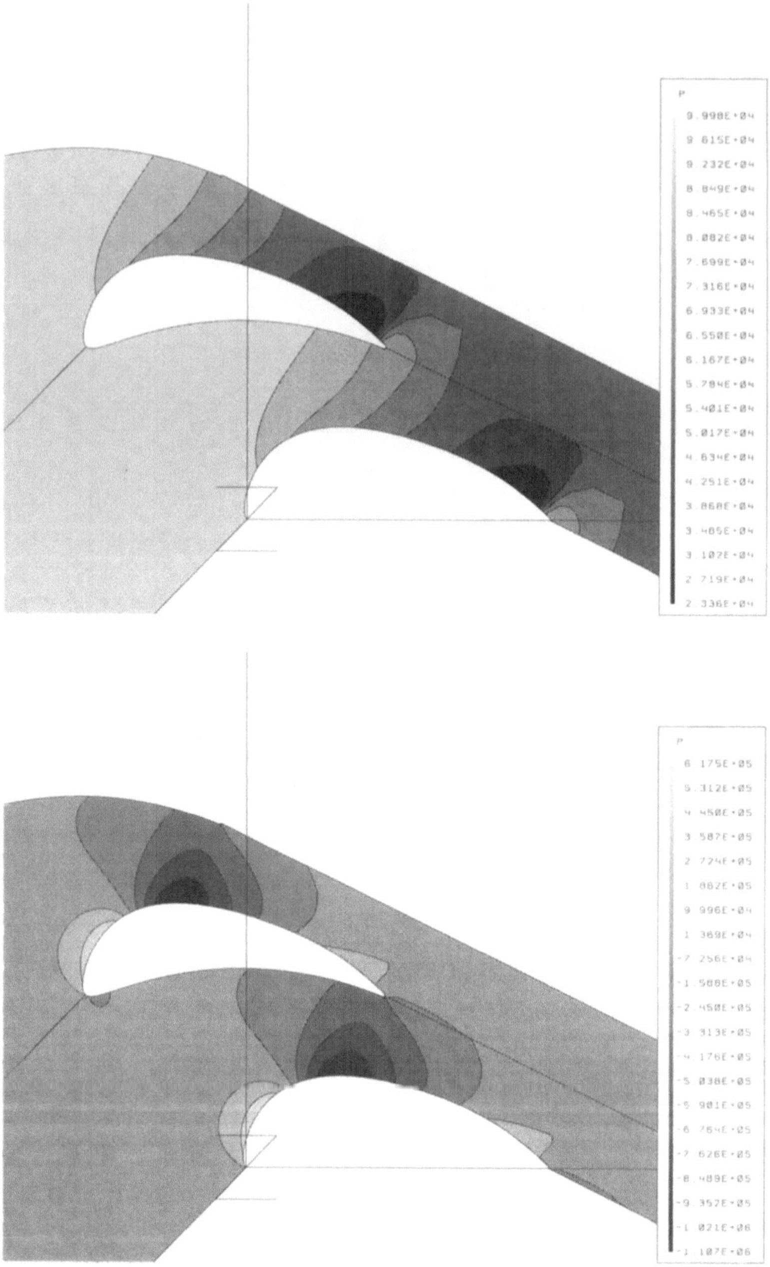

FIGURE 6. The pressure distribution of a turbine profile (upper figure) and the distribution of pressure sensitivity with respect to one of the Bézier coefficients at the suction side

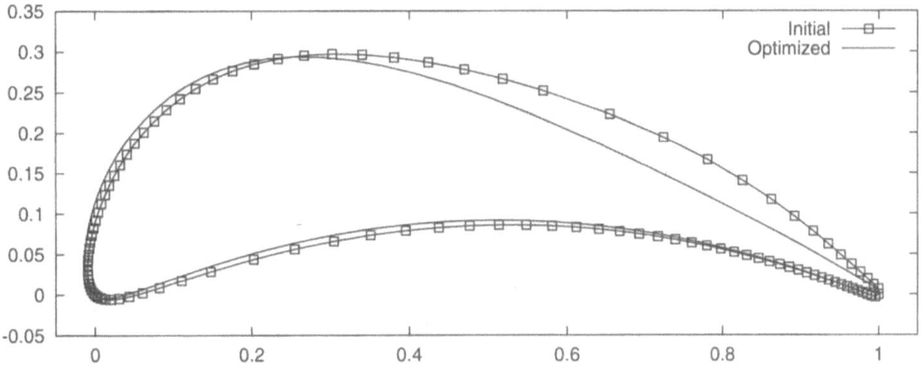

FIGURE 7. Initial and optimized profile

limited. We also tried difference quotients, but their use is not advisable because they are very unprecise and the choice of a suitable pertubation is very difficult.

A new algorithm for the calculation of the sensitivities is developed within this paper. The algorithm is tested on the shock-tube problem for the Euler equations, as well as on the evaluation of the sensitivities of the flow field between two turbine blades with respect to geometry changes. The calculation of sensitivities is not only useful for the computation of the gradients in optimization problems but also as an informational tool in design processes.

Acknowledgement

Sven-Olaf Stoll is indebted to the Siemens AG due to a PhD grant.

References

[AG96] J.R. Appel and M.D. Gunzburger. Sensitivity calculations in flows with discontinuities. In *Proceedings 14th AIAA Applied Aerodynamics Conference*, New Orleans, LA., June 17–20 1996.

[App97] J.R. Appel. *Sensitivity Calculations for Conservation Laws with Application to Discontinuous Fluid Flows.* PhD thesis, Virginia Polytechnic Institute and State University, 1997.

[BB95] J. Borggaard and J. Burns. A sensitivity equation approach to shape optimization in fluid flows. In M. Gunzburger, editor, *Flow Control*, pages 49–78. Springer, 1995.

[BB97] J. Borggaard and J. Burns. A pde sensitivity equation method for optimal aerodynamic design. *J. Comput. Phys.*, 136(2):366–384, 1997.

[BWZ] R. Bell, U. Wever, and Q. Zheng. Profile optimization for turbine blades. to appear in Survey on Mathematics in Industry, 2000.

[HL92] J. Hoschek and D. Lasser. *Grundlagen der geometrischen Datenverarbeitung.* Teubner, 1992.

[LeV92] R. J. LeVeque. *Numerical Methods for Conservation Laws.* Birkhäuser, 1992.

[LeV95] R.J. LeVeque. *CLAWPACK User Notes*, 1995. URL `ftp://amath.washington.edu/pub/rjl/programs/claw/doc`.

[Roe81] P.L. Roe. Approximate riemann solvers, parameter vectors, and difference schemes. *J. Comput. Phys.*, 43:357–372, 1981.

[Shu97] C.-W. Shu. Essentially non-oscillatory and weighted essentially non-oscillatory schemes for hyperbolic conservation laws. Technical report, Brown University, NASA, 1997.

[Spe93] P. Spellucci. *Numerische Verfahren der nichtlinearen Optimierung.* Birkhäuser, 1993.

[Sto98] S.-O. Stoll. Diskretisierung von Sensitivitätengleichungen. Diploma thesis, Darmstadt University of Technology, Faculty of Mathematics, 1998.

[Tra82] W. Traupel. *Thermische Turbomaschinen.* Springer, 1982.

[Ulb00] S. Ulbrich. A sensitivity and adjoint calculus for discontinuous solutions of hyperbolic conservation laws with source terms. Technical Report TR00-10, Department of Computational and Applied Mathematics, Rice University, Houston, Texas 77005-1892, 2000.

Peter Rentrop
Sven-Olaf Stoll
Institut für Wissenschaftliches Rechnen und Mathematische Modellbildung
Universität Karlsruhe (TH)
Engesserstr. 6
D-76128 Karlsruhe, Germany
E-mail addresses: `rentrop@iwrmm.math.uni-karlsruhe.de`
`stoll@iwrmm.math.uni-karlsruhe.de`

Utz Wever
Siemens AG
Corporate Technology
D-81739 München, Germany
E-Mail address: `utz.wever@mchp.siemens.de`

International Series of Numerical Mathematics, Vol. 139, 217–230

A Gauss-Newton Method for the Identification of Nonlinear Heat Transfer Laws

Arnd Rösch

Abstract. A fast and stable algorithm for efficient numerical identification of nonlinear heat transfer laws is introduced basing on a Gauss-Newton method. In this paper the theoretical background is investigated and numerical examples are discussed. The numerical experiences show that the algorithms proposed in the paper are suitable for problems having strongly perturbed data. Using stability estimates a posteriori estimates of the error are derived.

1. Introduction

In this paper we propose a Gauss-Newton Algorithm for the identification of a heat exchange coefficient which depends on temperature. We formulate this problem as an optimal control problem where the unknown heat exchange coefficient plays the part of the control:

$$\min \ \Phi(\alpha) = \int_0^T \int_\Gamma (u(t,x) - q(t,x))^2 dS_x dt, \tag{1}$$

subject to

$$\begin{aligned} \frac{\partial u}{\partial t}(t,x) &= \Delta_x u(t,x) && \text{on } (0,\mathrm{T}] \times \Omega \\ u(0,x) &= u^0(x) && \text{on } \Omega \\ \frac{\partial u}{\partial n}(t,x) &= \alpha(u(t,x))(\vartheta - u(t,x)) && \text{on } (0,\mathrm{T}] \times \Gamma \end{aligned} \tag{2}$$

where the *control* α is taken from the set

$$\begin{aligned} U_{ad} \ := \ & \{\alpha \in C^{1,\nu}[\vartheta_1,\vartheta_2], 0 < m_1 \le \alpha(u) \le M_1, m_2 \le \alpha'(u) \le M_2, \\ & \forall u \in [\vartheta_1,\vartheta_2], \quad \sup_{u_1,u_2\in[\vartheta_1,\vartheta_2]} \frac{|\alpha'(u_1) - \alpha'(u_2)|}{|u_1 - u_2|^\nu} \le C\}. \end{aligned}$$

In this setting, $\Omega \subset \mathbb{R}^m$ is a bounded domain with C^∞-boundary Γ, $T > 0$ a fixed time, $\vartheta \in \mathbb{R}$ a fixed temperature and $q \in L_2((0,T) \times \Omega)$ is a given function of "measurements". The constants ϑ_1 and ϑ_2 are defined by

$$\vartheta_1 = \min(\vartheta, \inf_{x\in\Omega} u^0(x)), \qquad \vartheta_2 = \max(\vartheta, \sup_{x\in\Omega} u^0(x)).$$

Interpreting the problem as a heating process, the variable u means the temperature of the material, u^0 the initial temperature, ϑ the constant temperature of the surrounding medium, and α the unknown heat transfer function, playing the part of the control.

In literature we find different approaches to attack this problem. If the aim consists in the identification of the dependence of an unknown function on the time and on the space we can find many papers, for instance Ito/Kunisch [4], Kunisch/Peichl [6]. Another and also interesting case is to suppose the unknown function as an function of the temperature. We find a standard method to solve such problems in the papers of Beck, we refer only to Beck/Blackwell/Clair [1]. The main idea is to identify first a linear boundary condition with least square methods and to calculate the original function in a second step. If the structure of the heat exchange function is known, results can obtained by methods of quadratic programming, see Kaiser/Tröltzsch [5]. A completely different approach is the direct determination of the unknown nonlinear law. This way is gone in a paper of Chavent/Lemonnier [2] for the identification of the nonlinearity for the quasilinear heat equation.

The technical background consists in the cooling of hot steel or glass in fluids or gases. The heat exchange coefficient depends on the boundary temperature and this dependence has a complicate and a priori unknown structure. From a technical point of view fast cooling processes with only few opportunities to measure the boundary temperature and relatively large errors in the measurements are of high interest. The above mentioned methods have large difficulties to generate sufficient good results for this uncomfortable situation. A suitable approach to overcome the problems are given in Rösch [7] in which a gradient type method is proposed and tested for a very simple example. The application of this method to this complicate practical situation delivers good and stable results but the convergence of this algorithm is very slow and this is the main cause to look for a better algorithm. Therefore, here we propose a Gauss-Newton method to overcome the depicted problems. Unfortunately the usual assumptions which ensure linear or quadratic convergence are not fulfilled even in the best case of correct data. Nevertheless, in our numerical tests we have observed a quite fast convergence to good and stable results. We resolve the problem of the missing convergence analysis by a posteriori estimates, which are derived in the last section.

2. Basic Properties

Let us recall some basic properties of the optimal control problem obtained in [11],[7],[8],[9].

Theorem 2.1. [11] *Let $u^0 \in W_p^{2\hat{\sigma}}(\Omega)$ with $\frac{m}{p} < 2\hat{\sigma}$. Then (2) has a unique solution $u \in C^{0,\delta}([0,T], W_p^{2\sigma}(\Omega))$ with a certain $\delta > 0$ and $\frac{m}{p} < 2\sigma < 2\hat{\sigma}$.*

In that paper this result is proved by means of the theory of analytic semigroups of linear continuous operators. It is possible to show that this solution is a weak solution in the sense of Ladyzhenskaya. For more regular solutions we have to require a compatibility assumption, see [9].

Theorem 2.2. [11] *The optimal control problem (1)–(2) has at least one solution.*

This follows from the compactness of the set U_{ad} in C^1 and the continuity of the control-state mapping and the continuity of the objective.

By Theorem 2.1, a control-state mapping $\alpha \mapsto u$ is defined. It has the following properties.

Theorem 2.3. [8] *The mapping $F: \alpha \mapsto u|_\Sigma$ is Fréchet differentiable from $C^1[\vartheta_1, \vartheta_2]$ to $C(\Sigma)$ with $\Sigma := [0,T] \times \Gamma$. Its derivative F' at $\alpha_o \in C^1[\vartheta_1, \vartheta_2]$ is given $F'(\alpha_o)\beta = v|_\Sigma$, where v is the unique solution of the initial-boundary value problem*

$$\begin{aligned} \frac{\partial v}{\partial t}(t,x) &= \Delta_x v(t,x) && \text{on } (0,T] \times \Omega \\ v(0,x) &= 0 && \text{on } \Omega \qquad (3) \\ \frac{\partial v}{\partial n}(t,x) &= (\alpha_o'(u_o(t,x))(\vartheta - u_o(t,x)) - \alpha_o(u_o(t,x)))v && \\ &\quad + \beta(u_o(t,x))(\vartheta - u_o(t,x)) && \text{on } (0,T] \times \Gamma. \end{aligned}$$

The formula for $F'(\alpha_o)\beta$ is the main tool we need for the definition and the implementation of the Gauss-Newton algorithm. In a numerical approach it is convenient to work with piecewise linear functions. Therefore it is desirable to extend this theory to $C^{0,1}$-functions. The reader is referred to [8].

Next we discuss the properties of the optimal control problem as an inverse problem. Naturally, inverse aspects are very important for the identification. Let us first discuss the case of error-free data q, i.e. the optimal solution α^* generates a state u^* and we have $u^*|_\Gamma = q$. The first question is as follows: In which sense α^* is the unique solution of the identification problem? The function α appears only in the boundary condition of (2). The argument of this function which are the boundary values $u(t,x)$ of the state. Consequently, the function α^* can only be determined on the union of all boundary values of u^*. Therefore we define the *reference set* of a control α.

Definition: Let $\alpha \in U_{ad}$ and u the corresponding solution of (2). The set

$$M(\alpha) := \bigcup_{t \in [0,T]\ x \in \Gamma} \{u(t,x)\}$$

is called *reference set* associated with α.

Each reference set is a subset of $[\vartheta_1, \vartheta_2]$. The crux of the matter is as follows. Every admissible control α generates its own reference set. According to the above discussion, the optimal control α^* can only be determined on its reference set $M(\alpha^*)$. In the case of error-free data we know the optimal reference set, since $u^*|_\Gamma = q$ and q is given. But in general we have no exact a priori information about the optimal reference set $M(\alpha^*)$.

Theorem 2.4. [7] *Suppose two solutions of (2) $u(\alpha_1)=u_1(x,t)$ and $u(\alpha_2)=u_2(x,t)$ to α_1, α_2 are equal, i.e. $u_1 = u_2 = u$ in the sense of the space $C([0,T], W_p^{2\sigma}(\Omega))$. Define $M = \{u(t,x) : t \in [0,T], x \in \Gamma\}$ and assume $\vartheta \notin M$. Then the controls are equal on the reference set M, i.e. $\alpha_1 = \alpha_2$.*

In the case of the 1D heat equation we were able to prove stability estimates. Therefore we regard instead of (2) the following one–dimensional problem (4):

$$\begin{aligned}
\frac{\partial u}{\partial t}(t,x) &= \frac{\partial^2 u}{\partial x^2}(t,x) && \text{on } (0,\mathrm{T}] \times (0,1) \\
u(0,x) &= u^0(x) && \text{on } (0,1) \\
\frac{\partial u}{\partial n}(t,0) &= 0 && \text{on } (0,\mathrm{T}] \\
\frac{\partial u}{\partial n}(t,1) &= \alpha(u(t,1))(\vartheta - u(t,1)) && \text{on } (0,\mathrm{T}].
\end{aligned} \tag{4}$$

Theorem 2.5. [9] *Let α_1, α_2 be admissible controls with associated states u_1 and u_2. Denote by M_1 the reference set of α_1. Then the stability estimate*

$$\|\alpha_1 - \alpha_2\|_{C(M_1)} \leq c\|u_1(.,1) - u_2(.,1)\|_{C[0,T]}^{\frac{1}{3}} \tag{5}$$

holds with a certain $c > 0$. Moreover, if the control α_1 and the initial data $u^0 \in C^{2,\nu'}[0,1]$ $(\nu' > 0)$ fulfil the compatibility conditions

$$\frac{du^0}{dx}(1) = \alpha_1(u^0(1))(\vartheta - u^0(1)), \qquad \frac{du^0}{dx}(0) = 0, \tag{6}$$

then

$$\|\alpha_1 - \alpha_2\|_{C(M_1)} \leq c\|u_1(.,1) - u_2(.,1)\|_{C[0,T]}^{\frac{1}{2}}. \tag{7}$$

In [9] also L_2-estimates of Hölder type are given. Let us short explain the inequalities. We assume that the boundary temperature $u(t,1)$ is Hölder continuous with exponent δ with respect to time. Then the exponent in the stability estimate is $\delta/(1+\delta)$, see [9]. For the 1-D heat equation we get the exponents $1/2$ for no compatibility and 1 for compatibility. Therefore we obtain the exponents $1/3$ and $1/2$ in (6) and (7).

3. A Gauss-Newton Algorithm

First we motivate the choice of the Gauss-Newton-algorithm. Assume for simplicity the best case of exact data $\Phi(\alpha^*) = 0$ and no restriction of U_{ad} is active for α^*. Then we might set up, for instance, a SQP-method. It is known that the SQP-method converges quadratically under certain assumptions. Among them a second order sufficient optimality condition (SSC) is most essential. It ensures local quadratically growth of the objective. In [10] it is shown that (SSC) implies Lipschitz stability of the control-state mapping. However, we do not have even in the best case Lipschitz stability, see [9]. Therefore we cannot expect quadratic convergence for a SQP-method for error-free data. Moreover, the implementation of a SQP-method has two additional drawbacks. The algorithm needs second derivatives of the control α. Therefore we cannot use piecewise linear functions for the discretization of α. Furthermore the calculation of the second order term is expensive. In the Gauss-Newton-method such problems do not occur. Under certain assumptions (which include $\Phi(\alpha^*) = 0$) the Gauss-Newton-algorithm converges quadratically as well, see Dennis/Schnabel [3] or Schwetlick [12]. Similar to the SQP-method one of its main assumptions is not fulfilled. The operator F' is a compact operator. Therefore the inverse is not bounded, which would be necessary for the proof of convergence. Nevertheless, we applied the Gauss-Newton-method. The result was very encouraging - we observed a fast convergence to stable results. In the last section we will justify the method by a posteriori estimates.

Let us introduce the Gauss-Newton principle first for the simple example to minimize

$$\min\ g(x) = \frac{1}{2}\|G(x)\|^2$$

where $G : \mathbb{R}^n \to \mathbb{R}^m$ is a given nonlinear function. The iteration of the Gauss-Newton method is defined by

$$x^{k+1} = x^k - [G'(x^k)^T G'(x^k)]^{-1} G'(x^k)^T G(x^k).$$

Equivalently x^{k+1} can be defined as solution of the quadratic problem

$$x^{k+1} = \operatorname{argmin}\{\frac{1}{2}\|G'(x^k)(x - x^k) + G(x^k)\|^2\}.$$

This formulation can be transferred to constrained problems while the formulation above is restricted to free minimization problems. Now let us return to the identification problem (1)–(2). We write our objective in the form

$$\Phi(\alpha) = \|u(t,x) - q(t,x)\|^2_{L_2(\Sigma)} = \|F(\alpha) - q\|^2_{L_2(\Sigma)} =: \|\varphi(\alpha)\|^2_{L_2(\Sigma)} \tag{8}$$

Now the formulation of the Gauss-Newton algorithm seems to be easy: Let α_n be the current iterate. Find β from

$$\min\ \|\varphi(\alpha_n) + \varphi'(\alpha_n)\beta\|^2_{L_2(\Sigma)} \tag{9}$$

subject to $\alpha_n + \beta \in U_{\text{ad}}$, where β plays the role of $\alpha_{n+1} - \alpha_n$. Then update $\alpha_{n+1} := \alpha_n + \beta$.

A critical view on this scheme shows different problems. We have to compute

$$\begin{array}{lcll} \varphi(\alpha_n) & = & F(\alpha_n) - q & \text{by solving (2)}, \\ \varphi'(\alpha_n)\beta & = & F'(\alpha_n)\beta & \text{by solving (3)}. \end{array}$$

We minimize an objective over Σ which is a m-dimensional manifold. The function β is defined on $[\vartheta_1, \vartheta_2]$ and depends only indirectly on t and x. Therefore the minimization problem has the structure

$$\min \|F'(\alpha_n)\beta(u_n(t,x)) - g(t,x)\|^2_{L_2(\Sigma)}. \tag{10}$$

Here we look for an optimal function β by given data u_n,g and known mapping F'. Thus we want to approximate the m-dimensional manifold $g(t,x)$ by the 1D-function $\beta(u_n)$. In general it is not clear how to solve such a problem.

The second problem is more difficult to recognize. It is connected with the reference sets of the controls α_n and α_{n+1}. Let us see what happens in such a iteration.
a) We start with α_n, $u_n = F(\alpha_n)$ and the reference set $M(\alpha_n)$
b) We solve

$$\min \ \|\varphi(\alpha_n) + \varphi'(\alpha_n)[\beta]\|_2^2 \quad \text{subject to} \quad \alpha_n + \beta \in U_{\text{ad}}.$$

and set $\alpha_{n+1} := \alpha_n + \beta$. Consequently, α_{n+1} is reasonably defined only on $M(\alpha_n)$.
c) We want to compute $u_{n+1} = F(\alpha_{n+1})$ by solving (2). The new state u_{n+1} may have values outside of $M(\alpha_n)$. Therefore we need for the computation of u_{n+1} values of α_{n+1} outside of $M(\alpha_n)$, but α_{n+1} is only reasonable defined on $M(\alpha_n)$. Therefore the value $\alpha_{n+1}(u_{n+1}(t,x))$ is in general *not declared.*

There are different ways to overcome this problem. One way is to start the iteration process on a larger "safety" set for instance on the interval $[\vartheta_1, \vartheta_2]$. Then we have no problems with the update. Naturally, for the larger interval we need more variables to get the same accuracy, but this additional effort is only a small problem. More critical is the fact that the control α_n has a part on its reference set $M(\alpha_n)$ and a part outside of it, i.e. on $[\vartheta_1, \vartheta_2] \setminus M(\alpha_n)$. Therefore we can change the control α_n on $[\vartheta_1, \vartheta_2] \setminus M(\alpha_n)$ without effects in the state or in the objective. This property deteriorates the behaviour of the minimization problem of the numerical algorithm. More in detail parts outside of the reference set lead to undefined optimization variables. A second way is to change the reference set in each iteration step. If the iteration process is successful then the variance of the reference sets becomes small. The advantages of this method is the smaller effort for the discretization and the knowledge of the reference sets. In this approach the update of the control requires an additional step. In this step the control is expanded or shrunken to the new reference set.

Let us now explain the implementation of the method which is connected with discretization. For the discretization we propose piecewise linear functions α. Clearly, such functions do not belong to $C^{1,\nu}$. We need this property only for the proof of Theorem 2.2. For the finite dimensional problem we do not need such an assumption. Therefore we omit this condition. Let $M(\alpha) = [u_A, u_B]$ be the current reference set. We define $u_i := i \cdot (u_B - u_A)/N$ (i=0..N) and piecewise linear basis functions

$$e_i(u) := \begin{cases} u - u_{i-1} & \text{if } u \in [u_{i-1}, u_i] \\ u_{i+1} - u & \text{if } u \in [u_i, u_{i+1}] \\ 0 & \text{otherwise.} \end{cases}$$

The function α is represented by

$$\alpha(u) = \sum_{i=0}^{N} \alpha_i e_i(u).$$

In this way we identify the function α by a vector $\underline{\alpha}$ having components $\alpha_i = \alpha(u_i)$. We describe U_{ad} by

$$\begin{array}{rcccl} m_1 & \leq & \alpha_i & \leq & M_1 \\ m_2 \cdot \dfrac{u_B - u_A}{N} & \leq & \alpha_{i+1} - \alpha_i & \leq & M_2 \cdot \dfrac{u_B - u_A}{N}. \end{array} \tag{11}$$

Analogously the increment β is represented. Next we define the *response functions*

$$v_i(t,x) = (F'(\alpha)e_i)(t,x) \qquad (i = 0..N).$$

Therefore we obtain

$$F'(\alpha)\beta = \sum_{i=0}^{N} \beta_i v_i.$$

In the numerical realization we have also to discretize the state space. Otherwise in practical situation the measurements often only given on a finite set Q of points (t_j, x_k) on the boundary. Hence we have to approximate the norm in (9) by a sum over Q. Therefore we need only the restriction on Q and define $A\beta := F'(\alpha)[\beta]|_Q$. The operator A can be represented by a matrix with $N+1$ columns and card(Q) rows. Therefore we obtain for the norm in the Gauss-Newton algorithm

$$\|F(\alpha_n)|_Q - q|_Q + A\beta\|_Q^2$$

which is now a simple quadratic functional. We describe the condition $\alpha_n + \beta \in U_{ad}$ by (11) and obtain a finite dimensional linear quadratic optimization problem which can be solved by a standard quadratic solver. We remark that the quadratic part of the objective is given by $A^T A$ and hence positive semidefinite.

Gauss-Newton Algorithm

1. Initialization: Choose $\underline{\alpha}^0$ and set $n := 0$.
2. Calculate $\underline{u}^n = F(\alpha)$ by solving (2), respectively, (4). Determine the reference set $M(\underline{\alpha}^n)$.
3. Approximate $\underline{\alpha}^n$ on the set $M(\underline{\alpha}^n)$.
4. Set up the matrix A by solving problem (3) for all functions e_j (j=0..N).
5. Establish and solve the quadratic problem

$$\min \|F(\alpha_n)|_Q - q|_Q + A\beta\|_Q^2 \text{ subject to } \quad \underline{\alpha}^n + \underline{\beta} \in \underline{U_{ad}}$$

6. Update $\underline{\alpha}^{n+1} := \underline{\alpha}^n + \underline{\beta}$, set $n := n+1$, and goto 2.

We terminated the algorithm after a fixed number of iterations for our numerical tests. However, the a posteriori error estimates in the last section justify a discrepancy principle as stopping criterion.

From a practical point of view tests with strongly perturbed data are interesting. In this case, the update step 6 is often too optimistic. In a *Modified*

Gauss-Newton Algorithm we use an additional line search $\underline{\alpha}^n + \lambda\underline{\beta}$ for the update step 6', where $\lambda \in (0,1]$. The choice of λ in step 6' is specified in the next section.

4. Numerical Results

Practical background of our example is a very fast cooling process. The cooling starts at a constant initial distribution $u^0 = 700^oC$. The cooling time is 10 seconds and the boundary temperature after the cooling process is about 300^oC. For the numerical results synthetic data are given. That means we use realistic material properties but we generate the "measurements" by a direct solver. The goal of our numerical examples is to identify the given function $\alpha(u) = \frac{367500}{1225-u}$. The one-dimensional domain Ω is a small interval (length 5mm). In this case the compatibility condition (6) is not fulfilled. For that reason the boundary temperature $u(t)$ behaves like the square root, see [9].

In the first example we apply the Gauss-Newton algorithm for error-free data. For the control a discretization of 20 intervals having length of about 20 degrees is taken. We start our iteration far away from the optimal point. The average of the error of the start iteration is about 80 degrees. Figures 1 and 2 show the objective value $\Phi^n = \Phi(\alpha^n)$ with respect to the iteration number n during the iteration process. We scaled the objective such that it describes the averaged discrepancy in degrees. Allthough, globally, Φ^n is decreasing, some iterations with increasing objective value occur. The average of the error after the 99th iteration is below 0.1 degree. It is easy to see that the convergence is not quadratic.

In this example this seems to be the expected result. We are surprised by a view on the control α^{99} (Figure 3). The functions α^* and α^{99} are nearly the same, except the part between 630 degrees and 700 degrees. What is the cause of this discrepancy? In Figure 4 we can see how many measurements are available in different intervals. In our critical region we do not have measurements. Here we cannot expect convergence. The quadratic solver applied on step 5 gives us the information that these variables do not influence the objective. For the identification of this part we would need additional measurements.

In the second example an error to the measurements is added. The largest error is 10 degrees at the time $t = 0.2s$. The error in the data reflects the practical circumstances. We computed the heat flow $Q_H = \frac{\partial u}{\partial n}$ with an unregularized identification method to each instant of time in a first step. In a second step, we calculated $\alpha(t) = Q_H(t)/(\vartheta - u(t))$. In a last step we reconstructed the dependence of α with respect to u. For the plots a linear interpolation of the sequence $(u(t_i), \alpha(t_i))$ is used. Because of the large error in the data, the boundary temperature is not monotone decreasing with respect to time. This effect is reflected by the loops in the Figure 5. The next result (Figure 6) is generated by a future regularizing method due to Beck. We got the best result for five time points with different weights. The error is less than the error of the unregularized problem, but the result is also useless in this case.

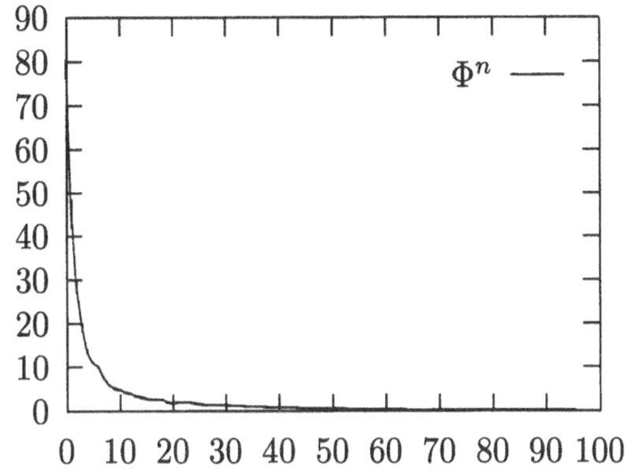

Figure 1

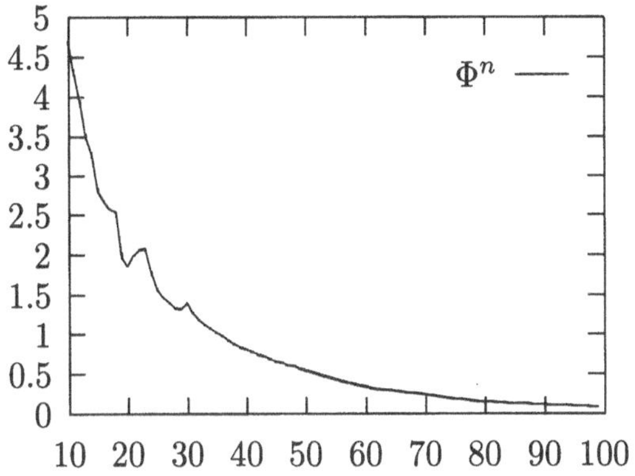

Figure 2

In this case the Gauss-Newton method exhibits a descent behaviour in the first 20 iterations. After these iterations, the objective value jumps in an interval between 2.3 and 3 degrees. For that reason we started different calculations by the modified Gauss-Newton algorithm. For the stepsize λ the bisection strategy was applied in our first test. In practice, this choice has a good convergence behaviour but we get a worse objective value which is about 2.5. Many local minima seem to exist near the best one. We computed the best results for a strategy which allows iteration steps where the objective increases. For this strategy we got an objective value of about 1.3. In view of the 10-degrees error in the data, all these values are acceptable. Figure 7 shows the optimal control α^* for error-free data and the controls generated by the Gauss-Newton algorithm and the modified Gauss-Newton algorithm for strongly perturbed data.

In the Figures 3 and 7, we plotted numerical results of the proposed method. In the iteration processes the bounds m_1 and M_1 were not active. In the invalid part in Figure 3 we can see a constant part of the numerical solution. Here the constraint $\alpha' = m_2 = 0$ is active. There is also a strong increasing part near 700 where the constraint $\alpha' = M_2$ is active. In Figure 7 for both numerical solutions we

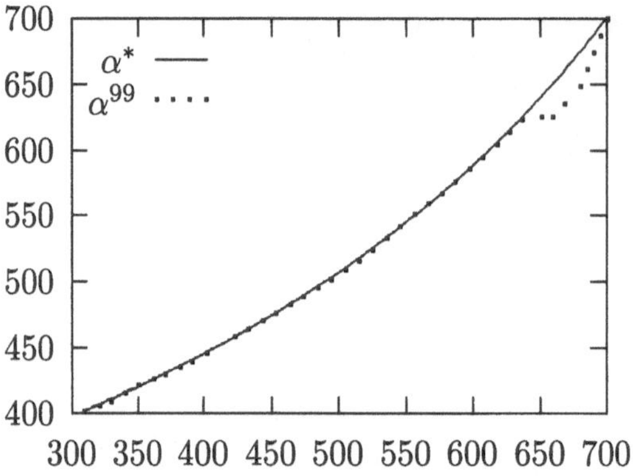

Figure 3

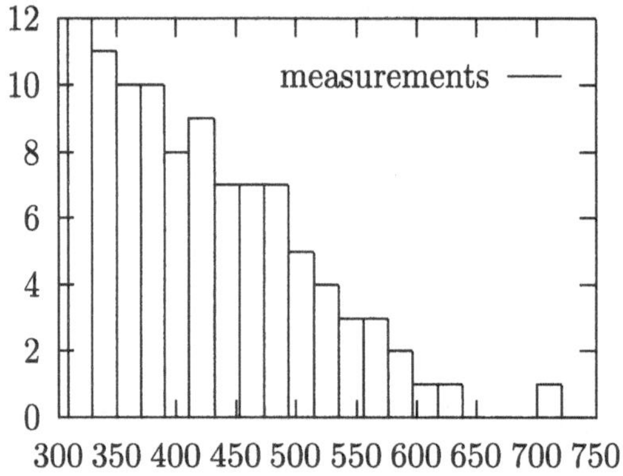

Figure 4

have parts where the restrictions $\alpha' = m_2 = 0$ and $\alpha' = M_2$ are active. Because of this fact a good a priori knowledge of m_2 and M_2 improves the numerical results.

For error-free data we observed convergence of the controls and their reference sets. In the case of strongly perturbed data we might enforce the convergence of the controls and their reference sets by a suitable step size rule. In our numerical tests we did not proceed this way. In general in our numerical tests we observed large variations of the reference sets during the first iterations (because we started far away from the optimal point). After about 10 iterations only small variations of the reference sets are obtained. In Figure 7 we can see that the reference sets of the numerical solutions are slightly different. Let us summarize our numerical experiences. The Gauss-Newton algorithm and the modified Gauss-Newton algorithm are fast and stable methods. In contrast to other methods, these algorithms generate good results also for strongly perturbed data. A good choice of step size λ can improve the performance of the modified Gauss-Newton algorithm.

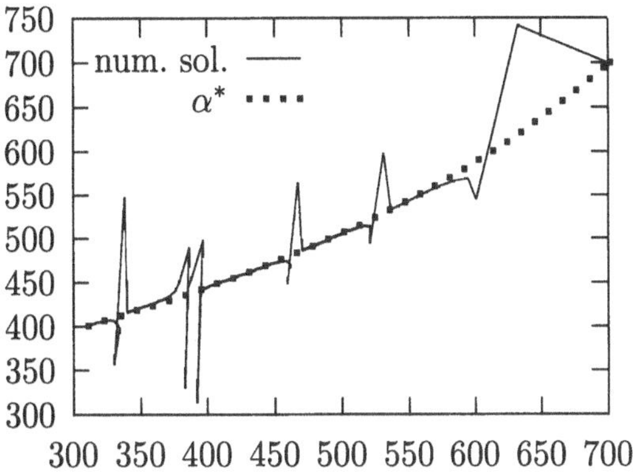

Figure 5

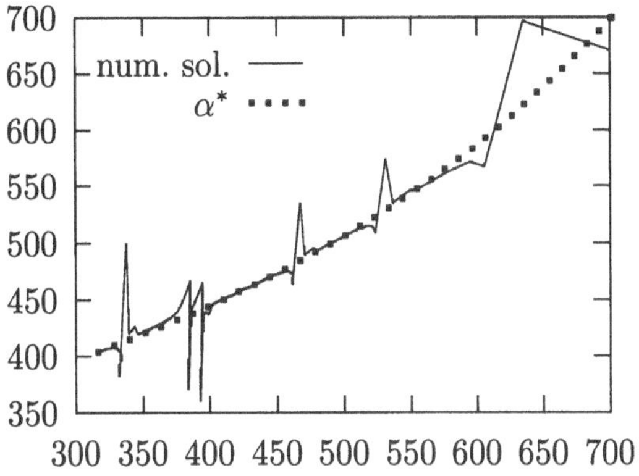

Figure 6

5. A Posteriori Estimates

In the last section we presented numerical results. We already noticed that we do not have a convergence proof of our algorithm. Therefore we aim to justify the results after the calculation. Using our stability results, a posteriori error estimates are derived. First we estimate the difference of the numerical solution α_n and the solution α^* of (1)–(2). This error is a criterion for the quality of the numerical algorithm. However, in practice, noisy data q^δ and an error level δ are given. We estimate the difference between the numerical solution α_n^δ of the perturbed problem and the solution α^* of (1)-(2) for exact data. This error estimate is a criterion of the quality of the analytical approach via the optimal control problem.

<u>A posteriori estimate with respect to the solution of the optimal control problem</u>

Let us discuss the following situation. Our numerical method terminated with the control $\alpha_n \in U_{ad}$ and we calculated the defect $F(\alpha_n) - q$. First we investigate a control $\alpha^* \in U_{ad}$ which is optimal in the following slightly general sense:

$$\|F(\alpha^*) - q\|_{X(\Sigma)} \leq \|F(\alpha) - q\|_{X(\Sigma)} \quad \forall \alpha \in U_{ad} \tag{12}$$

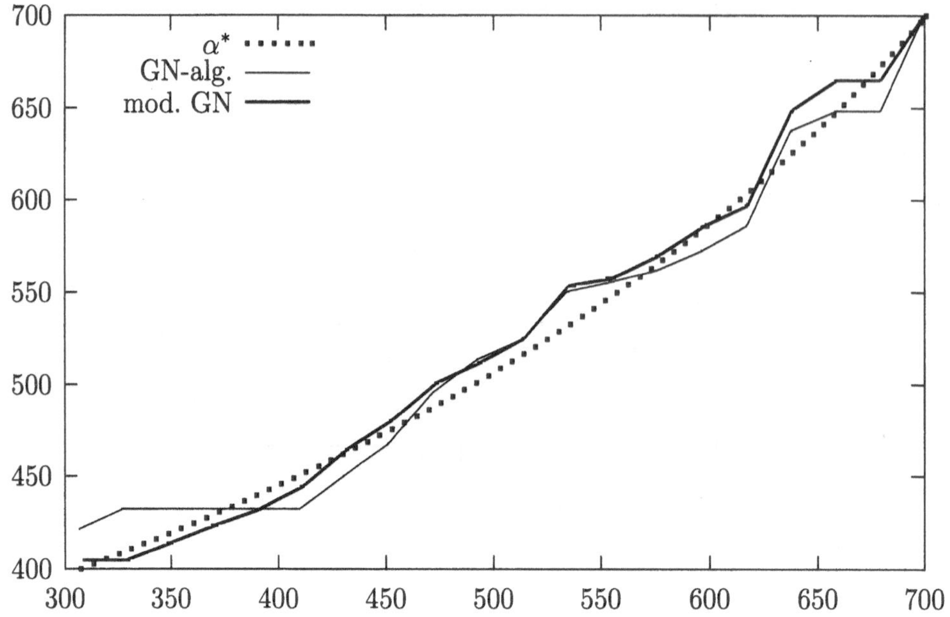

Figure 7

In our optimal control problem we minimized the L_2-norm of this expression, that means $X = L_2$. But we are also interested in estimates with respect to other norms for instance the C-norm, i.e. $X = C$. Furthermore we assume that a stability estimate of Hölder type holds. In this estimate we have to fix one control with its reference set. We know only the reference set $M(\alpha_n)$ of our numerical result.

Theorem 5.1. *Assume that a stability estimate holds in the form*

$$\|\alpha_n - \alpha^*\|_{Y(M(\alpha_n))} \leq c\|u_n - u^*\|^{\nu}_{X(\Sigma)}, \tag{13}$$

where X and Y are normed spaces. Then the a posteriori-estimate

$$\|\alpha_n - \alpha^*\|_{Y(M(\alpha_n))} \leq c(2\|u_n - q\|_{X(\Sigma)})^{\nu}. \tag{14}$$

is satisfied.

Proof. Note that (5) represents such an inequality with $X = C$, $Y = C$, $\nu = 1/3$. From optimality condition (12) we obtain the error estimate

$$\begin{aligned}\|\alpha_n - \alpha^*\|_{Y(M(\alpha_n))} &\leq c(\|u_n - q\|_{X(\Sigma)} + \|q - u^*\|_{X(\Sigma)})^{\nu} \\ &\leq c(2\|u_n - q\|_{X(\Sigma)})^{\nu}.\end{aligned}$$

The right-hand side of this inequality is given by the result of the iteration process. Therefore (14) is an a posteriori estimate of the error of the numerical solution with respect to the solution of the optimal control problem (1)–(2). □

A posteriori estimate with respect to the solution of the identification problem

Now we have exact data q^* and perturbed data q^δ where δ is the error level

$$\|q^\delta - q^*\|_{X(\Sigma)} \leq \delta \tag{15}$$

For our a posteriori estimate we can only use the knowledge about q^δ and δ. The iteration process is terminated with an iterate α_n^δ and the corresponding state u_n^δ.

Theorem 5.2. *Let stability estimate (13) be fulfilled. Then the a posteriori estimate*

$$\|\alpha_n^\delta - \alpha^*\|_{Y(M(\alpha_n^\delta))} \leq c(\|u_n - q^\delta\|_{X(\Sigma)} + \delta)^\nu \tag{16}$$

holds true.

Proof. Since α^* is the exact heat exchange coefficient we can replace the state u^* by the exact data q^*. We proceed as follows

$$\begin{aligned}\|\alpha_n^\delta - \alpha^*\|_{Y(M(\alpha_n^\delta))} &\leq c\|u_n - q^*\|_{X(\Sigma)}^\nu \\ &\leq c(\|u_n - q^\delta\|_{X(\Sigma)} + \|q^\delta - q^*\|_{X(\Sigma)})^\nu \\ &\leq c(\|u_n - q^\delta\|_{X(\Sigma)} + \delta)^\nu.\end{aligned}$$

Once again, the right-hand side of this inequality is known from the iteration process. □

The a posteriori estimate (16) justifies a discrepancy principle as stopping rule for the Gauss-Newton algorithm. Therefore the iteration process should be finished if $\|u_n - q^\delta\|_{X(\Sigma)} \leq \delta$. Hence we should stop the algorithm in our numerical example after 10–20 iterations.

References

[1] J. V. Beck, B. Blackwell, and Ch. R. St. Clair Jr., *Inverse heat conduction Ill-posed Problems,* A Wiley – Interscience Publication, New York, **1985**.

[2] G. Chavent and P. Lemonnier, *Identification de la Non-linearité d'une Équation Parabolique Quasilinéaire,* App. Math. and Opt., **1(2)** (1974), 121–161.

[3] J.E. Dennis and R.B. Schnabel. *Numerical Methods for Unconstrained Optimization and Nonlinear Equations,* SIAM, Philadelphia, **1996**.

[4] K. Ito and K. Kunisch. *The augmented Lagrangian method for parameter estimation in elliptic systems,* SIAM J.Control Opt., **28** (1990), 113–136.

[5] T. Kaiser and F. Tröltzsch. *An inverse problem arising in the steel cooling process,* Wissenschaftliche Zeitung TU Karl-Marx-Stadt, **29** (1987), 212–218.

[6] K. Kunisch and G. Peichl. *Estimation of a temporally and spatially varying diffusion coefficient in a parabolic system by an augmented Lagrangian technique,* Numerische Math., **59** (1991), 473–509.

[7] A. Rösch. *Identification of nonlinear heat transfer laws by optimal control,* Num. Funct. Analysis and Optimization, **15(3&4)** (1994), 417–434.

[8] A. Rösch. *Fréchet differentiability of the solution of the heat equation with respect to a nonlinear boundary condition,* Z. Anal. u. Anw., **15(3)** (1996), 603–618.

[9] A. Rösch. *Stability estimates for the identification of nonlinear heat transfer laws,* Inverse Problems, **12** (1996) 743–756.

[10] A. Rösch. *Second order optimality conditions and stability estimates for the identification of nonlinear heat transfer laws,* in: Control and estimation of distributed parameter systems, International conference in Vorau, Austria, July 14–20, 1996, Edited by W. Desch, number 126 in ISNM, Birkhäuser, **1998**, 237–246.

[11] A. Rösch and F. Tröltzsch. *An optimal control problem arising from the identification of nonlinear heat transfer laws,* Archives of Control Sciences, **1(3–4)** (1992), 183–195.

[12] H. Schwetlick. *Numerische Lösung nichtlinearer Gleichungen,* Deutscher Verlag der Wissenschaften, Berlin, **1979**.

Faculty of Mathematics
Chemnitz University of Technology
D-09107 Chemnitz, Germany
E-mail address: a.roesch@mathematik.tu-chemnitz.de

International Series of Numerical Mathematics, Vol. 139, 231–244

Topological Derivatives of Shape Functionals for Elasticity Systems

Jan Sokołowski, Antoni Żochowski

Abstract. The exact form of topological derivative (TD) and the computational procedure for its calculation is derived for a class of shape functionals in 3D elasticity. The derivation is based on asymptotic expansion of solutions. TD is used numerical shape optimization and for solving shape inverse problems.

1. Introduction

The topological derivative $\mathcal{T}_\Omega$ of a shape functional $\mathcal{J}(\Omega)$ is introduced in [14] in order to characterize the variation of $\mathcal{J}(\Omega)$ resulting from discontinuous deformation of of the domain Ω changing its topological properties, namely emerging of finite number of small ball–like voids. When applied in shape optimization, it allows to obtain more general, in comparison to varying only the external boundary, necessary optimality conditions. They should assure that joint effect of both topology and external shape changes does not lead to improvement of $\mathcal{T}_\Omega$ (in the class of admissible shapes).

Such a methodology constitutes an intermediate step between continuous shape optimization and material design, where an infinite number of infinitesimal voids is assumed.

The other use of the topological derivative is connected with approximating the influence of the holes in the domain on the values of integral functionals of solutions, what allows us to solve a class of shape inverse problems [7]. We refer the reader to [1], [12] for numerical results of computations for compliance optimization problems using the so–called bubble method. On the other hand, the topological derivatives are used to obtain numerical results for shape optimization problems reported in [2].

In general terms the notion of the *topological* derivative (TD) has the following meaning. Assume that $\Omega \subset \mathbb{R}^N$ is an open set and that there is given a shape

Partially supported by the grant 8T1100815 of the State Committee for the Scientific Research of the Republic of Poland and by the French–Polish research programme POLONIUM between INRIA–Lorraine and Systems Research Institute of the Polish Academy of Sciences.

functional

$$\mathcal{J} : \Omega \setminus K \to \mathbb{R}$$

for any compact subset $K \subset \overline{\Omega}$. We denote by $B_\rho(x), x \in \Omega$, the ball of radius $\rho > 0$, $B_\rho(x) = \{y \in \mathbb{R}^N | \|y - x\| < \rho\}$, $\overline{B_\rho(x)}$ is the closure of $B_\rho(x)$, and assume that there exists the following limit

$$\mathfrak{T}(x) = \lim_{\rho \downarrow 0} \frac{\mathcal{J}(\Omega \setminus \overline{B_\rho(x)}) - \mathcal{J}(\Omega)}{|\overline{B_\rho(x)}|}$$

which can be defined in an equivalent way by

$$\tilde{\mathfrak{T}}(x) = \lim_{\rho \downarrow 0} \frac{\mathcal{J}(\Omega \setminus \overline{B_\rho(x)}) - \mathcal{J}(\Omega)}{\rho^N}$$

The function $\mathfrak{T}(x), x \in \Omega$, is called the topological derivative of $\mathcal{J}(\Omega)$, and provides the information on the infinitesimal variation of the shape functional $\mathcal{J}$ if a small hole is created at $x \in \Omega$.

In several cases this characterization is constructive, i.e. TD can be evaluated for shape functionals depending on solutions of partial differential equations defined in the domain Ω.

For instance, in [16] TD is computed for the 3D elliptic Laplace type equation. In [14] the cases of 2D elliptic equation and of the 2D elasticity system is treated. In [15] TD is obtained for extremal values of cost functionals for a class of optimal control problems. All these examples have one common feature: the expression for TD may be calculated in the closed functional form.

As we shall see below, the 3D elasticity case is more difficult, since it requires evaluation of integrals on the unit sphere with the integrands which can be computed at any point, but the resulting functions have no explicit functional form. The results reported here are derived in the report [17].

In Section 5 we compare the results of the present paper with the formulae for 2D elasticity.

The main contribution of the present paper is the procedure for computations of the topological derivatives of shape functionals depending on the solutions of 3D elasticity systems. Therefore it constitutes an essential extension of the results given in [14] for the 2D case.

2. Elasticity Equations and Goal Functionals

We introduce 3D elasticity system in the form convenient for the evaluation of topological derivatives. Let us consider the elasticity equations in $\mathbb{R}^3$,

$$A^T D A u = f \quad \text{in} \quad \Omega, \tag{1}$$

$$u = g \quad \text{on} \quad \Gamma_1, \qquad B^T D A u = h \quad \text{on} \quad \Gamma_2,$$

and the same system in the domain with the spherical cavity $B_\rho(x_0) \subset \Omega$ centered at $x_0 \in \Omega$, $\Omega_\rho = \Omega \setminus \overline{B_\rho(x_0)}$,

$$\begin{aligned} A^T D A u^\rho &= f \quad \text{in} \quad \Omega_\rho, \\ u^\rho &= g \quad \text{on} \quad \Gamma_1, \end{aligned} \tag{2}$$
$$B^T D A u^\rho = h \quad \text{on} \quad \Gamma_2, \qquad B^T D A u^\rho = 0 \quad \text{on} \quad S_\rho(x_0) = \partial B_\rho(x_0).$$

Assuming that $0 \in \Omega$, we can consider the case $x_0 = 0$.
Here $u = (u_1, u_2, u_3)^T$ denotes the displacement field, g is a given displacement field on the fixed part Γ_1 of the boundary, h is a traction prescribed on the loaded part Γ_2 of the boundary. Finally, the volume forces are denoted by f. In addition, the following matrix differential operator and the matrix of material (Lame) coefficients are introduced,

$$A = \begin{bmatrix} \frac{\partial}{\partial x_1} & 0 & 0 \\ 0 & \frac{\partial}{\partial x_2} & 0 \\ 0 & 0 & \frac{\partial}{\partial x_3} \\ \frac{\partial}{\partial x_2} & \frac{\partial}{\partial x_1} & 0 \\ 0 & \frac{\partial}{\partial x_3} & \frac{\partial}{\partial x_2} \\ \frac{\partial}{\partial x_3} & 0 & \frac{\partial}{\partial x_1} \end{bmatrix}, \quad D = \begin{bmatrix} \lambda+2\mu & \lambda & \lambda & 0 & 0 & 0 \\ \lambda & \lambda+2\mu & \lambda & 0 & 0 & 0 \\ \lambda & \lambda & \lambda+2\mu & 0 & 0 & 0 \\ 0 & 0 & 0 & \mu & 0 & 0 \\ 0 & 0 & 0 & 0 & \mu & 0 \\ 0 & 0 & 0 & 0 & 0 & \mu \end{bmatrix}.$$

The following matrix is used for the Neumann boundary conditions

$$B = \begin{bmatrix} n_1 & 0 & 0 & n_2 & 0 & n_3 \\ 0 & n_2 & 0 & n_1 & n_3 & 0 \\ 0 & 0 & n_3 & 0 & n_2 & n_1 \end{bmatrix}^T,$$

where $n = [n_1, n_2, n_3]^T$ is the unit outward normal vector on $\partial\Omega_\rho$. In this notation the stress tensor is replaced by the vector $\sigma = [\sigma_{11}, \sigma_{22}, \sigma_{33}, \sigma_{12}\sigma_{23}, \sigma_{31}]^T$, strain tensor is given by the vector $\varepsilon = [\varepsilon_{11}, \varepsilon_{22}, \varepsilon_{33}, \gamma_{12}, \gamma_{23}, \gamma_{13}]^T$ (observe, that $\gamma_{12} = 2\varepsilon_{12}$) and the surface tractions are defined by the following formulae

$$\varepsilon = A \cdot u, \quad \sigma = D \cdot \varepsilon, \quad t = B \cdot \sigma. \tag{3}$$

The first shape functional under consideration depends on the displacement field,

$$J_u(\rho) = \int_{\Omega_\rho} F(u^\rho)\, d\Omega, \qquad F(u^\rho) = (u^\rho \cdot H \cdot u^\rho)^p = ((u^\rho)^T H u^\rho)^p , \tag{4}$$

where F is a C^2 function. It is also useful for further applications in the framework of elasticity to introduce the yield functional of the form

$$J_\sigma(\rho) = \int_{\Omega_\rho} [\sigma(u^\rho) \cdot S \cdot \sigma(u^\rho)]\, d\Omega = \int_{\Omega_\rho} [\sigma(u^\rho)^T S \sigma(u^\rho)]\, d\Omega, \tag{5}$$

where S is an isotropic matrix. Isotropicity means here, that S may be expressed as follows

$$S = [s_{ij}] = \begin{bmatrix} l+2m & l & l & 0 & 0 & 0 \\ l & l+2m & l & 0 & 0 & 0 \\ l & l & l+2m & 0 & 0 & 0 \\ 0 & 0 & 0 & 4m & 0 & 0 \\ 0 & 0 & 0 & 0 & 4m & 0 \\ 0 & 0 & 0 & 0 & 0 & 4m \end{bmatrix},$$

where l, m are real constants. Their values may vary for particular yield criteria. The following assumption assures, that J_u, J_σ are well defined for solutions of the elastity system.

(A) The domain Ω has piecewise smooth boundary, which may have reentrant corners with $\alpha < 2\pi$ created by the intersection of two planes. In addition, g, h must be compatible with $u \in H^1(\Omega; \mathbb{R}^3)$.

The interior regularity of u in Ω is determined by the regularity of the right hand side f of the elasticity system. For simplicity the following notation is used for functional spaces,

$$H_g^1(\Omega_\rho) = \{\psi \in H^1(\Omega_\rho; \mathbb{R}^3) \mid \psi = g \text{ on } \Gamma_1\},$$
$$H_{\Gamma_1}^1(\Omega_\rho) = \{\psi \in H^1(\Omega_\rho; \mathbb{R}^3) \mid \psi = 0 \text{ on } \Gamma_1\},$$
$$H_{\Gamma_1}^1(\Omega) = \{\psi \in H^1(\Omega; \mathbb{R}^3) \mid \psi = 0 \text{ on } \Gamma_1\}.$$

The weak solutions to the elasticity systems are defined in the standard way.

Find $u^\rho \in H_g^1(\Omega_\rho)$ such that, for every $\phi \in H_{\Gamma_1}^1(\Omega)$,

$$-\int_{\Omega_\rho} (Au^\rho)^T DA\phi \, d\Omega + \int_{\Gamma_2} h^T \phi \, dS = \int_{\Omega_\rho} f^T \phi \, d\Omega. \tag{6}$$

We introduce the adjoint state equations in order to simplify the form of shape derivatives of functionals J_u, J_σ. For functionals J_u, J_σ they take on the form:

Find $w^\rho, v^\rho \in H_{\Gamma_1}^1(\Omega_\rho)$ such that, for every $\phi \in H_{\Gamma_1}^1(\Omega)$,

$$-\int_{\Omega_\rho} (Aw^\rho)^T DA\phi \, d\Omega = \int_{\Omega_\rho} F_u'(u^\rho)^T \phi \, d\Omega, \tag{7}$$

$$-\int_{\Omega_\rho} (Av^\rho)^T DA\phi \, d\Omega = 2\int_{\Omega_\rho} \sigma(u^\rho)^T SDA\phi \, d\Omega. \tag{8}$$

3. Topological Derivative

We shall define the topological derivative of the functionals J_u, J_σ at the point x_0 as:

$$\mathcal{T} J_u(x_0) = \lim_{\rho \downarrow 0} \frac{dJ_u(\rho)}{d(|B_\rho(x_0)|)}, \quad \mathcal{T} J_\sigma(x_0) = \lim_{\rho \downarrow 0} \frac{dJ_\sigma(\rho)}{d(|B_\rho(x_0)|)}. \tag{9}$$

Now we may formulate the following result, giving the constructive method for computing the topological derivatives:

Theorem 1. *Assume that the distributed force is sufficiently regular, $f \in C^1(\Omega;\mathbb{R}^3)$, and (A) is satisfied, then*

$$\mathcal{T}J_u(x_0) = -\frac{1}{4\pi}\left[4\pi F(u) + 4\pi f^T w + K(D^{-1};\sigma(u),\sigma(w))\right]_{x=x_0}, \tag{10}$$

$$\mathcal{T}J_\sigma(x_0) = -\frac{1}{4\pi}\left[K(S;\sigma(u),\sigma(u)) + 4\pi f^T v + K(D^{-1};\sigma(u),\sigma(v))\right]_{x=x_0}, \tag{11}$$

where w, $v \in H^1_{\Gamma_1}(\Omega)$ are adjoint variables satisfying the integral identities (7) and (8) for $\rho = 0$, i.e. in the whole domain Ω instead of Ω_ρ, that is

$$\begin{aligned} -\int_\Omega (Aw)^T DA\phi \, d\Omega &= \int_\Omega F_u'(u)^T \phi \, d\Omega, \\ -\int_\Omega (Av)^T DA\phi \, d\Omega &= 2\int_\Omega \sigma(u)^T S\sigma(\phi) \, d\Omega \end{aligned} \tag{12}$$

for all test functions $\phi \in H^1_{\Gamma_1}(\Omega)$.

Some of the terms in (10), (11) require explanation. Given the 3×3 matrix M the function K is defined as an integral over the unit sphere $S_1(0) = \{x \in \mathbb{R}^3 \mid \|x\| = 1\}$ of the following functions:

$$K(M;\sigma(u(x_0)),\sigma(v(x_0))) = \int_{S_1(0)} \sigma^\infty(u(x_0);x)^T \cdot M \cdot \sigma^\infty(v(x_0);x)\, dS$$

The symbol $\sigma^\infty(u(x_0);x)$ denotes the stresses for the solution of the elasticity system in the infinite domain $\mathbb{R}^3 \setminus \overline{B_1(0)}$ with the following boundary conditions:

- no tractions are applied on the surface of the ball, $S_1(0) = \partial B_1(0)$;
- the stresses $\sigma^\infty(u(x_0);x)$ tends to the constant value $\sigma(u(x_0))$ as $\|x\| \to \infty$.

In this notation $\sigma^\infty(u(x_0);x)$ is a function of space variables depending on the functional parameter $u(x_0)$, while $\sigma(u(x_0))$ is a value of the stress tensor computed in the point x_0 for the solution u.

In order to derive the above formulae (10),(11) we calculate the derivatives of the functional $J_u(\rho)$ with respect to the parameter ρ, which determines the size of the hole $B_\rho(x_0)$, by using the material derivative method [13]. Then we pass to the limit $\rho \downarrow 0$ using the asymptotic expansions of u_ρ with respect to ρ. For the functional J_u the shape derivative with respect to ρ is given by

$$J_u'(\rho) = \int_{\Omega_\rho} F_u'(u^\rho)^T u^{\rho\prime} \, d\Omega - \int_{S_\rho(x_0)} F(u^\rho)\, dS, \tag{13}$$

and in the same way for the state equation:

$$-\int_{\Omega_\rho} (Au^{\rho\prime})^T DA\phi \, d\Omega + \int_{S_\rho(x_0)} (Au^\rho)^T DA\phi \, dS = -\int_{S_\rho(x_0)} f^T \phi \, dS, \tag{14}$$

where $u^{\rho\prime}$ is the shape derivative, i.e. the derivative of u_ρ with respect to ρ, [13].

After substitution of the test functions $\phi = w^\rho$ in the state equation, $\phi = u^{\rho\prime}$ in the adjoint state equation, we get

$$J_u'(\rho) = -\int_{S_\rho(x_0)} [F(u^\rho) + f^T w^\rho + \sigma(u_\rho)^T D^{-1} \sigma(w^\rho)]\, dS, \tag{15}$$

and similarly for J_σ

$$J_\sigma'(\rho) = -\int_{S_\rho(x_0)} [\sigma(u_\rho)^T S \sigma(u^\rho) + f^T v^\rho + \sigma(u_\rho)^T D^{-1} \sigma(v^\rho)]\, dS. \tag{16}$$

Observe, that both matrices D^{-1} and S are isotropic, and therefore the corresponding bilinear forms in terms of stresses are invariant with respect to the rotations of the coordinate system.

Now we exploit the fact, that

$$\frac{dJ_u(\rho)}{d(|B_\rho(x_0)|)} = \frac{1}{4\pi\rho^2} \cdot \frac{dJ_u}{d\rho},$$

and use the existence of the asymptotic expansions for u^ρ in the neighbourhood of $B_\rho(x_0)$, namely

$$u^\rho = u(x_0) + u^\infty + O(\rho^2). \tag{17}$$

In addition, u^∞ is proportional to ρ, $\|u^\infty\|_{\mathbb{R}^3} = O(\rho)$, on the surface $S_\rho(x_0)$ of the ball. The expansion of $\sigma(u^\rho)$ corresponding to (17) has the form

$$\sigma(u^\rho) = \sigma^\infty(u(x_0); x) + O(\rho). \tag{18}$$

It may be proved, that w^ρ and v^ρ have similar expansions.

Using the formulae (17),(18) we may justify the following passages to the limit:

$$\lim_{\rho\downarrow 0} \frac{1}{\rho^2} \int_{S_\rho(x_0)} (Au^\rho)^T D A v^\rho \, dS = K(D^{-1}; \sigma(u(x_0)), \sigma(v(x_0))),$$

$$\lim_{\rho\downarrow 0} \frac{1}{\rho^2} \int_{S_\rho(x_0)} (Au^\rho)^T D A w^\rho \, dS = K(D^{-1}; \sigma(u(x_0)), \sigma(w(x_0))),$$

$$\lim_{\rho\downarrow 0} \frac{1}{\rho^2} \int_{S_\rho(x_0)} \sigma(u_\rho)^T S \sigma(u^\rho)\, dS = K(S; \sigma(u(x_0)), \sigma(u(x_0))),$$

$$\lim_{\rho\downarrow 0} \frac{1}{\rho^2} \int_{S_\rho(x_0)} F(u^\rho)\, dS = 4\pi F(u(x_0)),$$

$$\lim_{\rho\downarrow 0} \frac{1}{\rho^2} \int_{S_\rho(x_0)} f^T \cdot w^\rho \, dS = 4\pi f^T(x_0) \cdot w(x_0),$$

$$\lim_{\rho\downarrow 0} \frac{1}{\rho^2} \int_{S_\rho(x_0)} f^T \cdot v^\rho \, dS = 4\pi f^T(x_0) \cdot v(x_0).$$

This completes the proof of the theorem.

The functions denoted above as

$$K(S;\sigma(u(x_0)),\sigma(u(x_0))) \quad \text{and} \quad K(D^{-1};\sigma(u(x_0)),\sigma(w(x_0)))$$

cannot be obtained in the closed form, in contrast with the two dimensional case. Therefore, we must approximate them using numerical quadrature. It is possible, because we may calculate the values of integrands at any point on the sphere.

4. Superposition of Asymptotic Solutions

Let us suppose, that the infinite elastic medium is stretched uniformly in the x_3 direction, so that at infinity we have $\sigma_{33} = const \neq 0$, while $\sigma_{11} = \sigma_{22} = 0$. This medium has a ball-shaped cavity, and we put the origin of the coordinate system in its centre. Then we select an arbitrary point on the surface of the ball and construct a plane Π tangent to the surface. The spherical coordinate system $\{\theta, \phi, r\}$ induces local orthogonal coordinates with the basis $\{\mathbf{e}_\theta, \mathbf{e}_\phi, \mathbf{e}_r\}$. Observe, that $\mathbf{e}_\theta, \mathbf{e}_\phi$ lie on the plane Π, see Fig. 1..

The local behaviour of the stresses around such a cavity is known (the so-called Leon solution, see e.g. [4]). Let the radius of the ball be ρ. Then in the defined above tangent coordinate system $\{\mathbf{e}_\theta, \mathbf{e}_\phi, \mathbf{e}_r\}$

$$\begin{aligned}
\sigma_{rr} &= \frac{\sigma_{33}}{7-5\nu}\{6[(\frac{\rho}{r})^3 - (\frac{\rho}{r})^5] + +[(7-5\nu) - 5(5-\nu)(\frac{\rho}{r})^3 + 18(\frac{\rho}{r})^5]\cos^2\theta\},\\
\sigma_{\theta\theta} &= \frac{\sigma_{33}}{2(7-5\nu)}\{2(7-5\nu) + (4-5\nu)(\frac{\rho}{r})^3 + 9(\frac{\rho}{r})^5 +\\
&+[-2(7-5\nu) + 5(1-2\nu)(\frac{\rho}{r})^3 - 21(\frac{\rho}{r})^5]\cos^2\theta\},\\
\sigma_{\phi\phi} &= \frac{3\sigma_{33}}{2(7-5\nu)}\{-(2-5\nu)(\frac{\rho}{r})^3 + (\frac{\rho}{r})^5 + +5[(1-2\nu)(\frac{\rho}{r})^3 - (\frac{\rho}{r})^5]\cos^2\theta\},\\
\sigma_{r\theta} &= \frac{\sigma_{33}}{7-5\nu}[-(7-5\nu) - 5(1+\nu)(\frac{\rho}{r})^3 + 12(\frac{\rho}{r})^5]\sin\theta\cos\theta.
\end{aligned}$$

Other components of the stress tensor vanish due to the symmetry.

The explicit expressions for displacements, corresponding to the stresses listed above, exist and are known, but we do not need them here. The displacements satisfy conditions, which may described as follows:

- the points on x_3 axis move only up or down;
- the equator of the sphere laying in the x_1, x_2 plane remains in this plane after deformation and does not rotate.

In addition, these displacements are proportional to ρ on the surface of the sphere.

The Leon solution given above corresponds to the situation, when very far from the origin (at infinity) the state of the material may be characterized by the following conditions:

- the axis Ox_1, Ox_2, Ox_3 are the principal stress directions;
- only σ_{33} does not vanish at infinity.

However, the solutions denoted in the previous section as σ^∞, u^∞ do not satisfy these conditions. But nevertheless they may be constructed by superposition of three Leon solutions. The procedure consists of the following steps:

- find principal stress directions of $\sigma(u(x_0))$;
- solve the Leon problem corresponding to the stretching in every of these principal directions separately;
- use superposition to add these three solutions.

Now let us consider the surface $S_\rho(0)$ of the sphere. If we substitute $r = \rho$ in the Leon solution, then there remain only two nonvanishing components of the stress tensor,

$$\sigma_{\theta\theta} = \frac{\sigma_{33}}{2(7-5\nu)}[27 - 15\nu - 30\cos^2\theta], \quad \sigma_{\phi\phi} = \frac{3\sigma_{33}}{2(7-5\nu)}[5\nu - 1 - 10\nu\cos^2\theta].$$

This means, that

- on the sphere $S_\rho(0)$ we have a plane stress state;
- the versors $\mathbf{e}_\theta, \mathbf{e}_\phi$ define principal stress directions for this plane stress state.

To simplify the above formulae, let us introduce the notations:

$$p_\theta = \frac{27-15\nu}{2(7-5\nu)}, \quad q_\theta = \frac{-30}{2(7-5\nu)}, \quad p_\phi = \frac{3(5\nu-1)}{2(7-5\nu)}, \quad q_\phi = \frac{-10\nu}{2(7-5\nu)}.$$

This leads to the expressions

$$\sigma_{\theta\theta} = \sigma_{33}(p_\theta + q_\theta\cos^2\theta), \qquad \sigma_{\phi\phi} = \sigma_{33}(p_\phi + q_\phi\cos^2\theta),$$

Observe, that we could exchange roles of axes Ox_1, Ox_2, Ox_3 and stretch the medium for example in the direction Ox_1. Then the corresponding local tangent coordinate system is rotated with respect to the system shown in Fig. 1. Moreover, the rotation angles δ_1, δ_2 are different at every point on the sphere. To be precise: the direction of $\mathbf{e}_\phi(\sigma_{33})$ is the same as given by the vector $\mathbf{e}_3 \times \mathbf{e}_r$, while $\mathbf{e}_\phi(\sigma_{11})$ lies along $\mathbf{e}_1 \times \mathbf{e}_r$. Therefore, if we want to superimpose the stresses at the selected point on the sphere due to stretchings in three directions, it is necessary to perform the following steps:

Step 1: select the reference coordinate system, say $\{\mathbf{e}_\theta, \mathbf{e}_\phi\}$ associated with the stretching in the direction Ox_3;

Step 2: compute the rotation angles δ_1, δ_2 of the coordinate systems $\{\mathbf{e}_{\theta_1}, \mathbf{e}_{\phi_1}\}$, $\{\mathbf{e}_{\theta_2}, \mathbf{e}_{\phi_2}\}$ (corresponding to stretchings in directions Ox_1 and Ox_2 respectively) with respect to the reference system, see Fig. 2;

Step 3: add all the stress states, transforming the stresses obtained in Step 2 to the reference coordinates (e.g. using Mohr's circle).

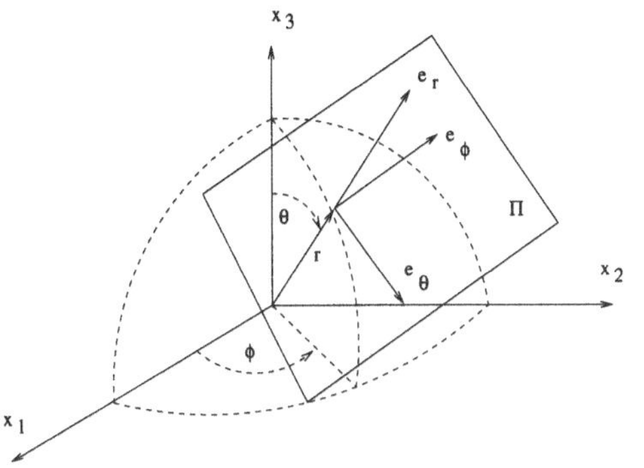

FIGURE 1. The spherical coordinate system and the plane Π tangent to the surface of the sphere. The local basis $\{\mathbf{e}_\theta, \mathbf{e}_\phi, \mathbf{e}_r\}$ is also shown.

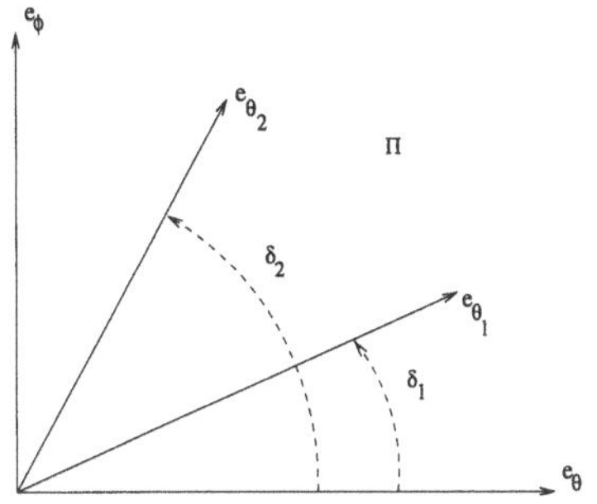

FIGURE 2. The relation between the reference coordinate system $\{\mathbf{e}_\theta, \mathbf{e}_\phi\}$ on the plane Π corresponding to the stretching in the direction Ox_3, and $\{\mathbf{e}_{\theta_1}, \mathbf{e}_{\phi_1}\}$, $\{\mathbf{e}_{\theta_2}, \mathbf{e}_{\phi_2}\}$, obtained in the Step 3 and represented by versors $\mathbf{e}_{\theta_1}$ and $\mathbf{e}_{\theta_2}$.

If we denote by $\theta_1, \theta_2, \theta_3$ the angles between the vectors $\mathbf{r}$ and $\mathbf{e}_1, \mathbf{e}_2, \mathbf{e}_3$ respectively, then the plane stresses on Π take on the forms (observe, that now we denote by θ_3 the spherical coordinate θ, in order to obtain symmetrical formulae):

$$\sigma_{\theta_1\theta_1} = \sigma_{11}(p_\theta + q_\theta \cos^2\theta_1), \qquad \sigma_{\phi_1\phi_1} = \sigma_{11}(p_\phi + q_\phi \cos^2\theta_1),$$

in the $\mathbf{e}_{\theta_1}, \mathbf{e}_{\phi_1}$ coordinate system,

$$\sigma_{\theta_2\theta_2} = \sigma_{22}(p_\theta + q_\theta \cos^2\theta_2), \qquad \sigma_{\phi_2\phi_2} = \sigma_{22}(p_\phi + q_\phi \cos^2\theta_2),$$

in the $\mathbf{e}_{\theta_2}, \mathbf{e}_{\phi_2}$ coordinate system, and finally

$$\sigma_{\theta_3\theta_3} = \sigma_{33}(p_\theta + q_\theta \cos^2\theta_3), \qquad \sigma_{\phi_3\phi_3} = \sigma_{33}(p_\phi + q_\phi \cos^2\theta_3),$$

in the $\mathbf{e}_{\theta_3}, \mathbf{e}_{\phi_3}$ coordinate system. All these three stress states must be superimposed in Step 3.

Let us repeat here for clarity, that the solution obtained in this way corresponds to the situation, when at infinity the principal stress directions coincide with the axis Ox_1, Ox_2, Ox_3 and the principal stresses have values $\{\sigma_{11}, \sigma_{22}, \sigma_{33}\}$.

In our situation we have a given displacement field u defined by the state equation and the corresponding stress $\sigma(u)$ which changes from point to point in the domain. Therefore, at every point $x \in \Omega$ the directions of principal stresses are also different. Hence we use one global reference coordinates, the same in which

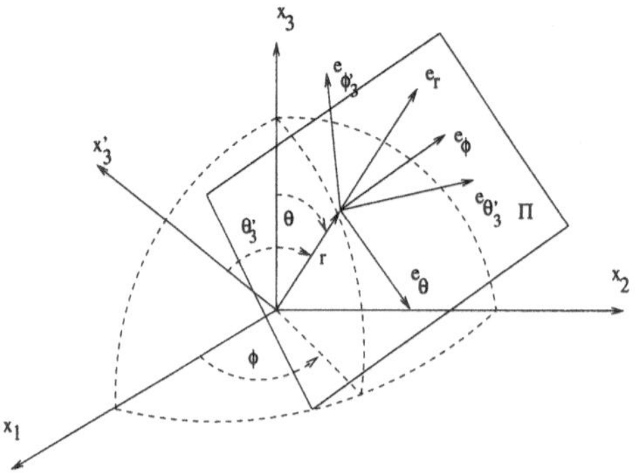

FIGURE 3. One of the principal stress axes Ox'_3 in the original spherical coordinates. The basis $\{\mathbf{e}_{\theta'_3}, \mathbf{e}_{\phi'_3}\}$ shows principal directions on Π corresponding to stretching along Ox'_3.

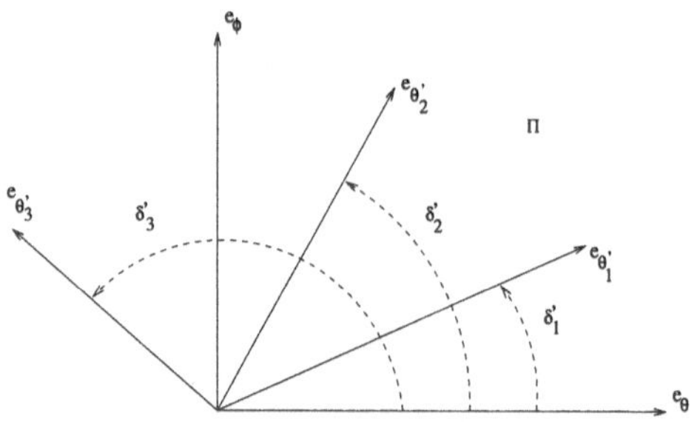

FIGURE 4. The original coordinates on Π and the rotated principal directions $\mathbf{e}_{\theta'_1}, \mathbf{e}_{\theta'_2}, \mathbf{e}_{\theta'_3}$ corresponding to stretching along Ox'_1, Ox'_2, Ox'_3 accordingly.

our original domain is defined, and treat the principal stress directions as rotated with respect to this system, see Fig. 3.

Let us fix the point $x_0 \in \Omega$. We define the principal stress directions at this point by versors $\{\mathbf{e}'_1, \mathbf{e}'_2, \mathbf{e}'_3\}$ forming with the vector $\mathbf{r}$ the angles $\theta'_1, \theta'_2, \theta'_3$. For clarity, we show again the formulae for the corresponding stresses on Π:

$$
\begin{aligned}
\sigma_{\theta'_1\theta'_1} &= \sigma'_{11}(p_\theta + q_\theta \cos^2\theta'_1), & \sigma_{\phi'_1\phi'_1} &= \sigma'_{11}(p_\phi + q_\phi \cos^2\theta'_1),\\
\sigma_{\theta'_2\theta'_2} &= \sigma'_{22}(p_\theta + q_\theta \cos^2\theta'_2), & \sigma_{\phi'_2\phi'_2} &= \sigma'_{22}(p_\phi + q_\phi \cos^2\theta'_2),\\
\sigma_{\theta'_3\theta'_3} &= \sigma'_{33}(p_\theta + q_\theta \cos^2\theta'_3), & \sigma_{\phi'_3\phi'_3} &= \sigma'_{33}(p_\phi + q_\phi \cos^2\theta'_3),
\end{aligned}
$$

where $\sigma'_{11}, \sigma'_{22}, \sigma'_{33}$ are the local values at the point x_0 of principal stresses, computed for the solution u of the elasticity problem. The axes $\{\mathbf{e}_{\theta'_1}, \mathbf{e}_{\phi'_1}\}$, $\{\mathbf{e}_{\theta'_2}, \mathbf{e}_{\phi'_2}\}$, $\{\mathbf{e}_{\theta'_3}, \mathbf{e}_{\phi'_3}\}$, are all rotated with respect to the axes induced by the global coordinate system $\{\mathbf{e}_\theta, \mathbf{e}_\phi\}$ by the angles $\delta'_1, \delta'_2, \delta'_3$, see Fig. 4. If these angle are known, we may transform all the stresses listed above to the $\{\mathbf{e}_\theta, \mathbf{e}_\phi\}$ coordinates. According to the well-known formulae,

$$\sigma_{\theta\theta}(\sigma'_{11}) = \frac{1}{2}(\sigma_{\theta'_1\theta'_1} + \sigma_{\phi'_1\phi'_1}) + \frac{1}{2}(\sigma_{\theta'_1\theta'_1} - \sigma_{\phi'_1\phi'_1})\cos 2\delta'_1,$$
$$\sigma_{\phi\phi}(\sigma'_{11}) = \frac{1}{2}(\sigma_{\theta'_1\theta'_1} + \sigma_{\phi'_1\phi'_1}) - \frac{1}{2}(\sigma_{\theta'_1\theta'_1} - \sigma_{\phi'_1\phi'_1})\cos 2\delta'_1,$$
$$\sigma_{\theta\phi}(\sigma'_{11}) = -\frac{1}{2}(\sigma_{\theta'_1\theta'_1} - \sigma_{\phi'_1\phi'_1})\sin 2\delta'_1.$$

The same relations hold for $\sigma_{\theta\theta}(\sigma'_{22}), \sigma_{\theta\phi}(\sigma'_{33})$. To obtain the resulting stresses, we add the individual contributions,

$$\sigma_{\theta\theta} = \sigma_{\theta\theta}(\sigma'_{11}) + \sigma_{\theta\theta}(\sigma'_{22}) + \sigma_{\theta\theta}(\sigma'_{33}),$$

and similarly for $\sigma_{\phi\phi}, \sigma_{\theta\phi}$. To make the above formulae complete, we have to calculate the angles δ'_i, θ'_i, $i = 1, 2, 3$. We use the following geometrical relations,

$$\mathbf{e}_\phi = \frac{\mathbf{e}_3 \times \mathbf{e}_r}{\|\mathbf{e}_3 \times \mathbf{e}_r\|}, \qquad \mathbf{e}_{\phi'_i} = \frac{\mathbf{e}'_i \times \mathbf{e}_r}{\|\mathbf{e}'_i \times \mathbf{e}_r\|}, \qquad i = 1, 2, 3.$$

This allows us to obtain the required trigonometric functions:

$$\cos\theta'_i = \mathbf{e}'_i \cdot \mathbf{e}_r, \qquad \cos\delta'_i = \mathbf{e}_{\phi'_i} \cdot \mathbf{e}_\phi, \qquad \sin\delta'_i = \|\mathbf{e}_{\phi'_i} \times \mathbf{e}_\phi\|, \qquad i = 1, 2, 3.$$

Before we collect all the formulae for stresses in the form convenient for numerical computations, it is necessary to introduce some additional notations.

$$\begin{aligned}
\boldsymbol{\sigma}_\Pi &= [\sigma_{\theta\theta}, \sigma_{\phi\phi}, \sigma_{\theta\phi}]^T & \boldsymbol{\sigma}'_p &= [\sigma'_{11}, \sigma'_{22}, \sigma'_{33}]^T \\
\mathbf{c} &= [\cos 2\delta'_1, \cos 2\delta'_2, \cos 2\delta'_3]^T & \mathbf{s} &= [\sin 2\delta'_1, \sin 2\delta'_2, \sin 2\delta'_3]^T \\
p_p &= \frac{1}{2}(p_\theta + p_\phi) & p_m &= \frac{1}{2}(p_\theta - p_\phi) \\
q_p &= \frac{1}{2}(q_\theta + q_\phi) & q_m &= \frac{1}{2}(q_\theta - q_\phi) \\
C &= \operatorname{diag}\{\cos^2\theta'_1, \cos^2\theta'_2, \cos^2\theta'_3\} & \mathbf{1} &= [1, 1, 1]^T.
\end{aligned}$$

Then the stress state on the plane Π induced by the displacement field $\mathbf{u}$ may be obtained through a simple matrix multiplication

$$\boldsymbol{\sigma}_\Pi(\mathbf{u}) = T \cdot \boldsymbol{\sigma}'_p,$$

where the matrix C has the form

$$T = \begin{bmatrix} p_p \cdot \mathbf{1}^T + q_p \cdot \mathbf{1}^T \cdot C + p_m \cdot \mathbf{c}^T + q_m \cdot \mathbf{c}^T \cdot C \\ p_p \cdot \mathbf{1}^T + q_p \cdot \mathbf{1}^T \cdot C - p_m \cdot \mathbf{c}^T - q_m \cdot \mathbf{c}^T \cdot C \\ -p_m \cdot \mathbf{s}^T - q_m \cdot \mathbf{s}^T \cdot C \end{bmatrix}.$$

In our case we have to do with two displacement fields, representing the state $\mathbf{u}$ and the adjoint state $\mathbf{v}$ of the system, so these formulae are used twice, in order to obtain $\boldsymbol{\sigma}_\Pi(\mathbf{u})$ and $\boldsymbol{\sigma}_\Pi(\mathbf{v})$.

The derivation described above is valid almost everywhere on the sphere, with the exception of 6 points, where the axes Ox'_1, Ox'_2, Ox'_3 pierce its surface.

However, there are no singularities there, and these particular cases may be treated easily, but we shall not describe it here.

Remark 1. *In the case of the energy functional the form of the topological derivative is well known in the literature in the different context than the shape optimization. We refer the reader to the work in progress [10] on the comparison of the different approaches to the derivation of the topological derivatives and the relation of the topological derivatives of the energy functional to the existing formulae in the literature in mechanics. Here we recall the formula given e.g. in [8].*
Let ω denote the ball in $\mathbb{R}^3$ of radius R and the center at the origin and let ω_ε be the ball of radius εR. Denote $\Omega_\varepsilon = \Omega \backslash \omega_\varepsilon$ and consider the elastic energy functional

$$\mathcal{E}(\Omega_\varepsilon) = \int_{\Omega_\varepsilon} \sigma(u_\varepsilon) : \epsilon(u_\varepsilon)$$

in the domain $\Omega_\varepsilon \subset \mathbb{R}^3$. It follows that the following expansion is obtained for the isotropic elasticity, see formula (5.16) in [8],

$$\mathcal{E}(\Omega_\varepsilon) = \mathcal{E}(\Omega) + \varepsilon^2 \sigma : H : \sigma + o(\varepsilon^2) ,$$

where $\sigma = A\epsilon$ denotes stress tensor in the body without cavity ω_ε, $A = 3K\Lambda_1 + 2\mu\Lambda_2$, $3K = \dfrac{E}{1-2\nu}$ and $2\mu = \dfrac{E}{1-\nu}$, finally

$$H = \frac{2(1-\nu)}{E} \pi R^3 (\Lambda_1 + \frac{10(1+\nu)}{7-5\nu} \Lambda_2)$$

and we use standard notation of isotropic elasticity.

This is much simplified in comparison to the general situation. The reason lies in the fact, that the adjoint variable coincides with the state, so, using isotropy, we may write down the integrand in the definition of the function $K(M; \sigma(u), \sigma(v))$ explicitely in terms of spherical coordinates. In general case of $u \neq v$ the tensors $\sigma(u)$ and $\sigma(v)$ are computed in different coordinate systems and therefore the principal directions of their projections on the plane tangent to the sphere are are rotated with respect to each other by the variable angle.

5. The Case of 2D Elasticity

For the convenience of the reader we recall here the results derived in [14] for the 2D case. Let us consider the elasticity equations in the plane,

$$A^T D A u = f \quad \text{in} \quad \Omega, \qquad u = g \quad \text{on} \quad \Gamma_1, \qquad B^T D A u = h \quad \text{on} \quad \Gamma_2, \tag{19}$$

and the same system for u^ρ in the domain with the circular hole $B_\rho(x_0) \subset \Omega$ centered at $x_0 \in \Omega$, $\Omega_\rho = \Omega \setminus \overline{B_\rho(x_0)}$, with additional boundary condition

$$B^T D A u^\rho = 0 \quad \text{on} \quad \Gamma_\rho. \tag{20}$$

Assuming that $0 \in \Omega$, we can consider the case $x_0 = 0$.

Here $u = (u_1, u_2)^T$ denotes the displacement field, and the operator A and matrices D, B are the 2D counterparts of their full versions, so that if the stress tensor is replaced by the vector $\sigma = [\sigma_{11}, \sigma_{22}, \sigma_{12}]^T$, the strain is given by $\varepsilon = [\varepsilon_{11}, \varepsilon_{22}, \gamma_{12}]^T$, then the formulae (3) hold.

The principal stresses associated with the displacement field u are denoted by $\sigma_I(u)$, $\sigma_{II}(u)$, the trace of the stress tensor $\sigma(u)$ is denoted by $\mathrm{tr}\sigma(u) = \sigma_I(u) + \sigma_{II}(u)$.

The shape functionals J_u,J_σ are defined in the same way as in the 3D problem, with the matrix S isotropic (that is similar to D). Only J_σ is slightly more general:

$$J_\sigma(\rho) = \int_{\Omega_\rho} [\sigma(u^\rho) \cdot S \cdot \sigma(u^\rho)]^p \, d\Omega = \int_{\Omega_\rho} [\sigma(u^\rho)^T S \sigma(u^\rho)]^p \, d\Omega, \tag{21}$$

The weak solutions to the elasticity system as well as adjoint equations are defined also analoguosly to the 3D case. Then we may formulate the following result [14]:

Theorem 2. *The expressions for the topological derivatives of the functionals J_u,J_σ have the form*

$$\mathcal{T} J_u(0) = -[\,F(u) + f^T w + \frac{1}{E}(\,a_u a_w + 2b_u b_w \cos 2\delta\,)\,]_{x=x_0}, \tag{22}$$

$$\mathcal{T} J_\sigma(0) = -[s_{22}^p K_p(a_u, b_u) + f^T v + \frac{1}{E}(\,a_u a_v + 2b_u b_v \cos 2\delta\,)\,]_{x=x_0}. \tag{23}$$

Here the function K_p is defined by the expression

$$K_p(a,b) = \frac{1}{2\pi} \int_0^{2\pi} (a - 2b\cos 2\theta)^{2p} \, d\theta = \begin{cases} a^2 + 2b^2 & \text{for } p = 1 \\ a^4 + 6b^4 + 12a^2 b^2 & \text{for } p = 2 \end{cases}$$

and in addition

$$a_u = \mathrm{tr}\sigma(u), \qquad a_w = \mathrm{tr}\sigma(w), \qquad a_v = \mathrm{tr}\sigma(v),$$
$$b_u = \sigma_I(u) - \sigma_{II}(u), \qquad b_w = \sigma_I(w) - \sigma_{II}(w), \qquad b_v = \sigma_I(v) - \sigma_{II}(v).$$

Finally, the angle δ denotes the angle between principal stress directions for displacement fields u and w in (22), and for displacement fields u and v in (23).

In comparison to the 3D case the expressions are much simpler, due to the fact, that on the plane the rotation of one coordinate system with respect to the other is defined by the single value of the angle (here δ). This is a purely 2D phenomenon and it allows us to compute the value of K_p in the closed form.

References

[1] H.A. ESCHENAUER, V.V. KOBELEV, A. SCHUMACHER *Bubble method for topology and shape optimization of structures* Struct. Optimiz. **8**(1994), pp. 42–51.

[2] S. GARREAU, PH. GUILLAUME, M. MASMOUDI, *The topological asymptotic for PDE systems: the elasticity case* (2000) to appear.

[3] D. GÖHDE, *Singuläre Störung von Randwertproblemen durch ein kleines Loch im Gebiet*, Zeitschrift für Analysis und ihre Anwendungen Vol.4(5)(1985), pp. 467–477.

[4] H. G. Hahn, *Elastizitätstheorie*, B.G. Teubner Stuttgart, 1985.

[5] A. Herwig, *Elliptische Randwertprobleme zweiter Ordnung in Gebieten mit einer Fehlstelle*, Zeitschrift für Analysis und ihre Anwendungen No.8(2)(1989), pp. 153–161.

[6] A. M. Il'in, *Matching of Asymptotic Expansions of Solutions of Boundary Value Problems*, Translations of Mathematical Monographs, Vol. 102, AMS 1992.

[7] L. Jackowska-Strumiłło, J. Sokołowski, A. Żochowski *The topological derivative method and artificial neural networks for numerical solution of shape inverse problems*, Rapport de Recherche No. 3739, 1999, INRIA-Lorraine.

[8] M. Kachanov, I. Tsukrov, B. Shafiro, *Effective moduli of solids with cavities of various shapes* Appl. Mech. Rev. Volume 47, Number 1, Part 2, pp. S151–S174.

[9] T. Lewinski, J. Sokolowski, *Optimal Shells Formed on a Sphere. The Topological Derivative Method.* Rapport de Recherche No. 3495, 1998, INRIA-Lorraine

[10] T. Lewinski, J. Sokolowski, *Topological derivative for nucleation of non-circular voids* Contemporary Mathematics, American Math. Soc. vol 268, 2000, pp. 341–361.

[11] T. Lewinski, J. Sokolowski, *Energy change due to appearing of cavities in elastic solids* in preparation.

[12] A. Shumacher, *Topologieoptimierung von Bauteilstrukturen unter Verwendung von Lochpositionierungkriterien*, Ph.D. Thesis, Universität–Gesamthochschule–Siegen, Siegen, 1995.

[13] J. Sokołowski, J-P. Zolesio, *Introduction to Shape Optimization. Shape Sensitivity Analysis*, Springer Verlag, 1992.

[14] J. Sokołowski, A. Żochowski *On topological derivative in shape optimization*, SIAM Journal on Control and Optimization. Volume 37, Number 4, 1999, pp. 1251–1272.

[15] J. Sokołowski, A. Żochowski *Topological derivative for optimal control problems*, Proceedings of Fifth International Symposium on Methods and Models in Automation and Robotics, Międzyzdroje, Poland, August, 1998, pp. 111–116; special issue of Control and Cybernetics *Recent Advances in Control of PDEs* **28**(1999) pp. 611–626.

[16] J. Sokołowski, A. Żochowski, *Topological derivatives for elliptic problems*, Inverse Problems, Volume 15, Number 1, 1999, pp. 123–134.

[17] J. Sokołowski, A. Żochowski, *Topological derivatives of shape functionals for elasticity systems* Les prépublications de l'Institut Elie Cartan 35/99.

[18] J. Sokołowski, A. Żochowski, *Optimality conditions for simultaneous topology and shape optimization* Les prépublications de l'Institut Elie Cartan 7/2001

E-mail address: `sokolows@iecn.u-nancy.fr`
E-mail address: `zochowsk@ibspan.waw.pl`

Institut Elie Cartan, Laboratoire de Mathématiques
Université Henri Poincaré Nancy I, B.P. 239
54506 Vandoeuvre lès Nancy Cedex, France

Systems Research Institute of the Polish Academy of Sciences
ul. Newelska 6, 01-447 Warszawa, Poland

International Series of Numerical Mathematics, Vol. 139, 245–257

Control Variational Methods for Differential Equations

Jürgen Sprekels, Dan Tiba

Abstract. We review recent results established in the literature via the optimal control approach to differential equations, and we show that a systematic study of general variational inequalities associated to fourth-order operators can be performed by similar methods.

1. Introduction

Optimal control approaches associated to domain decomposition methods or to fictitious domains methods are well-known in the scientific literature devoted to numerical methods for differential equations. They may be viewed as applications of the general least squares minimization procedure, and we quote the works of Lions and Pironneau [11], Glowinski, Lions and Pironneau [6], Neittaanmäki and Tiba [13], for recent advances in this area.

It turns out that in certain important examples, arising for instance in mechanics, standard variational formulations based on the minimization of energy can be advantageously replaced by appropriate optimal control formulations that yield the existence and the uniqueness of the solution under low regularity conditions on the coefficients, i.e. on the geometric parameters of the problem. Other useful consequences of this new approach concern general results on the continuous/differentiable dependence of the solution on these parameters and, even, explicit solutions (in the case of arches) obtained via duality theory in optimal control. These theoretical developments are important in the setting of shape optimization problems in structural mechanics.

In Section 2, we shall give a brief presentation of some recently obtained results along these lines, following the works of Sprekels and Tiba [14], [15], [16], Ignat, Sprekels and Tiba [10], Arnăutu, Langmach, Sprekels and Tiba [1]. Although complete proofs are not included for the sake of brevity, most relevant arguments are carefully described, and precise quotations of the literature are indicated.

In Section 3, we shall study variational inequalities associated to fourth-order differential operators, emphasizing the applications to obstacle-type problems for clamped arches and plates. We underline that our approach is constructive and easily implemented using piecewise linear finite elements in the computations, since it involves just second order differential equations.

Finally, we notice that our optimal control variational formulation for differential systems provides, via the corresponding Pontryagin maximum principle, a nonstandard decomposition of the original equations, which is at the core of our argument.

2. The Optimal Control Approach

We start with a simplified model (Bendsoe [3]) of a clamped plate, with variable thickness $u \in L^\infty(\Omega)_+$, and with normalized mechanical constants,

$$\Delta(u^3\,\Delta y) = f \quad \text{in } \Omega\,, \tag{2.1}$$

$$y = \frac{\partial y}{\partial n} = 0 \quad \text{on } \partial\Omega\,. \tag{2.2}$$

Here, Ω is a bounded Lipschitzian domain in $I\!R^N$ (for $N = 2$, the plate model is obtained), $f \in L^2(\Omega)$ denotes the load and $y \in H_0^2(\Omega)$ the deflection.

We also consider the distributed control problem ($\varepsilon > 0$ is a "small" parameter),

$$\operatorname{Min}\left\{\frac{1}{2\varepsilon}\int_{\partial\Omega}\left(\frac{\partial y}{\partial n}\right)^2 d\sigma + \frac{1}{2}\int_{\Omega}\ell\,h^2\,dx\right\}, \tag{2.3}$$

subject to

$$\Delta y = \ell\,g + \ell\,h \quad \text{in } \Omega\,, \tag{2.4}$$

$$y = 0 \quad \text{on } \partial\Omega\,. \tag{2.5}$$

Here $\ell = u^{-3} \in L^\infty(\Omega)_+$, and g is defined by $\Delta g = f$ in Ω, $g = 0$ in $\partial\Omega$.

If $0 < m \le u \le M$ a.e. in Ω then $\ell \ge M^{-3}$ a.e. in Ω, and the coercivity of (2.3) gives the existence of a unique optimal pair $[y_\varepsilon\,,\, h_\varepsilon] \in [H^2(\Omega) \cap H_0^1(\Omega)] \times L^2(\Omega)$. It is unique by the strict convexity.

The Pontryagin maximum principle for the unconstrained optimal control problem (2.3)–(2.5) is given by (2.4), (2.5), and, for some $p_\varepsilon \in H^1(\Omega)$, by:

$$\Delta p_\varepsilon = 0 \quad \text{in } \Omega\,, \tag{2.6}$$

$$p_\varepsilon = \frac{1}{\varepsilon}\frac{\partial y_\varepsilon}{\partial n} \quad \text{on } \partial\Omega\,, \tag{2.7}$$

$$p_\varepsilon + h_\varepsilon = 0 \quad \text{a.e. in } \Omega\,. \tag{2.8}$$

The pair $[0\,, -g]$ gives the cost $\frac{1}{2}\int_\Omega \ell\,g^2\,dx$ in (2.3), independently of $\varepsilon > 0$. This shows that $[y_\varepsilon\,,\, h_\varepsilon]$ are bounded with respect to $\varepsilon > 0$ since $\ell \ge M^{-3} > 0$ a.e. in Ω, as noticed before. Moreover, again due to (2.3), $\frac{\partial y_\varepsilon}{\partial n} \to 0$ strongly in $L^2(\partial\Omega)$. From (2.6), (2.8), we get that h_ε is harmonic, which is preserved by passing to the limit in the weak topology of $L^2(\Omega)$. A simple limitting argument in (2.3)–(2.5), and the definitions of ℓ, g, give:

Theorem 2.1. *The solution of* (2.1), (2.2) *is the limit of the optimal states* y_ε *in* $H^2(\Omega) \cap H_0^1(\Omega)$ *weak, for* $\varepsilon \to 0$.

This result was established in the paper of Arnăutu, Langmach, Sprekels and Tiba [1]. It is also valid for simply supported plates, i.e. with (2.2) replaced by

$$y = \Delta y = 0 \quad \text{on } \partial\Omega. \tag{2.2'}$$

The above discussion shows that the original fourth-order boundary value problem (2.1), (2.2) is equivalent with (2.4), (2.5), with h some harmonic mapping in $L^2(\Omega)$, and with the extra condition $\frac{\partial y}{\partial n} = 0$ on $\partial\Omega$.

Assume that $\ell_n \to \ell$ weakly* in $L^\infty(\Omega)$, and denote by y_n the solution of (2.1), (2.2) associated to $u_n = \ell_n^{-\frac{1}{3}}$, and by h_n the corresponding harmonic mappings appearing in (2.4). It is a standard argument to see that $\{y_n\}$ is bounded in $H_0^2(\Omega)$ and that $\{h_n\}$ is bounded in $L^2(\Omega)$; moreover, they weakly converge, on a subsequence, to the limits $y \in H_0^2(\Omega)$, respectively $h \in L^2(\Omega)$. The difficulty to pass to the limit in the equations is related to the products $u_n^3 \Delta y_n$ or $\ell_n h_n$ appearing in (2.1), respectively (2.4), and to the weak convergence. However, as h_n, h are harmonic, the solid mean property gives that $h_n(x) \to h(x)$, $\forall x \in \Omega$, and the Egorov theorem shows that $h_n \to h$ strongly in $L^s(\Omega)$, $\forall s < 2$. Then, we clearly get that $\ell_n h_n \to \ell h$ weakly in $L^2(\Omega)$, and we can pass to the limit in (2.4). Notice that the weak limit of u_n is in general different from $\ell^{-\frac{1}{3}}$, but we have:

Theorem 2.2. *Assume that* $\ell_n \to \ell$ *weakly* in* $L^\infty(\Omega)$. *If* $y = \lim y_n$ *in* $H_0^2(\Omega)$ *weak, then it satisfies the equation*

$$\Delta(\ell^{-1} \Delta y) = f \quad \text{in } \Omega. \tag{2.1'}$$

This result gives the "continuous" dependence of the solution on the coefficient in (2.1), in the weak* topology of $L^\infty(\Omega)$. It was established in Sprekels and Tiba [15] and has important consequences in the existence theory for shape optimization problems or in homogenization problems for plates.

We now consider the differentiability with respect to the coefficient ℓ:

Theorem 2.3. *The mappings* $\ell \mapsto y$ *and* $\ell \mapsto h$ *are Gâteaux differentiable from* $L^\infty(\Omega)$ *into* $H^2(\Omega)$ *and* $L^2(\Omega)$, *respectively, and the directional derivatives at* ℓ *in the direction* $v \in L^\infty(\Omega)$ *satisfy:*

$$\Delta \bar{y} = \ell \bar{h} + v(h + g) \quad \text{in } \Omega, \tag{2.9}$$

$$\bar{y} = \frac{\partial \bar{y}}{\partial n} = 0 \quad \text{on } \partial\Omega, \tag{2.10}$$

$$\Delta \bar{h} = 0 \quad \text{in } \Omega. \tag{2.11}$$

The solution $[\bar{y}, \bar{h}]$ *of* (2.9)–(2.11) *is unique in* $H_0^2(\Omega) \times L^2(\Omega)$.

This result was established in Ignat, Sprekels and Tiba [10], and an essential ingredient in the necessary estimates is the observation that the decomposition of (2.1) provided by (2.4) has the orthogonality property $\Delta y \perp h$, in the $L^2(\Omega)$-inner product. **Theorem 2.3.** allows the writing of the optimality conditions in shape optimization problems for plates, without differentiability assumptions on the coefficients. The obtained gradient can be used in numerical experiments. We also stress that the approximation of (2.1), (2.2) via (2.3)–(2.5) is a simple and efficient method for the computation of solutions to plate equations. Numerical examples related to **Theorems 2.1.–2.3.** can be found in the work of Arnăutu, Langmach, Sprekels and Tiba [1].

Remark. The variant of the control variational approach given by (2.3) includes the penalization in the cost of one of the boundary conditions (2.2). In the sequel, we briefly describe another variant based on the use of constrained control problems.

With this aim, we now consider the Kirchhoff–Love model for clamped arches:

$$\int_0^1 \left[\frac{1}{\delta}(v_1' - c\,v_2)\,(u_1' - c\,u_2) + (v_2' + c\,v_1)'\,(u_2' + c\,u_1)'\right] ds$$

$$= \int_0^1 (f_1\,u_1 + f_2\,u_2)\,ds\,, \quad \forall\, u_1 \in H_0^1(0,1)\,, \quad \forall\, u_2 \in H_0^2(0,1)\,. \quad (2.12)$$

Above, $\varphi : [0,1] \to I\!R^2$ is the parametrization of a smooth clamped arch with the curvature denoted by c, and with the (constant) thickness given by $\sqrt{\delta}$. The mappings $v_1 \in H_0^1(0,1)$, $v_2 \in H_0^2(0,1)$ are the tangential and the normal components of the deformation, while $[f_1\,,\, f_2]$ is a similar notation for the load, in the local system of axes. A thorough presentation of the model for $\varphi \in C^3(0,1)$ via Dirichlet's principle and Korn's inequality may be found in Ciarlet [5].

Let $\theta : [0,1] \to I\!R$ denote the angle between the tangent vector to the arch (given by φ') and the horizontal axis. If φ is smooth, then $\theta' = c$. We also consider the orthogonal matrix

$$W(t) = \begin{pmatrix} \cos\theta(t) & \sin\theta(t) \\ -\sin\theta(t) & \cos\theta(t) \end{pmatrix} \quad (2.13)$$

and the functions $\ell\,, h\,, g_1\,, g_2$ constructed from $f_1, f_2 \in L^2(0,1)$ as follows:

$$g_1 = \delta\,\ell\,, \quad -g_2'' = h\,, \quad g_2(0) = g_2(1) = 0\,, \quad (2.14)$$

$$\begin{bmatrix} \ell \\ h \end{bmatrix}(t) = -\int_0^t W(t)\,W^{-1}(s) \begin{bmatrix} f_1(s) \\ f_2(s) \end{bmatrix} ds\,. \quad (2.15)$$

We define the constrained control problem

$$\operatorname{Min}\left\{\frac{1}{2\,\delta}\int_0^1 u^2\,ds + \frac{1}{2}\int_0^1 (z')^2\,ds\right\}, \quad (2.16)$$

subject to $u \in L^2(0,1)$, $z \in H_0^1(0,1)$, such that the mappings $[v_1\,,\,v_2]$ given by

$$\begin{bmatrix} v_1 \\ v_2 \end{bmatrix}(t) = \int_0^t W(t)\,W^{-1}(s)\begin{bmatrix} u+g_1 \\ z+g_2 \end{bmatrix}(s)\,ds \tag{2.17}$$

satisfy $v_1(1) = v_2(1) = 0$ in the sense that

$$\int_0^1 W^{-1}(s)\begin{bmatrix} u(s)+g_1(s) \\ z(s)+g_2(s) \end{bmatrix} ds = \begin{bmatrix} 0 \\ 0 \end{bmatrix}. \tag{2.18}$$

We underline that relations (2.13)–(2.18) are meaningful under the mere assumption that $\theta \in L^\infty(0,1)$. Then, $[v_1\,,\,v_2] \in L^\infty(0,1)^2$ represent the mild solution of the Cauchy problem (written formally)

$$v_1' - c\,v_2 = u + g_1 \quad \text{in } [0,1], \tag{2.19}$$

$$v_2' + c\,v_1 = z + g_2 \quad \text{in } [0,1], \tag{2.20}$$

$$v_1(0) = v_2(0) = 0. \tag{2.21}$$

The constraint (2.18) is a terminal state constraint, expressed as a control constraint, since the state system (2.17) is in explicit form and the matrix $W(t)$ is nonsingular.

We denote by $[u_\delta\,,\,z_\delta] \in L^2(0,1)\times H_0^1(0,1)$ the unique optimal control associated to (2.16)–(2.18). It exists due to the coercivity of the cost functional and since the pair $[-g_1\,,\,-g_2]$ is clearly admissible. We also denote by $[v_1^\delta\,,\,v_2^\delta] \in L^\infty(0,1)^2$ the optimal state corresponding to $[u_\delta\,,\,z_\delta]$ via (2.17).

Theorem 2.4. *If* $\theta \in W^{2,\infty}(0,1)$, *then* $[v_1^\delta\,,\,v_2^\delta]$ *is the solution of* (2.12).

We briefly indicate the argument:

We get $c \in W^{1,\infty}(0,1)$, and (2.17) can be written in the strong form (2.19)–(2.21). The same holds for (2.15).

The Euler equation associated to (2.16)–(2.18) is

$$\frac{1}{\delta}\int_0^1 u_\delta\,\mu\,ds + \int_0^1 z_\delta'\,\zeta'\,ds = 0, \tag{2.22}$$

for any $[\mu\,,\,\zeta] \in L^2(0,1) \times H_0^1(0,1)$ such that

$$\int_0^1 W^{-1}(s)\begin{bmatrix} \mu(s) \\ \zeta(s) \end{bmatrix} ds = \begin{bmatrix} 0 \\ 0 \end{bmatrix}. \tag{2.23}$$

For any $u_1 \in H_0^1(0,1)$, $u_2 \in H_0^2(0,1)$, we introduce

$$\tilde{\mu} = u_1' - c\,u_2 \in L^2(0,1), \tag{2.24}$$

$$\tilde{\zeta} = u_2' + c\,u_1 \in H_0^1(0,1), \tag{2.25}$$

and we have, consequently, that

$$\begin{bmatrix} u_1 \\ u_2 \end{bmatrix}(t) = \int_0^t W(t)\,W^{-1}(s) \begin{bmatrix} \tilde{\mu}(s) \\ \tilde{\zeta}(s) \end{bmatrix} ds\,. \tag{2.26}$$

As $u_1(1) = u_2(1) = 0$, we see that $[\tilde{\mu}\,,\,\tilde{\zeta}]$ given by (2.24), (2.25), satisfy (2.23) and may be used in (2.22), whence

$$\begin{aligned} 0 \;&=\; \frac{1}{\delta}\int_0^1 \Big((v_1^\delta)' - c\,v_2^\delta - g_1\Big)\Big(u_1' - cu_2\Big)\,ds \\ &\quad + \int_0^1 \Big((v_2^\delta)' + c\,v_1^\delta - g_2\Big)'\Big(u_2' + c\,u_1\Big)'\,ds \\ &= \frac{1}{\delta}\int_0^1 \Big((v_1^\delta)' - c\,v_2^\delta\Big)\Big(u_1' - c\,u_2\Big)\,ds + \int_0^1 \Big((v_2^\delta)' + c\,v_1^\delta\Big)'\Big(u_2' + c\,u_1\Big)'\,ds \\ &\quad - \int_0^1 \ell\Big(u_1' - c\,u_2\Big)\,ds - \int_0^1 h\Big(u_2' + c\,u_1\Big)\,ds\,, \end{aligned} \tag{2.27}$$

where we have used (2.19), (2.20) and (2.14). By partial integration in the last two terms in (2.27), and by (2.15), we recover from (2.27) the equation (2.12).

Remark. **Theorem 2.4.** shows that the constrained control problem (2.16)–(2.18) is a weak formulation of the Kirchhoff–Love model under very low geometric regularity assumptions. Other arguments along these lines can be found in Sprekels and Tiba [16], Ignat, Sprekels and Tiba [10].

We also notice that the constraint (2.18) is affine and finite dimensional. This allows a complete solution of the control problem via duality theory, Barbu and Precupanu [2]. In the work of Ignat, Sprekels and Tiba [10] the dual control problem giving the (two-dimensional) Lagrange multiplier is explicitly derived, and the results are used for numerical experiments with Lipschitzian arches, i.e. for $\varphi \in W^{1,\infty}(0,1)^2$. Moreover, by writing a Pontryagin-type maximum principle for the problem (2.16)–(2.18), continuity and differentiability of the solution $[v_1^\delta\,,\,v_2^\delta]$ with respect to the parameter $\theta \in L^\infty(0,1)$ can be studied. In this way, in Ignat, Sprekels and Tiba [10], a complete theoretical and numerical analysis of shape optimization problems associated with Kirchhoff–Love arches is performed.

Remark. For the plate equation (2.1), (2.2), the corresponding constrained control problem is

$$\mathrm{Min}\left\{\frac{1}{2}\int_\Omega \ell\,h^2\,dx\right\},$$

subject to (2.4), (2.5) and to the constraint

$$\frac{\partial y}{\partial n} = 0 \quad \text{on } \partial\Omega\,. \tag{2.28}$$

We notice that (2.28) is affine, but infinite dimensional. A dual control problem (unconstrained!) can be obtained, but it remains an infinite dimensional optimization problem. Consequently, it is not possible to find an explicit solution, in general, and a standard treatment is to employ the penalization of (2.28) in the cost, as in (2.3).

3. Variational Inequalities for Fourth-order Differential Operators

We shall show that the technique presented in the previous section can also be applied to establish existence results for general variational inequalities.

We examine first the case of clamped plates subjected to unilateral conditions, since it is more intuitive. We define the control problem with state constraints,

$$\text{Min}\left\{\frac{1}{2\varepsilon}\int_{\partial\Omega}\left(\frac{\partial y}{\partial n}\right)^2 d\sigma + \frac{1}{2}\int_{\Omega}\ell\, h^2\, dx\right\}, \tag{3.1}$$

subject to

$$\Delta y = \ell g + \ell h \quad \text{in } \Omega\,, \tag{3.2}$$

$$y = 0 \quad \text{on } \partial\Omega\,, \tag{3.3}$$

$$y \in \mathcal{K}\,. \tag{3.4}$$

The notations are the same as in Section 2, and $\mathcal{K} \subset H^2(\Omega)$ is a closed convex subset such that $\mathcal{K} \cap H_0^2(\Omega) \neq \emptyset$. Then, the pair $[\hat{y}\,,\, \ell^{-1}\Delta\hat{y} - g]$, with $\hat{y} \in \mathcal{K} \cap H_0^2(\Omega)$, is clearly admissible, and the corresponding cost is independent of $\varepsilon > 0$.

We denote by $[y_\varepsilon\,,\, h_\varepsilon] \in H^2(\Omega) \times L^2(\Omega)$, the unique optimal pair of (3.1)–(3.4) (recall that $\ell \geq M^{-3} > 0$ in Ω). We have:

$$\frac{1}{2\varepsilon}\int_{\partial\Omega}\left(\frac{\partial y_\varepsilon}{\partial n}\right)^2 d\sigma + \frac{1}{2}\int_{\Omega}\ell\, h_\varepsilon^2\, dx \leq \frac{1}{2}\int_{\Omega}\ell\,(\ell^{-1}\Delta\hat{y} - g)^2\, dx\,. \tag{3.5}$$

Let $[z\,,\, v] \in H^2(\Omega) \times L^2(\Omega)$ be another admissible pair, i.e. satisfying (3.2)–(3.4). We consider admissible variations of the type

$$[y_\varepsilon\,,\, h_\varepsilon] + \lambda\,[z - y_\varepsilon\,,\, v - h_\varepsilon] \in \mathcal{K} \tag{3.6}$$

with $\lambda \in [0,1]$. By comparing the optimal cost with that associated to (3.6), we get the inequality

$$0 \leq \frac{1}{\varepsilon}\int_{\partial\Omega}\frac{\partial y_\varepsilon}{\partial n}\left(\frac{\partial z}{\partial n} - \frac{\partial y_\varepsilon}{\partial n}\right) d\sigma + \int_{\Omega}\ell\, h_\varepsilon\,(v - h_\varepsilon)\, dx\,. \tag{3.7}$$

We introduce again the auxiliary function $p_\varepsilon \in H^1(\Omega)$ given by (2.6), (2.7), and we underline that this is not the adjoint mapping from control theory, since it does not take into account the state constraint (3.4). A general discussion about this approach in state-constrained control problems, in a different setting, may be found in Bergounioux and Tiba [4].

Inequality (3.7) may be rewritten as

$$0 \leq \int_{\partial\Omega} p_\varepsilon \left(\frac{\partial z}{\partial n} - \frac{\partial y_\varepsilon}{\partial n} \right) d\sigma + \int_\Omega \ell\, h_\varepsilon \,(v - h_\varepsilon)\, dx\,. \tag{3.8}$$

Multiplying by p_ε in the equation for $z - y_\varepsilon$ and integrating by parts, we obtain

$$-\int_\Omega p_\varepsilon\, \ell\,(v - h_\varepsilon)\, dx = -\int_\Omega p_\varepsilon\, \Delta(z - y_\varepsilon)\, dx = -\int_{\partial\Omega} p_\varepsilon \frac{\partial}{\partial n}(z - y_\varepsilon)\, d\sigma\,. \tag{3.9}$$

Combining (3.8) and (3.9), we can infer that

$$0 \leq \int_\Omega \ell\,(p_\varepsilon + h_\varepsilon)\,(v - h_\varepsilon)\, dx \tag{3.10}$$

for any control v admissible for (3.1)–(3.4). Relation (3.10) corresponds to the Pontryagin maximum principle and can be reformulated as

$$0 \leq \int_\Omega (p_\varepsilon + h_\varepsilon)\,(\Delta z - \Delta y_\varepsilon)\, dx\,. \tag{3.11}$$

Consider now the special case $z \in H_0^2(\Omega)$. We notice:

$$\begin{aligned}
-\int_\Omega p_\varepsilon \Delta(z - y_\varepsilon)\, dx &= \int_\Omega \nabla p_\varepsilon \nabla(z - y_\varepsilon)\, dx - \int_{\partial\Omega} p_\varepsilon \frac{\partial}{\partial n}(z - y_\varepsilon)\, d\sigma \\
&= \frac{1}{\varepsilon} \int_{\partial\Omega} \left(\frac{\partial y_\varepsilon}{\partial n} \right)^2 d\sigma - \int_\Omega \Delta p_\varepsilon (z - y_\varepsilon)\, dx + \int_{\partial\Omega} \frac{\partial p_\varepsilon}{\partial n}(z - y_\varepsilon)\, d\sigma \\
&= \frac{1}{\varepsilon} \int_{\partial\Omega} \left(\frac{\partial y_\varepsilon}{\partial n} \right)^2 d\sigma \geq 0\,.
\end{aligned} \tag{3.12}$$

By (3.11), (3.12), we infer that

$$0 \leq \int_\Omega h_\varepsilon (\Delta z - \Delta y_\varepsilon)\, dx \tag{3.13}$$

for $z \in H_0^2(\Omega)$ admissible. From (3.2), (3.13), and the definitions of ℓ, g, one easily obtains that

$$\int_\Omega u^3\, \Delta y_\varepsilon\, \Delta(y_\varepsilon - z)\, dx = \int_\Omega f\,(y_\varepsilon - z)\, dx \quad \forall\, z \in \mathcal{K} \cap H_0^2(\Omega)\,. \tag{3.14}$$

From (3.5), it is obvious that $\{h_\varepsilon\}$ is bounded in $L^2(\Omega)$, and by virtue of (3.2), $\{y_\varepsilon\}$ is bounded in $H^2(\Omega) \cap H_0^1(\Omega)$. Again (3.5) shows that

$$\frac{\partial y_\varepsilon}{\partial n} \to 0 \quad \text{strongly in } L^2(\partial\Omega). \tag{3.15}$$

Then, we have $y_\varepsilon \to y^*$ weakly in $H^2(\Omega)$, and $y^* \in \mathcal{K} \cap H_0^2(\Omega)$. By using the weak lower semicontinuity of quadratic forms, we can take $\varepsilon \to 0$ in (3.14) and finally arrive at the result:

Theorem 3.1. *The mapping $y^* \in \mathcal{K} \cap H_0^2(\Omega)$ is the unique solution to the variational inequality*

$$\int_\Omega u^3 \, \Delta y^* \, \Delta(y^* - z)\, dx \le \int_\Omega f\,(y^* - z)\, dx \quad \forall\, z \in \mathcal{K} \cap H_0^2(\Omega). \tag{3.16}$$

Remark. The above argument yields the existence of the solution to (3.16) and its approximation by the control problem (3.1)–(3.4). Uniqueness is obtained immediately, by contradiction.

Remark. Important examples entering into the formulation (3.16) are the obstacle problem, obtained for

$$\mathcal{K} = \{z \in H^2(\Omega) \cap H_0^1(\Omega)\,;\, a \le z \le b \text{ a.e. } \Omega\},$$

or the variational inequality studied by Glowinski et al. [7] via a direct method, corresponding to

$$\mathcal{K} = \{z \in H^2(\Omega) \cap H_0^1(\Omega)\,;\, a \le \Delta z \le b \text{ a.e. in } \Omega\}.$$

Here, a, b are some given mappings such that $\mathcal{K}$ is nonvoid. If the boundary conditions are changed, or if unilateral conditions on the boundary are considered, then other subspaces of $H^2(\Omega)$ have to be taken into account, and the argument proceeds similarly. A variational inequality for a partially clamped plate was studied by an ad-hoc method in Sprekels and Tiba [14].

Remark. Variational inequalities are obtained by imposing constraints in the variational formulation of the corresponding equation, Lions and Stampacchia [12]. By comparing **Theorem 3.1.** with **Theorem 2.1.**, we see that this remains valid for the control variational method, as well.

We now continue the study of variational inequalities associated to Kirchhoff–Love arches. We consider the state constrained control problem given by (2.16), (2.17), and

$$[v_1\,,\, v_2] \in \mathcal{C}\,, \tag{3.17}$$

where $\mathcal{C} \subset L^\infty(0,1)^2$ is a closed convex set, compatible with the null initial conditions. Notice that (2.18) is no longer imposed and that relations (2.16), (2.17) correspond to a partially clamped arch (in $t = 0$), while (3.17) will yield the unilateral conditions on the arch, as we shall see in the sequel.

All the notations have the same significance as in Section 2; however, the control space for z is $V = \{w \in H^1(0,1)\,;\, w(0) = 0\}$, and the definitions of g_1, g_2 are replaced by

$$g_1 = \delta\ell, \quad -g_2'' = h, \quad g_2(0) = g_2'(1) = 0, \tag{3.18}$$

$$\begin{bmatrix} \ell \\ h \end{bmatrix}(t) = \int_t^1 W(t)\,W^{-1}(s)\begin{bmatrix} f_1(s) \\ f_2(s) \end{bmatrix} ds. \tag{3.19}$$

As we have no constraints on the control variables u, z, admissibility may be assumed in connection with (3.17), and we obtain again the existence of a unique optimal quadruple denoted by u_δ, z_δ, v_1^δ, v_2^δ, in $L^2(0,1) \times V \times C$.

We take admissible control variations of the type

$$[u_\delta\,,\, z_\delta] + \lambda[u - u_\delta\,,\, z - z_\delta], \quad \lambda \in [0,1], \tag{3.20}$$

with $[u\,,\,z]$ any admissible control. A simple argument yields the Euler inequality

$$0 \le \frac{1}{\delta}\int_0^1 u_\delta(u - u_\delta)\,ds + \int_0^1 z_\delta'(z - z_\delta)'\,ds. \tag{3.21}$$

Under the regularity assumption $W \in W^{2,\infty}(0,1)^4$ (as in **Theorem 2.4.**), we shall show that v_1^δ, v_2^δ are the solutions of a general variational inequality.

Fix any $[w_1\,,\,w_2] \in C \cap [V \times U]$, with $U = \{z \in H^2(0,1)\,;\, z(0) = z'(0) = 0\}$. The corresponding controls, generating w_1, w_2 via (2.16), (2.17), are

$$\mu = w_1' - c\,w_2 - g_1 \in L^2(0,1), \tag{3.22}$$

$$\zeta = w_2' + c\,w_1 - g_2 \in V, \tag{3.23}$$

(due to the regularity of W). Moreover, v_1^δ, v_2^δ satisfy (2.19)–(2.21).

Using (3.22), (3.23), and (2.19)–(2.21), in (3.21), we obtain

$$\begin{aligned} 0 \;\le\; & \frac{1}{\delta}\int_0^1 \Big(-g_1 + (v_1^\delta)' - c\,v_2^\delta\Big)\Big(w_1' - c\,w_2 - (v_1^\delta)' + c\,v_2^\delta\Big)\,ds \\ & + \int_0^1 \Big(-g_2 + (v_2^\delta)' - c\,v_1^\delta\Big)'\Big(w_2' + c\,w_1 - (v_2^\delta)' - c\,v_1^\delta\Big)'\,ds. \end{aligned} \tag{3.24}$$

We compute first the terms:

$$-\frac{1}{\delta}\int_0^1 g_1\Big(w_1' - c\,w_2 - (v_1^\delta)' + c\,v_2^\delta\Big)\,ds + \int_0^1 g_2''\Big(w_2' + c\,w_1 - (v_2^\delta)' - c\,v_1^\delta\Big)\,ds$$

$$= -\int_0^1 \ell\Big(w_1' - c\,w_2 - (v_1^\delta)' + c\,v_2^\delta\Big)\,ds - \int_0^1 h\Big(w_2' + c\,w_1 - (v_2^\delta)' - c\,v_1^\delta\Big)\,ds$$

$$= -\int_0^1 f_1\,(w_1 - v_1^\delta)\,ds - \int_0^1 f_2(w_2 - v_2^\delta)\,ds\,. \tag{3.25}$$

In (3.25), we have repeatedly integrated by parts, and we have made use of (3.18), (3.19). Combining (3.24) and (3.25), we have proved the following result:

Theorem 3.2. *If* $W \in W^{2,\infty}(0,1)^4$, *then* v_1^δ, v_2^δ *given by* (2.16), (2.17), (3.17) *satisfy*

$$\frac{1}{\delta}\int_0^1 \Big((v_1^\delta)' - c\,v_2^\delta\Big)\Big((v_1^\delta)' - c\,v_2^\delta - w_1' + c\,w_2\Big)\,ds$$

$$+ \int_0^1 \Big((v_2^\delta)' + c\,v_1^\delta\Big)'\Big((v_2^\delta)' + c\,v_1^\delta - w_2' - c\,w_1\Big)'\,ds$$

$$\leq \int_0^1 f_1(v_1^\delta - w_1)\,ds + \int_0^1 f_2(v_2^\delta - w_2)\,ds\,, \tag{3.26}$$

for a.e. $[w_1\,,\,w_2] \in \mathcal{C} \cap [V \times U]$.

Remark. If the convex $\mathcal{C}$ includes null conditions at the point $t = 1$, then we obtain a variational inequality for a clamped arch. **Theorem 3.2.**, compared with **Theorem 2.4.**, is an example of how the spaces (for the state and for the control) should be adapted when different boundary conditions are imposed. The method introduced in this section allows for general unilateral conditions and various boundary conditions. We conjecture that it also allows the extension of **Theorem 2.2.** to the case of variational inequalities. Concerning the differentiability properties discussed in **Theorem 2.3.**, it is known that, generally, they are not valid for variational inequalities.

Remark. In the case of fourth-order ordinary differential equations, the works of Hlavacek, Bock and Lovisek [8], [9], Sprekels and Tiba [16] discussed variational inequalities associated to beam models. **Theorem 3.2.** seems to be a first result in the literature related to arches submitted to unilateral conditions. The problem (2.16), (2.17), (3.17) is a new weak formulation of the variational inequality (3.26), valid for $W \in L^\infty(0,1)^4$.

We close this presentation with some short examples. If C has the form:

$$C = L^\infty(0,1) \times \{v_2 \in L^\infty(0,1)\,;\, \alpha \le v_2 \le \beta \text{ a.e. in } (0,1)\}\,,$$

then we have an obstacle problem for the normal component of the deflection (α and β are some given mappings such that C allows null initial conditions for v_2). Obstacle problems for the tangential component or for both components are obtained similarly.

Under the smoothness hypothesis $W \in W^{2,\infty}(0,1)^4$, the solution $[v_1^\delta, v_2^\delta]$ is in $V \times U$, and we can impose from the beginning that C is a closed convex subset of $V \times U$. One situation of interest is given by:

$$C = \Big\{[v_1, v_2] \in V \times U\,;\, v_1(1) \ge r\Big\}$$

with $r \in I\!R$ a given constant. This represents a partially clamped arch with a unilateral condition on the tangential component in the end point $t = 1$. Similar formulations may easily be written for the normal component or for both, or in other points, and so on.

References

[1] V. Arnăutu, H. Langmach, J. Sprekels and D. Tiba, *On the approximation and the optimization of plates*, Numer. Funct. Anal. Optim., **21** (2000), no. 3–4, 337–354.

[2] V. Barbu and Th. Precupanu, *Convexity and optimization in Banach spaces*, Sijthoff and Noordhoff, Leyden, 1978.

[3] M. Bendsoe, *Optimization of structural topology, shape, and material*, Springer-Verlag, Berlin, 1995.

[4] M. Bergounioux and D. Tiba, *General optimality conditions for constrained convex control problems*, SIAM J. Control Optim., **34** (1996), no. 2, 698–711.

[5] Ph. Ciarlet, *The finite element method for elliptic problems*, North-Holland, Amsterdam, 1978.

[6] R. Glowinski, J.-L. Lions and O. Pironneau, *Decomposition of energy spaces and applications*, CRAS Paris, Série I, **330** (1999), 445–452.

[7] R. Glowinski, L. Marini and M. Vidrascu, *Finite element approximation and iterative solutions of a fourth order elliptic variational inequality*, IMA J. Numer. Anal., **4** (1984), 127–167.

[8] I. Hlavacek, I. Bock and J. Lovisek, *Optimal control of a variational inequality with applications to structural analysis, I. Optimal design of a beam with unilateral supports*, Appl. Math. Optimiz., **11** (1984), 111–143.

[9] I. Hlavacek, I. Bock and J. Lovisek, *Optimal control of a variational inequality with applications to structural analysis, II. Local optimization of stresses in a beam, III. Optimal design of an elastic plate*, Appl. Math. Optimiz., **13** (1985), 117–135.

[10] A. Ignat, J. Sprekels and D. Tiba, *Analysis and optimization of nonsmooth mechanical structures*, Preprint 581, Weierstrass Institute for Applied Analysis and Stochastics, Berlin, 2000, submitted to SIAM J. Control Optimiz.

[11] J.-L. Lions and O. Pironneau, *Virtual control, replicas and decomposition of operators*, CRAS Paris, Série I, **330** (2000), no. 1, 47–54.

[12] J.-L. Lions and G. Stampacchia, *Variational inequalities*, Comm. Pure Appl. Math., **20** (1967), 493–519.

[13] P. Neittaanmäki and D. Tiba, *An embedding of domains approach in free boundary problems and optimal design*, SIAM J. Control Optimiz., **33** (1995), no. 5, 1587–1602.

[14] J. Sprekels and D. Tiba, *On the approximation and optimization of fourth-order elliptic systems*, ISNM **133**, Birkhäuser Verlag, Basel, 1999, 277–286.

[15] J. Sprekels and D. Tiba, *A duality approach in the optimization of beams and plates*, SIAM J. Control Optimiz., **37** (1998–1999), no. 2, 486–501.

[16] J. Sprekels and D. Tiba, *Sur les arches lipschitziennes*, CRAS Paris, Série I, **331** (2000), 179–184.

Jürgen Sprekels, Dan Tiba
Weierstrass Institute
for Applied Analysis and Stochastics
Mohrenstrasse 39
D–10117 Berlin, Germany
E-mail address: `sprekels@wias-berlin.de, tiba@wias-berlin.de`

Dan Tiba
Institute of Mathematics, Romanian Academy
P. O. Box 1–764
RO–70700 Bucharest, Romania
E-mail address: `dtiba@imar.ro`

International Series of Numerical Mathematics, Vol. 139, 259–265

Control of Czochralski Crystal Growth

Axel Voigt, Karl-Heinz Hoffmann

Abstract. The goal of the paper is the development of proper numerical models for optimization and control of industrial crystal growth processes. To this end we concentrate on the numerical treatment, simulation and optimal control of techniques for growing high quality semiconductor crystals.

1. Introduction

The growth of high quality bulk single crystals is one of the key processes that determine the yield and the profitability in semiconductor device manufacturing. The electronic industry, which is based on that materials is today the largest industry in the world. In the 1999–2002 international technology roadmap for semiconductors the Semiconductor Industry Association (SIA) predict an annual growth of the industry of approximately 20% and more than 200 billion in sales by the year 2000. The growth is being driven by the surge of wired and wireless information appliances, as well as Internet infrastructure products. Wafer suitability for microelectronic devices is determined by several different measurements that include surface uniformity, lattice perfection, impurity concentration and crystal electrical properties such as resistivity. Increasing device size increases the requirements on the quality of the crystal. For a given concentration of defects, device failures are greater if fewer, more expensive devices are fabricated on the wafer. Crystal quality will therefore be the utmost importance in the coming years as these trends continue to decrease the tolerance on wafer specifications. It is widely accepted that today without an effective attendant process modeling no appropriate quality of the crystal can be achieved. Besides the help in understanding the complex processes in crystal growth the model can also be used in predicting optimal parameters to grow appropriate crystals. The main technique to grow Silicon is the Czochralski technique. In a simple form the Czochralski technique consists of a crucible which contains the charge material to be crystallized, a heater capable of melting the charge and a pull rod which is positioned axially above the crucible. A seed crystal is attached to the lower end of the pull rod and the pull rod is then lowered until the end of the seed crystal is dipped into the melt. The temperature of the melt is adjusted until a meniscus is supported by the end of the seed. If a thermal steady state has been achieved, the pull rod is slowly lifted and crystallization onto the end of the seed occurs where a single crystal grows in cylindrical ingot form. This article concentrates on the numerical modeling and optimal control of the growth process of Si, which include a large number of physical transport mechanisms, such as convection, mass transport, segregation, interface dynamics, defect formation and radiative heat transfer, which interact in complex ways, see Figure 1.

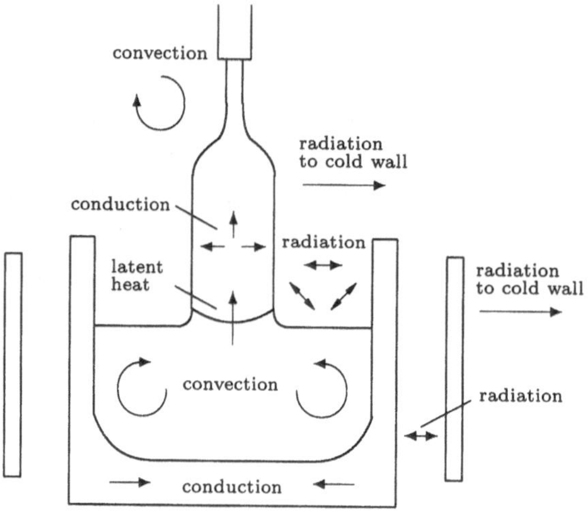

FIGURE 1. Schematic visualization of heat exchange processes

2. Mathematical Model

The governing equations can be derived from the basic principles of conservation of mass, momentum and energy. In the melt we have the Navier-Stokes equation

$$\frac{\partial \mathbf{u}_m}{\partial t} + (\mathbf{u}_m \cdot \nabla)\mathbf{u}_m - \nu\Delta\mathbf{u}_m + \nabla p = \beta(T_m - T_0)\mathbf{g}, \tag{1}$$

$$\nabla \cdot \mathbf{u}_m = 0, \tag{2}$$

with $\mathbf{u}_m$ velocity, p pressure and T_m temperature, where we assume the melt being Newtonian and variations of melt density are negligible, except in the calculation of the body force induced by thermal buoyancy, which is the onset of Grashof convection. This simplification is known as Boussinesq approximation.

Heat transfer in the Czochralski apparatus occurs by conduction, convection and radiation. The equations for the temperature are

$$\rho_{i_0} c_{i_p} \left(\frac{\partial T_i}{\partial t} + \mathbf{u}_i \cdot \nabla T_i \right) - \nabla \cdot (k_i \nabla T_i) = \rho_{i_0} P_i, \tag{3}$$

where ρ_{i_0} is the density of the component, c_{i_p} is the heat capacity, $\mathbf{u}_i$ is the velocity, k_i is the thermal conductivity and P_i is a heat source term. i indicates the different components of the apparatus, like melt, crystal, crucible, heaters, heat shields and insulators. The heat sources P_i are nonzero only in the heaters. We neglect the influence of a gas flow in our model. Further we assume that almost no radiation is emitted or absorbed by the gas and that radiation is only diffuse. Moreover, we consider that emission, absorption and reflection of radiating waves occure only at the surfaces. The heat balance at the surface relating the heat flux caused by conduction to the surface q, the radiosity and the irradiation then becomes $q = \mathbf{G}(\sigma T^4)$, where $\mathbf{G} = (\mathbf{I} - \epsilon\mathbf{K}(\mathbf{I} - (1-\epsilon)\mathbf{K})^{-1})\epsilon$ is the infinite dimensional equivalent of the so called Gebhart factor. ϵ is the emissivity, the operator $\mathbf{K}$ is defined as

$$\mathbf{K}\lambda(\mathbf{r}) = \int_\Gamma \omega(\mathbf{r},\mathbf{s})\Xi(\mathbf{r},\mathbf{s})\lambda(\mathbf{s})ds, \ \forall u \in \Gamma, \tag{4}$$

where $\Xi(\mathbf{r},\mathbf{s})$ is a visibility factor. This is mainly based on Tiihonen [4]. Analytic results concerning the heat equation with non-local radiation terms can be found in Tiihonen [5] and Metzger [2]. Along the free melt-crystal interface we have mass conservation and a no-slip boundary condition. Due to the rotation of the crystal this no-slip condition is the onset of forced convection by crystal rotation. Because the melt-crystal interface is a thermal free boundary the thermal conditions need to specify the temperature and the position and shape of the boundary, we set $T_m = T_c = T_{eq}$ at that boundary, where T_{eq} is the equilibrium melting temperature. The second thermal condition takes account of the heat flux balance between the two phases, the heat flux in the crystal is greater than in the melt by the amount of latent heat released during solidification, i.e. $k_m\nabla T_m \cdot \mathbf{n} - k_c\nabla T_c \cdot \mathbf{n} = \rho_{c_0} l(\mathbf{u}_c - \mathbf{u}_f)\cdot\mathbf{n}$, with l the latent heat, $\mathbf{n}$ the normal vector and $\mathbf{u}_f$ the speed of the phase-boundary. The meniscus separating the melt from the ambient gas is another free interface, whose shape is part of the problem unknowns. Here we again make a Boussinesq-type approximation, where we assume that the surface-tension coefficient γ equals a constant γ_0, except in the boundary condition for the tangential stress, where we assume a linear dependence $\gamma = \gamma_0(1 - \alpha(T_m - T_0))$ on temperature. The traction condition for tangential and normal stress reads $\boldsymbol{\sigma}\ \mathbf{n}\cdot\mathbf{t} = -\gamma_0(\alpha\nabla T_m \cdot \mathbf{t})$ and $\boldsymbol{\sigma}\ \mathbf{n}\cdot\mathbf{n} = \gamma_0 2H - p_g$, where p_g is the pressure in the gas phase and H is the mean curvature. The temperature differences at the interface have an influence on the transport of momentum and heat near the interface. The surface-tension gradient resulting from this differences acts like a shear stress on the melt-gas interface and thereby generates a surface flow. This phenomenon is known as Marangoni convection. We assume that the melt perfectly adheres to the crucible wall and specify a no-slip condition for the velocity which is the onset of forced convection due to rotation of the crucible. At the remaining boundaries we only need to specify a condition for the temperature. At the surfaces of the melt, the crystal, the crucible, the heaters, heat-shields, insolators and the inner surface of the enclosure we take

$$k_i\nabla T_i \cdot \mathbf{n} = -\mathbf{G}(\sigma T_i^4), \tag{5}$$

where i stands for the different components. We need to describe initial conditions for the temperature T_i in all components of the apparatus, conditions for the velocity $\mathbf{u}_m$ in the melt and an initial domain for melt and crystal.

3. Numerical Methods

Often in industrial crystal growth one can assume an axisymmetric furnace, so that we formulate our models in cylindrical coordinates and are able to reduce the complexity of the problem to 2D. Our simulation of Czochralski growth of Silicon models the full time-dependent problem with free capillary and phase boundaries. The capillary surface is calculated by minimizing the total energy, which include surface, coating and gravitational energy with the constraint of volume conservation. Therefore the capillary surface is approximated by a polygon $s_i = (r_i, h_i)$, $i = 1, \dots, n$. The solid bodies will be described by $s_i^* = (r_i^*, h_i^*)$, $i = 1, \dots, m$. The total energy E is approximated by the sum of E_{surf}, E_{coat} and E_{grav}

$$
\begin{aligned}
E_{surf} &= \gamma 2\pi \sum_{i=1}^{n-1} \frac{r_i + r_{i+1}}{2} \|s_{i+1} - s_i\|, \\
E_{coat} &= \gamma 2\pi\beta \sum_{i=1}^{,-1} \frac{r_i^* + r_{i+1}^*}{2} \|s_{i+1}^* - s_i^*\|, \\
E_{grav} &= \rho g 2\pi \sum_{i=1}^{n-1} \frac{r_i - r_{i+1}}{3} \left(\left(\frac{h_i + h_{i+1}}{2} \right)^2 \frac{r_i + r_{i+1}}{2} + \frac{h_i^2 r_i + h_{i+1}^2 r_{r+1}}{4} \right) \\
&\quad - \rho\omega_0^2 \pi \sum_{i=1}^{n-1} \frac{r_i - r_{i+1}}{5} \left(4\left(h_i + h_{i+1}\right) \left(\frac{r_i + r_{i+1}}{2} \right)^3 \right. \\
&\quad \left. + \frac{3h_i r_i^3 + 3h_{i+1} r_{i+1}^3 - h_i r_i r_{i+1}^2 - h_{i+1} r_{i+1} r_i^2}{4} \right).
\end{aligned}
$$

The volume V is approximated by

$$
V = \gamma\lambda 2\pi \sum_{i=1}^{n-1} \frac{r_{i+1} - r_i}{4} \left(\left(h_i + h_{i+1}\right)\left(r_i + r_{i+1}\right) + \frac{\left(h_{i+1} - h_i\right)\left(r_{i+1} - r_i\right)}{3} \right).
$$

The minimization problem now reads as follows

$$
\min_{s \text{ feasible}} E(s) \quad \text{s.t.} \quad V(s) = A. \tag{6}
$$

By use of a Lagrange multiplier λ the resulting necessary condition

$$
\begin{aligned}
E'(s) + \lambda V'(s) &= 0 \\
V(s) - A &= 0
\end{aligned}
$$

can be solved by the Newton-method by making use of the special structure of the iteration matrix [3]. The Navier-Stokes equation is discretized by the finite element ansatz of Bernardi and Rauge [1] with divergence free elements. This elements satisfies the Ladyzenskaja-Babuska-Brezzi (LBB) condition. For the discretization of the heat equation piecewise linear finite elements are used. A front-tracking method is applied in order to solve for the Stefan boundary and the time diskretisation is based on the method of characteristics. Because of the moving domain space-time basis functions are used

$$\phi_i(\mathbf{x},t) = \frac{t-t^n}{t^{n+1}-t^n}\phi_i^{n+1}(\mathbf{x}) + \left(1 - \frac{t-t^n}{t^{n+1}-t^n}\phi_i^n(\mathbf{x})\right). \tag{7}$$

The algorithm for one timestep reads as follows

1. Solve the Laplace-Young equation by minimizing the total energy and find a starting domain.
2. Find an initial solution of the transport equations in that domain.
3. From the heat flux difference at the phase boundary calculate the growth speed and a new Stefan boundary.
4. Move the new Stefan boundary by $\mathbf{u}_p \cdot (t^{n+1} - t^n)$
5. Calculate the crystallized volume δV.
6. With the new volume $V - \delta V$ calculate a new capillary boundary.
7. Solve the transport equations on the displaced domain.
8. goto 3.

where $\mathbf{u}_p$ is the pulling velocity and $(t^{n+1} - t^n)$ the time step.

4. Results

For the Cz-crystal growth of Silicon we use a control strategy in order to grow crystals with constant diameter. This is done by observing the triple point and adjusting the pulling velocity by

$$u_s^{n+1} = u_s^n + \alpha(t^n - t^{n+1})u_d^n, \tag{8}$$

with u_s^n pulling velocity during the n-th time step, u_d^n velocity of the triple point during the n-th time step in the radial direction and α a process parameter. The influence of the convection on the temperature distribution and the shape of the phase boundary is analyzed. Figure 1 display the stream function and the temperature distribution in the melt-crystal element.

At $t = 0$ the stationary solution of the transport equations in the initial domain is shown, the other results are the solutions at different time steps.

This calculations indicate that there is a strong dependence of the shape of the phase boundary and the temperature distribution in the crystal on the flow field in the melt. The adjusted pulling velocity in order to grow crystals with a constant diameter is shown in Figure 2. The velocity varies between 1 $\frac{\mathrm{cm}}{\mathrm{h}}$ and 7 $\frac{\mathrm{cm}}{\mathrm{h}}$ and is smaller if convection is taken into account. The control strategy used here is similar to the one used in industrial CZ growth processes, where one also observes an oscillation in the pull-velocity.

5. Conclusions

We numerically investigated the growth of Silicon single crystals by the Czochralski method. The influence of convection in the melt on the temperature field in the

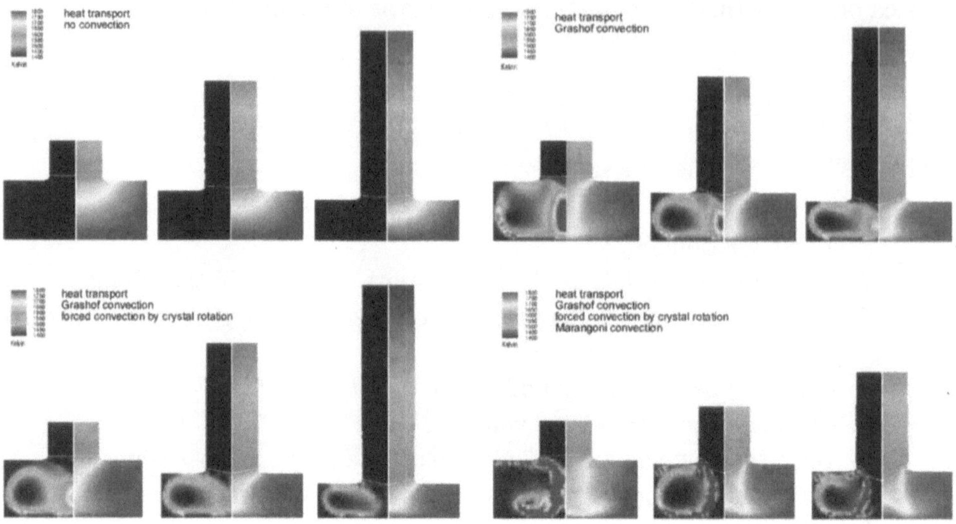

Figure 1: Temperature field and stream function at $t = 0, 40min, 80min$

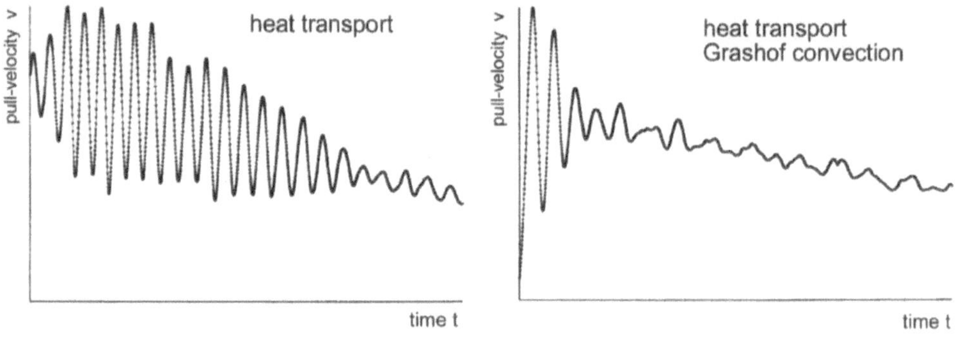

Figure 2: Pulling velocity as function of time

crystal was analyzed. The process was controlled in order to grow crystals with a constant diameter. The observed variations in the pull-velocity are in qualitative agreement with experimental observation.

6. Acknowledgements

This research was partly funded by the *Bundesministerium für Bildung, Wissenschaft, Forschung und Technologie (BMBF)*, number 03$HO7TM$20 of *Mathematische Verfahren zur Lösung von Problemstellungen in Industrie und Wirschaft.*

References

[1] C. Bernardi and G. Raugel, *Analysis of some finite elements for the Stokes problem,* Math. of Comp., **44** (1985), 71–79.

[2] M. Metzger, *Existence for a time-dependent heat equation with non-local radiation terms,* Math. Meth. Appl. Sci. **22** (1999) 1101–1119.

[3] W. Seifert, *Numerische Behandlung freier Ränder beim Kristallziehen,* PhD-Thesis at the Technical University Munich, (1996).

[4] T. Tiihonen, J. Järvinen and R. Nieminen *Time-dependent simulation of Czochralski silicon crystal growth,* J. Crystal Growth, **180** (1997) 468–476.

[5] T. Tiihonen, *Stefan-Bolzmann radiation on non-convex surfaces,* Math. Meth. Appl. Sci., **20** (1997) 47–57.

research center caesar
Friedensplatz 16
53111 Bonn, Germany
E-mail address: `voigt@caesar.de`, `hoffmann@caesar.de`

International Series of Numerical Mathematics, Vol. 139, 267–278

Boundary Control of the Burgers Equation: Optimality Conditions and Reduced-order Approach

Stefan Volkwein

Abstract. In this article bilaterally control constrained optimal control problems with boundary control are considered. First- and second-order optimality conditions are analyzed. The method of proper orthogonal decomposition is used to solve the control problem numerically.

1. Introduction

In this paper we consider control constrained optimal control problems for the Burgers equation:

$$\min J(y,u,v) = \frac{1}{2}\int_Q \alpha_Q |y - z_Q|^2 \, dxdt + \frac{1}{2}\int_0^T \beta|u|^2 + \gamma|v|^2 \, dt \tag{1a}$$

subject to

$$\begin{aligned} y_t - \nu y_{xx} + y y_x &= f \quad \text{in } Q = (0,T)\times\Omega, && \text{(1b)}\\ \nu y_x(\cdot,0) + \sigma_0 y(\cdot,0) &= u \quad \text{in } (0,T) && \text{(1c)}\\ \nu y_x(\cdot,1) + \sigma_1 y(\cdot,1) &= v \quad \text{in } (0,T), && \text{(1d)}\\ y(0,\cdot) &= y_0 \quad \text{in } \Omega = (0,1) \subset \mathbb{R}, && \text{(1e)} \end{aligned}$$

and

$$(u,v) \in U_{\mathsf{ad}} \times V_{\mathsf{ad}} \subset L^2(0,T) \times L^2(0,T), \tag{1f}$$

where $T > 0$ is fixed and $\nu > 0$ denotes a viscosity parameter. We assume that $\alpha_Q \in L^\infty(Q)$ is a non-negative weight, $z_Q \in L^2(Q)$ denotes a given desired state, β, γ are positive constants and $\sigma_0, \sigma_1 \in L^\infty(0,T)$. Moreover, let $f \in L^2(Q)$, $y_0 \in L^\infty(\Omega)$ and $u_a, u_b, v_a, v_b \in L^\infty(0,T)$ with $u_a \le u_b$ and $v_a \le v_b$ almost everywhere (a.e.) in Q. The sets of admissible controls are given by

$$\begin{aligned} U_{\mathsf{ad}} &= \{u \in L^2(0,T) : u_a \le u \le u_b \text{ a.e. in } (0,T)\},\\ V_{\mathsf{ad}} &= \{v \in L^2(0,T) : v_a \le v \le v_b \text{ a.e. in } (0,T)\}. \end{aligned}$$

In this article we present first- and second-order optimality conditions for (1). The work extends the analysis done in [18] to bilaterally constrained optimal control problems. For the numerical realization we use the SQP (sequential quadratic programming) method combined with a primal-dual active set strategy. For the spatial discretization the method of proper orthogonal decomposition (POD) is applied, which is a method for deriving low order models of dynamical systems. It was successfully used in different fields including signal analysis and pattern recognition (see e.g. [10]), fluid dynamics and coherent structures (see e.g. [4, 16]) and more recently in control theory (see e.g. [2, 3, 11, 12]).

The article is organized as follows. Section 2 is devoted to review some results for problem (1). First-order optimality conditions are studied in Section 3. In Section 4 we analyze second-order optimality conditions. To solve (1) numerically we introduce the POD method in the fifth section and present a numerical example.

2. Preliminaries

By $L^2(0,T;H^1(\Omega))$ we denote the space of measurable functions from $[0,T]$ to $H^1(\Omega)$, which are square integrable. When t is fixed, the expression $\varphi(t)$ stands for the function $\varphi(t,\cdot)$ considered as a function in Ω only. The space $W(0,T)$ is defined by

$$W(0,T) = \{\varphi \in L^2(0,T;H^1(\Omega)) : \varphi_t \in L^2(0,T;H^1(\Omega)')\},$$

where $H^1(\Omega)'$ denotes the dual of $H^1(\Omega)$. The space $W(0,T)$ is a Hilbert space endowed with the common inner product, see [7], for instance. Recall that $W(0,T)$ is continuously embedded into $C([0,T];L^2(\Omega))$, the space of all continuous functions from $[0,T]$ into $L^2(\Omega)$. Thus, there exists an embedding constant $C_E > 0$ such that

$$\|\varphi\|_{C([0,T];L^2(\Omega))} \le C_E \|\varphi\|_{W(0,T)} \quad \text{for all } \varphi \in W(0,T). \tag{2}$$

Definition 2.1. *A function $y \in W(0,T)$ is called a* weak solution of (1b)–(1e) *if $y(0) = y_0$ in $L^2(\Omega)$ and*

$$\begin{aligned} &\langle y_t(t), \varphi\rangle_{(H^1)',H^1} + \sigma_1(t)y(t,1)\varphi(1) - \sigma_0(t)y(t,0)\varphi(0) \\ &+ \int_\Omega \nu y_x(t)\varphi' + y(t)y_x(t)\varphi\,dx = \int_\Omega f(t)\varphi\,dx + v(t)\varphi(1) - u(t)\varphi(0) \end{aligned}$$

for all $\varphi \in H^1(\Omega)$ and $t \in (0,T)$ a.e., where $\langle\cdot,\cdot\rangle_{(H^1)',H^1}$ denotes the dual pair associated with $H^1(\Omega)$ and its dual.

Now we proceed by writing (1) in an abstract form. Therefore, we define the Hilbert spaces

$$X = W(0,T) \times L^2(0,T) \times L^2(0,T), \quad Y = L^2(0,T;H^1(\Omega)) \times L^2(\Omega)$$

and introduce the subset

$$\emptyset \neq K_{\mathsf{ad}} = W(0,T) \times U_{\mathsf{ad}} \times V_{\mathsf{ad}} \subset X.$$

Moreover, let $\tilde{e} : X \to L^2(0,T;H^1(\Omega)')$ be defined by

$$\begin{aligned}\langle \tilde{e}(y,u,v), \lambda\rangle_{L^2(0,T;H^1(\Omega)'),L^2(0,T;H^1(\Omega))} \\ = \int_0^T \langle y_t(\cdot), \lambda(\cdot)\rangle_{(H^1)',H^1} + \Big(\int_\Omega \nu y_x\lambda_x + yy_x\lambda - f\lambda\,dx\Big)dt \\ + \int_0^T (\sigma_1 y(\cdot,1) - v)\lambda(\cdot,1) + (u - \sigma_0 y(\cdot,0))\lambda(\cdot,0)\,dt\end{aligned}$$

for $\lambda \in L^2(0,T;H^1(\Omega))$. Then we set

$$e : X \to Y, \quad (y,u,v) \mapsto \big((-\tfrac{d^2}{dx^2} + I)^{-1}\tilde{e}(y,u,v), y(0) - y_0\big),$$

where $(-\frac{d^2}{dx^2} + I)^{-1} : H^1(\Omega)' \to H^1(\Omega)$ is the Neumann solution operator associated with

$$\int_\Omega w'\varphi' + w\varphi\,dx = \langle g, \varphi\rangle_{(H^1)',H^1} \quad \text{for all } \varphi \in H^1(\Omega),$$

where $g \in H^1(\Omega)'$. Now we can express the optimal control problem (1) as:

$$\min J(x) \text{ subject to } x \in K_{\text{ad}} \text{ and } e(x) = 0. \tag{P}$$

Note that both J and e are twice continuously Fréchet-differentiable and their second Fréchet-derivatives are Lipschitz-continuous on X. Theorem 2.2 guarantees that the optimal control problem (P) has a solution. For the proof we refer to [20].

Theorem 2.2. *There exists an optimal solution $x^* = (y^*, u^*, v^*)$ of problem* (P).

The following result proved in [20] implies a standard constraint qualification condition.

Proposition 2.3. *For every $\bar{x} \in X$ the operator $e_y(\bar{x})$ is bijective. Here and in the following, the subscript denotes as usual the associated partial derivative.*

3. First-order Necessary Optimality Conditions

This section is devoted to present the first-order necessary optimality conditions for (P). Problem (P) is a non-convex programming problem so that there will be probably occur different local minima. Numerical methods will deliver a local minimum close to their starting point. Therefore, we do not restrict our investigations to global solutions of (P). We will assume that a fixed reference solution is given satisfying certain first- and second-order optimality conditions (ensuring local optimality of the solution). Let us define the active sets at $x^* = (y^*, u^*, v^*) \in K_{\text{ad}}$ by $\mathcal{U}^* = \mathcal{U}_a^* \cup \mathcal{U}_b^*$ and $\mathcal{V}^* = \mathcal{V}_a^* \cup \mathcal{V}_b^*$, where

$$\begin{aligned}\mathcal{U}_a^* &= \{t \in [0,T] : u^*(t) = u_a(t) \text{ a.e.}\}, & \mathcal{U}_b^* &= \{t \in [0,T] : u^*(t) = u_b(t) \text{ a.e.}\}, \\ \mathcal{V}_a^* &= \{t \in [0,T] : v^*(t) = v_a(t) \text{ a.e.}\}, & \mathcal{V}_b^* &= \{t \in [0,T] : v^*(t) = v_b(t) \text{ a.e.}\}.\end{aligned}$$

The corresponding inactive sets at x^* are given by $\mathcal{I}_{U_{\text{ad}}}^* = [0,T] \setminus \mathcal{U}^*$ and $\mathcal{I}_{V_{\text{ad}}}^* = [0,T] \setminus \mathcal{V}^*$. First-order necessary optimality conditions are presented in the next theorem.

Theorem 3.1. *Let $x^* = (y^*, u^*, v^*) \in K_{\mathsf{ad}}$ be a local solution to* (P). *Then there exists a unique pairs $p^* = (\lambda^*, \mu^*) \in W(0,T) \times L^2(\Omega)$ and $(\xi^*, \eta^*) \in L^2(0,T) \times L^2(0,T)$ satisfying*

$$
\begin{aligned}
-\lambda_t^* - \nu\lambda_{xx}^* - y^*\lambda_x^* &= -\alpha_Q(y^* - z_Q) && \text{in } Q, && \text{(3a)}\\
\left.\begin{aligned}\nu\lambda_x^*(\cdot,0) + (y^*(\cdot,0)+\sigma_0)\lambda^*(\cdot,0) &= 0\\ \nu\lambda_x^*(\cdot,1) + (y^*(\cdot,1)+\sigma_1)\lambda^*(\cdot,1) &= 0\end{aligned}\right\} && && \text{in } (0,T), && \text{(3b)}\\
\lambda^*(T) &= 0 && \text{in } \Omega, && \text{(3c)}\\
\mu^* &= \lambda^*(0) && \text{in } \Omega, && \text{(3d)}\\
e(x^*) &= 0, \quad x^* \in K_{\mathsf{ad}}, && && \text{(3e)}\\
\beta u^* + \lambda^*(\cdot,0) + \xi^* &= 0 && \text{in } (0,T), && \text{(3f)}\\
\gamma v^* - \lambda^*(\cdot,1) + \eta^* &= 0 && \text{in } (0,T), && \text{(3g)}\\
\xi^*|_{\mathcal{U}_a^*} &\le 0,\ \xi^*|_{\mathcal{U}_b^*} \ge 0,\ \xi^*|_{\mathcal{I}_{U_{\mathsf{ad}}}^*} = 0, && && \text{(3h)}\\
\eta^*|_{\mathcal{V}_a^*} &\le 0,\ \eta^*|_{\mathcal{V}_b^*} \ge 0,\ \eta^*|_{\mathcal{I}_{V_{\mathsf{ad}}}^*} = 0, && && \text{(3i)}
\end{aligned}
$$

where, for instance, $\xi^*|_{\mathcal{U}_a^*}$ denotes the restriction of the multiplier ξ^* on the subset $\mathcal{U}_a^*$ of $[0,T]$.

Proof. Let us introduce the Lagrangian associated with (P) by

$$L(x,p) = J(x) + (e(x), p)_Y.$$

Due to Proposition 2.3 there exists a unique $p^* = (\lambda^*, \mu^*) \in Y$ such that

$$L_x(x^*, p^*) = 0 \quad \text{in } W(0,T). \tag{4}$$

The proof that the Lagrange multiplier λ^* satisfies (3a)–(3c) can be done analogously to [19]. Condition (3e) denotes feasibility and is clearly satisfied. Due to the optimality of u^* the following variational inequality holds

$$L_u(x^*, p^*)(u - u^*) \ge 0 \quad \text{for all } u \in U_{\mathsf{ad}}.$$

Setting

$$(-\xi^*, u - u^*)_{L^2(0,T)} := L_u(x^*, p^*)(u - u^*) = (\beta u^* + \lambda^*(\cdot,0), u - u^*)_{L^2(0,T)} \tag{5}$$

for all $u \in U_{\mathsf{ad}}$ we obtain (3f). Analogously, we get (3g). For the proof of (3h) and (3i) we refer the reader to [9]. From Remark 4.7 the uniqueness of the pair (ξ^*, η^*) will follow. □

In the next lemma we provide an estimate for the Lagrange multiplier λ^*, which will be useful in the second-order analysis carried out in Section 4. The proof follows from variational techniques, see [20].

Lemma 3.2. *There exists a constant $C = C(\nu, T, y^*, \sigma_0, \sigma_1) > 0$ such that*

$$\|\lambda^*\|_{L^\infty(0,T;L^2(\Omega))} + \|\lambda^*\|_{L^2(0,T;H^1(\Omega))} \le C\, \|\alpha_Q(y^* - z)\|_{L^2(Q)}.$$

4. Second-order Optimality Conditions

Let us recall the following definition.

Definition 4.1. *Let K be a convex subset of a Hilbert space Z and $z \in K$. The set*

$$T_K(z) = \{\tilde{z} \in Z : \text{ there exists } z(\sigma) = z + \sigma\tilde{z} + o(\sigma) \in K,\ \sigma \geq 0\}$$

is called the tangent cone at the point z. *Moreover, the* normal cone N_K at the point z *is given by*

$$N_K(z) = \{\tilde{z} \in Z : (\tilde{z}, \hat{z} - z)_Z \leq 0 \text{ for all } \hat{z} \in K\}.$$

In case of $z \notin K$ these two cones are set equal to the empty set.

For $K = K_{\mathsf{ad}}$ we have the following characterizations.

Lemma 4.2. *Let $x = (y, u, v) \in K_{\mathsf{ad}}$.*

a) $T_{K_{\mathsf{ad}}}(x) = W(0,T) \times T_{U_{\mathsf{ad}}}(u) \times T_{V_{\mathsf{ad}}}(v)$, *where*

$$T_{U_{\mathsf{ad}}}(u) = \{\tilde{u} \in L^2(0,T) : \tilde{u}(t) \in T_{[u_a(t),u_b(t)]}(u(t)) \text{ for } t \in [0,T] \text{ a.e.}\}$$

and $T_{V_{\mathsf{ad}}}(v)$ accordingly, where for $a, b, s \in \mathbb{R}$ with $a \leq b$

$$T_{[a,b]}(s) = \begin{cases} \mathbb{R}^+ = \{t \in \mathbb{R} : t \geq 0\} & \text{if } s = a, \\ \mathbb{R}^- = \{t \in \mathbb{R} : t \leq 0\} & \text{if } s = b, \\ \mathbb{R} & \text{otherwise.} \end{cases}$$

b) $N_{K_{\mathsf{ad}}}(x) = \{0\} \times N_{U_{\mathsf{ad}}}(u) \times N_{V_{\mathsf{ad}}}(v)$, *where*

$$N_{U_{\mathsf{ad}}}(u) = \{\tilde{u} \in L^2(0,T) : \tilde{u}(t) \in N_{[u_a(t),u_b(t)]}(u(t)) \text{ for } t \in [0,T] \text{ a.e.}\}$$

and $N_{V_{\mathsf{ad}}}(v)$ accordingly.

c) *Moreover,*

$$\begin{aligned} &T_{U_{\mathsf{ad}}}(u^*) \cap \{\xi^*\}^\perp \\ &\quad = \{u \in L^2(0,T) : u \geq 0 \text{ on } \mathcal{U}_a^*,\ u \leq 0 \text{ on } \mathcal{U}_b^* \text{ and } u = 0 \text{ on } \mathcal{U}_\pm^*\} \end{aligned} \tag{6}$$

and $T_{V_{\mathsf{ad}}}(u^*) \cap \{\eta^*\}^\perp$ accordingly, where $(\xi^*, \eta^*) \in N_{U_{\mathsf{ad}}} \times N_{V_{\mathsf{ad}}}$ are the Lagrange multipliers introduced in Theorem 3.1, $S^\perp$ denotes the orthogonal complement of a set S, and $\mathcal{U}_\pm^* = \{t \in [0,T] : \xi^* > 0 \text{ or } \xi^* < 0 \text{ a.e.}\} \subset \mathcal{U}^*$.

Proof. The characterization of the tangent and normal cones is a classical result. For a proof we refer to [15]. What remains to show is (6). Due to (5) we have $\xi^* \in N_{V_{\mathsf{ad}}}(u^*)$ satisfying $\xi^* = 0$ on the set $[0,T] \setminus \mathcal{U}_\pm^*$. Suppose that $t \in \mathcal{U}_a^*$. We conclude $T_{[u_a(t),u_b(t)]}(u^*(t)) = \mathbb{R}^+$. Thus, $u \in T_{U_{\mathsf{ad}}}(u^*)$ implies $u \geq 0$ on $\mathcal{U}_a^*$ by part a). Analogously, $u \leq 0$ on $\mathcal{U}_b^*$ holds. Hence,

$$\begin{aligned} T_{U_{\mathsf{ad}}}(u^*) \cap \{\xi^*\}^\perp &= \{\tilde{u} \in L^2(0,T) : \tilde{u}(t) \in T_{[u_a(t),u_b(t)]}(u^*(t)) \text{ for } t \in [0,T] \text{ a.e.}\} \\ &\quad \cap \left\{u \in L^2(0,T) : \int_{\mathcal{U}_\pm^*} \xi^* u \, dt = 0\right\}. \end{aligned}$$

Since $\xi^* > 0$ and $u \geq 0$ on $\mathcal{U}_a^* \cap \mathcal{U}_\pm^*$ and $\xi^* < 0$ and $u \leq 0$ on $\mathcal{U}_b^* \cap \mathcal{U}_\pm^*$, (6) holds. ☐

Suppose that the point $\bar{x} = (\bar{y}, \bar{u}, \bar{v}) \in X$ satisfies the first-order necessary optimality conditions. By Proposition 2.3 there exists unique Lagrange multipliers $\bar{p} = (\lambda, \mu) \in Y$ and $(\bar{\xi}, \bar{\eta}) \in N_{U_{\mathsf{ad}}} \times N_{V_{\mathsf{ad}}}$ satisfying the first-order necessary optimality conditions

$$L_x(\bar{x}, \bar{p}) + (0, \bar{\xi}, \bar{\eta})^\mathsf{T} = 0, \;\; \bar{x} \in K_{\mathsf{ad}} \text{ and } e(\bar{x}) = 0. \tag{7}$$

Now we introduce the critical cone at $\bar{x}$, which is the set of directions of non increase of the cost that are tangent to the feasible set $\{x \in K_{\mathsf{ad}} : \; e(x) = 0\}$.

Definition 4.3. *The* critical cone at $\bar{x}$ *is defined by*

$$C(\bar{x}) = \{h \in T_{K_{\mathsf{ad}}}(\bar{x}) : J_x(\bar{x})h \leq 0 \text{ and } e_x(\bar{x})h = 0\}.$$

The critical cone at $\bar{x}$ can be characterized as follows.

Lemma 4.4. *Let* ker $e'(\bar{x})$ *denote the kernel of* $e_x(\bar{x})$. *Then we obtain* $J_x(\bar{x})h = 0$, *whenever* $h \in C(\bar{x})$, *and*

$$h = (h_1, h_2, h_3) \in C(\bar{x}) = \{h \in T_{K_{\mathsf{ad}}}(\bar{x}) \cap \{0, \bar{\xi}, \bar{\eta}\}^\mathsf{T} : h \in \ker e'(\bar{x})\}.$$

Proof. The proof follows the arguments in [6]. Let $h = (h_1, h_2, h_3) \in T_{K_{\mathsf{ad}}}(\bar{x}) \cap \ker e'(\bar{x})$. From (7) and $h \in \ker e'(\bar{x})$ we infer that

$$0 = \big(L_x(\bar{x}, \bar{p}) + (0, \bar{\xi}, \bar{\eta})^\mathsf{T}\big)h = J_x(\bar{x})h + (\bar{\xi}, h_2)_{L^2(0,T)} + (\bar{\eta}, h_3)_{L^2(0,T)}.$$

From $(\bar{\xi}, \bar{\eta}) \in N_{U_{\mathsf{ad}}} \times N_{V_{\mathsf{ad}}}$ we conclude that $(\bar{\xi}, h_2)_{L^2(0,T)} \leq 0$ and $(\bar{\eta}, h_3)_{L^2(0,T)} \leq 0$. Hence, $J_x(\bar{x})h \geq 0$. Moreover, $J_x(\bar{x})h \leq 0$ is equivalent with the fact that $(\bar{\xi}, h_2)_{L^2(0,T)} = 0$ and $(\bar{\eta}, h_3)_{L^2(0,T)} = 0$, i.e., $h_2 \in \{\bar{\xi}\}^\mathsf{T}$ and $h_3 \in \{\bar{\eta}\}^\mathsf{T}$. □

Now we turn to the second-order necessary optimality conditions. Let $h = (h_1, h_2, h_3) \in X$. We find

$$L_{xx}(\bar{x}, \bar{p})(h, h) = \int_Q \alpha_Q h_1^2 + 2h_1(h_1)_x\bar{\lambda}\, dxdt + \int_0^T \beta h_2^2 + \gamma h_3^2\, dt. \tag{8}$$

In Theorem 2.2 we have denoted by x^* the local solution to (P). The associated unique Lagrange multipliers are p^*, ξ^* and η^*, see Theorem 3.1.

Definition 4.5. *The* second-order necessary optimality conditions *are defined as*

$$L_{xx}(x^*, p^*)(h, h) \geq 0 \text{ for all } h \in C(x^*). \tag{9}$$

Now let $\bar{x} = x^*$ be a local solution to (P).

Theorem 4.6. *The point* (x^*, p^*) *satisfies the second-order necessary optimality condition* (9).

Proof. The equality constraints can be written as

$$e(x) \in K_Y = \{0\} \subset Y,$$

where, of course, K_Y is a closed convex set. The result follows from Theorem 2.7 in [6] if the following strict semi-linearized qualification condition

$$0 \in \operatorname{int}\,\{e'(x^*)((K_{\mathsf{ad}} - x^*) \cap \{0, \xi^*, \eta^*\}^\perp)\} \subset Y. \tag{CQA}$$

In our case we have

$$(K_{\mathsf{ad}} - x^*) \cap \{0, \xi^*, \eta^*\}^\perp = W(0,T) \times \left((U_{\mathsf{ad}} \times V_{\mathsf{ad}} - (u^*, v^*)) \cap \{\xi^*, \eta^*\}^\perp \right).$$

Let $z \in Y$ be arbitrary, close enough to zero. Then (CQA) follows if there exist an element $(y, u, v) \in W(0,T) \times ((U_{\mathsf{ad}} - u^*) \cap \{\xi^*\}^\perp) \times ((V_{\mathsf{ad}} - v^*) \cap \{\eta^*\}^\perp)$ satisfying

$$e'(x^*)(y, u, v) = z. \tag{10}$$

Due to Proposition 2.3 the operator $e_y(x^*)$ is bijective. Thus, there exists even a unique $y \in W(0,T)$ such that

$$e_y(x^*)y = z - e_u(x^*)u - e_v(x^*)v.$$

This gives (10) so that the claim follows. □

Remark 4.7. As it is proved in [6], condition (CQA) implies uniqueness of the Lagrange multipliers p^*, ξ^* and η^*.

To prove Theorem 4.10 below we make use of the following lemma. Recall that we have introduced the point $\bar{x}$ satisfying the first-order necessary optimality conditions (7). Let

$$\mathcal{U}_a = \{t \in [0,T] : \bar{u}(t) = u_a(t) \text{ a.e.}\} \text{ and } \mathcal{U}_b = \{t \in [0,T] : \bar{u}(t) = u_b(t) \text{ a.e.}\}$$

and set $\mathcal{U} = \mathcal{U}_a \cup \mathcal{U}_b$. For $\bar{v} \in V_{\mathsf{ad}}$ the active sets $\mathcal{V}_a$, $\mathcal{V}_b$, and $\mathcal{V}$ are defined analogously.

Lemma 4.8. *Let $h = (h_1, h_2, h_3) \in \ker e_x(\bar{x})$. Then there exists a constant $C_{\mathsf{ker}} > 0$ depending only on $\bar{x}$, ν, T, σ_0, and σ_1 but independent of (h_2, h_3) such that*

$$\|h_1\|^2_{W(0,T)} \le C_{\mathsf{ker}} \left(\|h_2\|^2_{L^2(0,T)} + \|h_3\|^2_{L^2(0,T)} \right). \tag{11}$$

Moreover, $h_2 \ge 0$ on $\mathcal{U}_a$, $h_2 \le 0$ on $\mathcal{U}_b$, $u = 0$ on $\mathcal{I}_{U_{\mathsf{ad}}} = [0,T] \setminus \mathcal{U}$ and $h_3 \ge 0$ on $\mathcal{V}_a$, $h_3 \le 0$ on $\mathcal{V}_b$, $h_3 = 0$ on $\mathcal{I}_{V_{\mathsf{ad}}} = [0,T] \setminus \mathcal{V}$.

Proof. Due to Lemma 4.2 it remains to prove the estimate (11). Suppose that $h = (h_1, h_2, h_3) \in \ker e_x(\bar{x})$. Then it follows that $h_1(0) = 0$ in Ω and

$$\begin{aligned} \int_0^T & \langle (h_1)_t(\cdot), \varphi(\cdot) \rangle_{(H^1)', H^1} + \sigma_1 h_1(\cdot, 1)\varphi(\cdot, 1) - \sigma_0 h_1(\cdot, 0)\varphi(\cdot, 0)\, dt \\ & + \int_Q \nu (h_1)_x \varphi_x + (\bar{y} h_1)_x \varphi \, dx dt = \int_0^T h_3 \varphi(\cdot, 1) - h_2 \varphi(\cdot, 0)\, dt = 0 \end{aligned} \tag{12}$$

for all $\varphi \in L^2(0,T; H^1(\Omega))$. Taking $\varphi = h_1$ as test function in (12), the estimate follows from variational techniques, see [20]. □

Definition 4.9. *Suppose that $\bar{x}$ satisfies the first-order necessary optimality conditions with associated unique Lagrange multipliers $\bar{p} \in Y$, $\bar{\xi} \in N_{U_{\mathsf{ad}}}(\bar{u})$, and $\bar{\eta} \in N_{V_{\mathsf{ad}}}(\bar{v})$. At $(\bar{x}, \bar{p})$ the* second-order sufficient optimality condition *holds if there exists a $\kappa > 0$ such that*

$$L_{xx}(\bar{x}, \bar{p})(h, h) \ge \kappa \, \|h\|^2_X \quad \text{for all } h \in C(\bar{x}).$$

Theorem 4.10. *If $\|\alpha_Q(y^* - z_Q)\|_{L^2(Q)}$ is sufficiently small, the second-order sufficient optimality condition is satisfied.*

Proof. Let $h = (h_1, h_2, h_3) \in C(x^*) \setminus \{0\}$. Using (2), (8), Lemma 4.8 and Proposition 2.3 we estimate

$$L_{xx}(x^*, p^*)(h,h) \geq \frac{\beta}{2} \|h_2\|^2_{L^2(0,T)} + \frac{\gamma}{2} \|h_3\|^2_{L^2(0,T)} + \left(\frac{\min(\beta,\gamma)}{2C_{\text{ker}}} - 2C_E C \|\alpha_Q(y^* - z_Q)\|_{L^2(Q)}\right) \|h_1\|^2_{W(0,T)}.$$

If $\|\alpha_Q(y^* - z_Q)\|_{L^2(Q)} < \min(\beta,\gamma)/(5CC_{\text{ker}}C_E)$, the claim follows. □

5. Reduced-order Approach

Since (P) is an infinite-dimensional optimal control problem, it requires discretization before it can be realized numerically. Our reduced-order approach to optimal control problems such as (P) is based on approximating the nonlinear dynamics by a Galerkin technique utilizing POD basis functions that contain characteristics of the expected flow.

5.1. The POD-method

In this section we introduce the POD method. For that purpose let X be a real Hilbert space endowed with inner product $(\cdot\,,\cdot)_X$ and norm $\|\cdot\|_X$. For $w_1, \ldots, w_n \in X$ we set $\mathcal{W} = \text{span}\,\{w_1, \ldots, w_n\}$, and refer to $\mathcal{W}$ as the ensemble consisting of the snapshots $\{w_j\}_{j=1}^n$, at least one of which is assumed to be non-zero. Let $\{\psi_i\}_{i=1}^d$ denote an orthonormal basis of $\mathcal{W}$ with $d = \dim \mathcal{W}$. Then each member of the ensemble can be expressed as

$$w_j = \sum_{i=1}^{d} (w_j, \psi_i)_X \psi_i \quad \text{for } j = 1, \ldots, n. \tag{13}$$

The method of POD consists in choosing the orthonormal basis such that for every $\ell \in \{1, \ldots, d\}$ the mean square error between the elements w_j, $1 \leq j \leq n$, and the corresponding ℓ-th partial sum of (13) is minimized on average:

$$\begin{gathered} \frac{1}{n} \min_{\{\psi_k\}_{k=1}^{\ell}} \sum_{j=1}^{n} \left\| w_j - \sum_{k=1}^{\ell} (w_j, \psi_k)_X \psi_k \right\|_X^2 \\ \text{subject to } (\psi_i, \psi_j)_X = \delta_{ij} \text{ for } 1 \leq i \leq \ell, 1 \leq j \leq i. \end{gathered} \tag{14}$$

A solution $\{\psi_i\}_{i=1}^{\ell}$ to (14) is called a POD basis of rank ℓ. We introduce the correlation matrix $K = ((K_{ij})) \in \mathbb{R}^{n \times n}$ corresponding to the snapshots $\{w_j\}_{j=1}^n$ by $K_{ij} = (w_j, w_i)_X/n$. The matrix K is positive semi-definite and has rank d. The solution of (14) can be found in [16], for instance.

Proposition 5.1. *Let $\lambda_1 \geq \ldots \geq \lambda_d > 0$ denote the non-zero eigenvalues of K and $v_1, \ldots, v_d \in \mathbb{R}^n$ the associated eigenvectors. Then a POD basis of rank $\ell \leq d$ is given by*

$$\psi_i = \frac{1}{\sqrt{\lambda_i}} \sum_{j=1}^{n} (v_i)_j w_j,$$

where $(v_i)_j$ is the j-th component of the eigenvector v_i. Moreover, we have the error formula

$$\frac{1}{n} \sum_{j=1}^{n} \left\| w_j - \sum_{i=1}^{\ell} (w_j, \psi_i)_X \psi_i \right\|_X^2 = \sum_{i=\ell+1}^{d} \lambda_i.$$

5.2. Numerical example

To solve (P) we apply the SQP method combined with a primal-dual active set strategy, see [17] in the case of a distributed control problem for the Burgers equation. The partial differential equations are discretized by utilizing ℓ POD basis functions for the spatial discretization and the implicit Euler method for the time integration. For more details we refer to [8], where a non-linear boundary control problem for the heat equation is solved by using the POD-method.

Let us present a numerical example. The programs were written in MATLAB Version 5.3 executed on a DIGITAL Alpha 21264 computer. We choose $T = 1$, $\nu = 0.25$, $f = z = 0$, $\beta = \gamma = -\sigma_0 = 0.01$, $\sigma_1 = 0$, and $y_0 = 1$ in $[0, 1/2]$ and $y_0 = 0$ in $(1/2, 1]$. The time grid is chosen to be $t_j = (j-1)T/40$, $j = 1, \ldots, 41$. The POD basis is determined by a dynamic strategy, see e.g. [1, 8]:

Algorithm 5.2. 1. *Choose $u_\ell^0, v_\ell^0 \in L^2(0,T)$, $i_{\max} > 0$ and set $i = 0$.*

2. *For $u = u_\ell^i$ and $v = v_\ell^i$ solve (1b)–(1e) by utilizing 40 piecewise linear finite elements to get y^i. Compute λ^i from (3a)–(3c) with $y^* = y^i$ and choose the snapshots $w_j = y^i(t_j)$ and $w_{j+m} = \lambda^i(t_j)$ for $j = 1, \ldots, 41$.*
3. *If $i \leq i_{\max}$:*
   (a) *If $i > 0$, then add $\psi_1^{i-1}, \ldots, \psi_k^{i-1}$ to the snapshots ensemble, i.e. $n = 2 \cdot 41 + k$ in the context of Section 5.1.*
   (b) *Compute a POD basis $\{\psi_j^i\}_{j=1}^{\ell}$ as proposed in Proposition 5.1.*
   (c) *Solve the reduced-order optimal control problem to get $(y_\ell^i, u_\ell^i, v_\ell^i)$.*
   (d) *If $i = 0$ or $i + 1 < i_{\max}$ or*

$$\mathtt{res}(i) = \frac{\|(y_\ell^i, u_\ell^i, v_\ell^i) - (y_\ell^{i-1}, u_\ell^{i-1}, v_\ell^{i-1})\|_X}{\|(y_\ell^i, u_\ell^i, v_\ell^i)\|_X} \geq \varepsilon > 0,$$

   then set $i = i + 1$ and goto 2.

Note that we also include information of the adjoint dynamics into the snapshots. This improves the numerics significantly.

We solve (P) for $U_{\mathsf{ad}} = V_{\mathsf{ad}} = L^2(0,T)$ (unconstrained case) with $\ell = 6$ and for $u_a = v_b = 0$, $u_b = 3$, $v_b = -0.5$ (constrained case) with $\ell = 5$. Moreover we choose $u_\ell^0 = v_\ell^0 = 0$, $k = 4$ and $i_{\max} = 2$ in Algorithm 5.2. The POD based solutions are compared with results from discretization using 40 piecewise linear

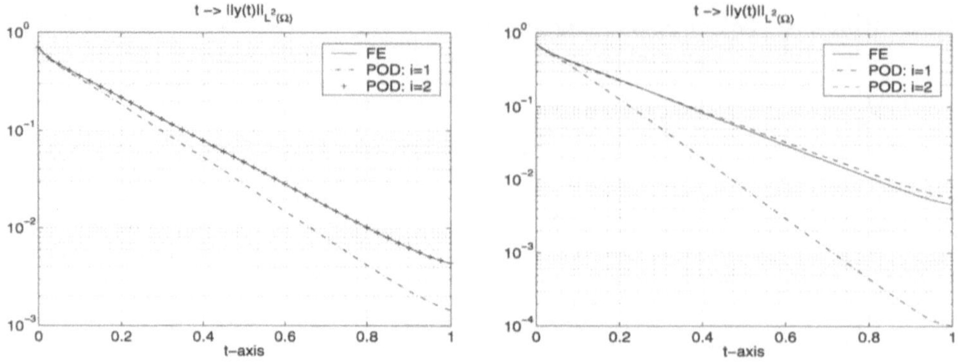

FIGURE 1. State: unconstrained (left) and constrained (right) case.

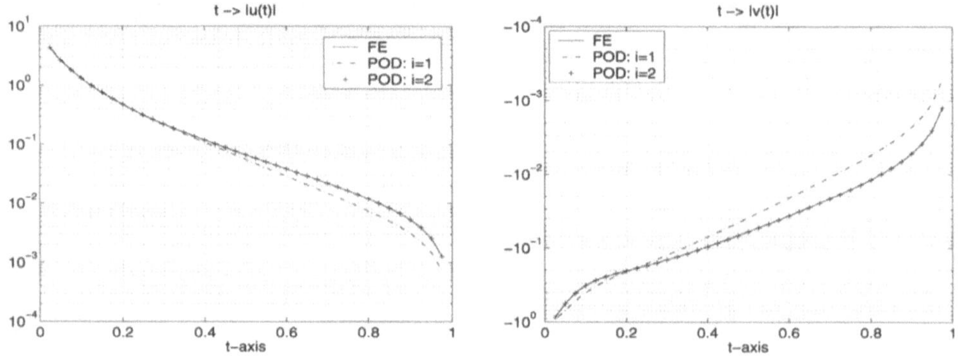

FIGURE 2. Controls: unconstrained case.

	J_{fe}	J_{pod}	$\lvert J_{fe} - J_{pod}\rvert / J_{fe}$
unconstrained	0.026697	0.026732	0.13 %
constrained	0.027230	0.026714	1.89 %

TABLE 1. Values of the cost functionals.

finite elements. From Figures 1–2 it turns out that for the unconstrained problem the FE- and POD-based solutions nearly coincides. The relative error of the costs is about 0.13%, see Table 1. For the reduction of the computing time and the computation effort we refer to Table 2.

In the constrained case the POD-based solution does not nearly coincide with the FE result, see Figures 1 and 3. But however, the relative error of the costs are less than 2% and the CPU time is reduced by a factor 3.2, see Tables 1–2. Note that the reduction of the number of M-flops is by a factor of 18.

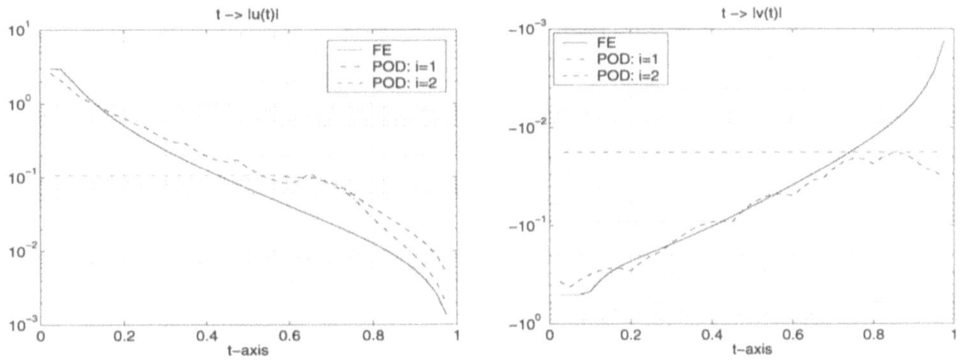

FIGURE 3. Controls: constrained case.

	FE	POD
unconstrained	12.5	3.5
constrained	179.4	55.0

	FE	POD
unconstrained	146	19
constrained	4732	259

TABLE 2. CPU times in seconds (right) and M-flops (left).

References

[1] K. Afanasiev and M. Hinze. Adaptive control of a wake flow using proper orthogonal decomposition. In *Shape Optimization & Optimal Design*, Lecture Notes in Pure and Applied Mathematics. Marcel Dekker, 2001.

[2] J. A. Atwell and B. B. King. Reduced order controllers for spatially distributed systems via proper orthogonal decomposition. *SIAM Journal Scientific Computation*, to appear.

[3] H. T. Banks, R. C. H. del Rosario, and R. C. Smith. Reduced order model feedback control design: Numerical implementation in a thin shell model. Technical report CRSC-TR98-27, North Carolina State University, 1998.

[4] G. Berkooz, P. Holmes, and J. L. Lumley. *Turbulence, Coherent Structures, Dynamical Systems and Symmetry*. Cambridge Monographs on Mechanics. Cambridge University Press, 1996.

[5] J. F. Bonnans. Second-order analysis for control constrained optimal control problems of semilinear elliptic systems. *Appl. Math. Optim.*, 38:303–325, 1998.

[6] J. F. Bonnans and H. Zidani. Optimal control problems with partially polyhedric constraints. *SIAM J. Control Optim.*, 37(6):1726–1741, 1999.

[7] R. Dautray and J.-L. Lions. *Mathematical Analysis and Numerical Methods for Science and Technology. Volume 5: Evolution Problems I.* Springer-Verlag, Berlin, 1992.

[8] F. Diwoky and S. Volkwein. Nonlinear boundary control for the heat equation utilizing proper orthogonal decomposition. Proceedings of the workshop *Fast Solution of Discretized Optimization Problems*, WIAS, Berlin (2000). To appear.

[9] M. Hintermüller. A primal-dual active set algorithm for bilaterally control constrained optimal control problems. Technical Report No. 146, Special Research Center F 003 *Optimization and Control*, Project area *Continuous Optimization and Control*, University of Graz & Technical University of Graz, 1998.

[10] K. Fukunaga. *Introduction to Statistical Recognition*. Academic Press, 1990.

[11] K. Kunisch and S. Volkwein. Control of Burgers' equation by a reduced order approach using proper orthogonal decomposition. *J. Optimization Theory and Applications*, 102(2):345–371, March 1999.

[12] H. V. Ly and H. T. Tran. Modelling and control of physical processes using proper orthogonal decomposition. *Mathematical and Computer Modeling*, 33:223–236, 2001.

[13] S. M. Robinson. Stability theorems for systems of inequalities, Part II: differentiable nonlinear systems. *SIAM J. Numer. Anal.*, 13:497–513, 1976.

[14] S. M. Robinson. Strongly regular generalized equations. *Math. of Operation Research*, 5:43–62, 1980.

[15] R. T. Rockafellar. Conjugate duality and optimization, 1974. Regional conference Series in Applied Mathematics.

[16] L. Sirovich. Turbulence and the dynamics of coherent structures, parts I-III. *Quart. Appl. Math., XLV*, pages 561–590, 1987.

[17] F. Tröltzsch and S. Volkwein. The SQP method for bilaterally control constrained optimal control of the Burgers equation. Technical Report No. 202, Special Research Center F 003 *Optimization and Control*, Project area *Continuous Optimization and Control*, University of Graz & Technical University of Graz, 2000.

[18] S. Volkwein. *Mesh-Independence of an Augmented Lagrangian-SQP Method in Hilbert Spaces and Control Problems for the Burgers Equation*. PhD thesis, Department of Mathematics, Technical University of Berlin, October 1997.

[19] S. Volkwein. Distributed control problems for the Burgers equation. *Computational Optimization and Applications*, 18:133–158, 2001.

[20] S. Volkwein. Second-order conditions for boundary control problems of the Burgers equation. Submitted to *Control and Cybernetics*, 2000.

[21] E. Zeidler. *Nonlinear Functional Analysis and its Application III. Variational Methods and Optimization*. Springer-Verlag, New York, 1985.

S. Volkwein
Institut für Mathematik
Karl-Franzens-Universität Graz
Heinrichstrasse 36
A–8010 Graz
Austria
E-mail address: `stefan.volkwein@uni-graz.at`

International Series of Numerical Mathematics, Vol. 139, 279–289

Flow Matching by Shape Design for the Navier-Stokes System

M. Gunzburger and S. Manservisi

Abstract. We consider a simple shape design problem for the Navier-Stokes system in two-dimensions. The shape of part of the boundary is determined so that flow matches, as well as possible, a given flow. An optimality system is derived and the adjoint equation method is used to determine the shape gradient of the design functional.

1. Introduction

Shape design problems associated with the Navier-Stokes system have wide application. Previous studies devoted to optimal shape design problems for the Stokes and Navier-Stokes equations can be found in [3, 4, 5, 7, 9, 13, 14, 15, 16, 18]. Some of these deal with the existence and regularity of solutions but generally lack a coherent first-order necessary condition; often the regularity assumed cannot be used in numerical algorithms. Other papers deal with reformulations of the problem, mainly to simplified situations. The embedding domain technique explored in [17] provides an equivalent formulation of the shape design problem on a fixed domain and an explicit formula for shape variations for the Navier-Stokes equation is proposed. We recover this result and extend it in the framework of a more general Lagrange multiplier technique that can easily take into account other constraints.

The main focus is to find an appropriate formulation of the optimal shape design problem that is attractive for consistent numerical computations. We consider the full Navier-Stokes case and recover the first-order necessary conditions. The resulting optimality condition is a system of equations and variational inequalities which express the problem in a compact and coherent mathematical formulation. Although we deal with a specific, flow matching problem, the approach used here is discussed in general terms and can be used for many other optimal control problems involving different objective functionals, classes of shape controls, and more complicated domains. For the case of drag minimization and for details concerning the results of this paper, see [10].

1.1. The Model Shape Control Problem

We consider the two-dimensional, Navier-Stokes flow through the channel Ω. The velocity $\vec{u}$ and pressure p satisfy the stationary Navier-Stokes system

$$-\nu\Delta\vec{u} + (\vec{u}\cdot\nabla)\vec{u} + \nabla p = \vec{f} \quad \text{in } \Omega \tag{1}$$

$$\nabla\cdot\vec{u} = 0 \qquad \text{in } \Omega \tag{2}$$

$$\vec{u} = \vec{g} = \begin{cases} \vec{g}_i & \text{on } \Gamma_i,\ i = 1,3 \\ \vec{0} & \text{on } \Gamma_2 \cup \Gamma_4 \end{cases} \tag{3}$$

$$\int_\Gamma \vec{g}\cdot\vec{n}\, ds = 0\,, \tag{4}$$

where $\vec{f}$ is the given body force, ν is the inverse of the Reynolds number whenever appropriate nondimensionalizations are used, and $\vec{g}_1$ and $\vec{g}_3$ are given velocities at the inflow Γ_1 and outflow Γ_3. Along the bottom Γ_4 and top Γ_2 of the channel the velocity vanishes. The set $\Gamma_\alpha = \{\vec{x} = (x,z) \in \Re^2 \,|\, x \in (a,b), z = \alpha(x)\} \subset \Gamma_2$ is the shape which is to be determined. We denote the interval (a,b) by I and the domain Ω by $\Omega(\alpha)$.

We define a set of allowable shapes in the following way. Let c_0, d_0, c_2, and d_2 be positive constants and z_1 and z_2 be the location of the controlled surface Γ_α at $x = a$ and $x = b$, respectively. Then, the set

$$\{\alpha(x) \in C^1(I) \quad | \quad c_i \le \alpha^{(i)} \le d_i \text{ for } i = 0,2,$$
$$\alpha(a) = z_1,\ \alpha(b) = z_2, \text{ and } \alpha^{(1)}(a) = \alpha^{(1)}(b) = 0\}$$

may be a suitable set of allowable shapes, where $\alpha^{(i)}$ denotes the i-th derivative of α.

In order to enforce the regularity of the boundary, we take α to be the solution of the Poisson equation

$$\frac{d^2\alpha}{dx^2} = q \ \text{ on } I\,, \quad \alpha(a) = z_1, \quad \alpha(b) = z_2\,, \tag{5}$$

where $c_2 \le q \le d_2$ for all $x \in I$ is an unknown function. The other boundary conditions on α, i.e., $\alpha'(a) = \alpha'(b) = 0$, impose constraints on allowable functions q. In fact, if α and q are related by (5), then $\alpha'(a) = \alpha'(b) = 0$ if and only if

$$\int_a^b q(x)\, dx = 0\,, \qquad \int_a^b \int_a^x q(\xi)\, d\xi dx = z_2 - z_1\,. \tag{6}$$

Since q is bounded, we have that $\alpha \in C^1(I)$ and Γ is piecewise $C^{1,1}$ with convex corners. We note that the lower bound for α, i.e., $\alpha(x) \ge c_0\ \forall x \in I$, is necessary to avoid the intersection of Γ_2 and Γ_4 and that the second derivative bounds are necessary to assure that the curvature on Γ_α can be computed.

We focus on the minimization of the cost functional

$$J(\vec{u},q,\alpha) = \frac{1}{2}\int_{\Omega(\alpha)} \nabla(\vec{u}-\vec{u}_0) : \nabla(\vec{u}-\vec{u}_0)\, d\vec{x} + \frac{\beta}{2}\int_I q^2\, dx\,, \tag{7}$$

where $\vec{u}_0$ is a given flow field and β is a nonnegative constant. For $\beta = 0$, the functional (7) represents the discrepenacy between the velocity field $\vec{u}$ and the velocity field of the given flow $\vec{u}_0$. The target flow field $\vec{u}_0$ is not required to be solenoidal nor to satisfy any specific boundary conditions; also, it is assumed that $\vec{u}_0$ is defined over all allowable domains $\Omega(\alpha)$.

Formally speaking, the design problem we consider is the *flow matching problem* of finding $\vec{u}$, α, and q such that the functional (7) is minimized subject to the Navier-Stokes system (1)–(4) and the relations (5) and (6) being satisfied.

1.2. Notations

Depending on the context, C and K denote generic constants whose values also depend on context. We denote by $H^s(\mathcal{O})$, $s \in \Re$, the standard Sobolev space of order s with respect to the set $\mathcal{O}$, which is either the flow domain Ω, or its boundary Γ, or part of its boundary. Whenever m is a nonnegative integer, the inner product over $H^m(\mathcal{O})$ is denoted by $(f,g)_m$ and (f,g) denotes the inner product over $H^0(\mathcal{O}) = L^2(\mathcal{O})$. Hence, we associate with $H^m(\mathcal{O})$ its natural norm $\|f\|_{m,\mathcal{O}} = \sqrt{(f,f)_m}$. Whenever possible, we will neglect the domain label in the norm. For vector-valued functions and spaces, we use boldface notation. For example, $\mathbf{H}^s(\Omega) = [H^s(\Omega)]^n$ denotes the space of $\Re^n$-valued functions such that each component belongs to $H^s(\Omega)$. Of special interest is the space

$$\mathbf{H}^1(\Omega) = \left\{ v_j \in L^2(\Omega) \;\middle|\; \frac{\partial v_j}{\partial x_k} \in L^2(\Omega) \quad \text{for } j,k = 1,2 \right\}$$

equipped with the norm $\|\vec{v}\|_1 = (\sum_{k=1}^2 \|v_k\|_1^2)^{1/2}$. We define the space

$$\mathbf{V}(\Omega) = \{ \vec{u} \in \mathbf{H}^1(\Omega) \mid \nabla \cdot \vec{u} = 0 \}$$

and the space of infinite differentiable solenoidal functions by

$$\mathcal{V}(\Omega) = \{ \vec{u} \in \mathbf{C}_0^\infty(\bar{\Omega}) \mid \nabla \cdot \vec{u} = 0 \}.$$

For $\Gamma_s \subset \Gamma$ with nonzero measure, we also consider the subspace

$$\mathbf{H}^1_{\Gamma_s}(\Omega) = \{ \vec{v} \in \mathbf{H}^1(\Omega) \mid \vec{v} = \vec{0} \quad \text{on } \Gamma_s \}.$$

Also, we write $\mathbf{H}^1_0(\Omega) = \mathbf{H}^1_\Gamma(\Omega)$. For any $\vec{v} \in \mathbf{H}^1(\Omega)$, we write $\|\nabla\vec{v}\|$ for the seminorm. Let $(\mathbf{H}^1_{\Gamma_s})^*$ denote the dual space of $\mathbf{H}^1_{\Gamma_s}$. Note that $(\mathbf{H}^1_{\Gamma_s})^*$ is a subspace of $\mathbf{H}^{-1}(\Omega)$, where the latter is the dual space of $\mathbf{H}^1_0(\Omega)$. The duality pairing between $\mathbf{H}^{-1}(\Omega)$ and $\mathbf{H}^1_0(\Omega)$ is denoted by $< \cdot, \cdot >$.

Let $\vec{g}$ be an element of $\mathbf{H}^{1/2}(\Gamma)$. It is well known that $\mathbf{H}^{1/2}(\Gamma)$ is a Hilbert space with norm

$$\|\vec{g}\|_{1/2,\Gamma} = \inf_{\vec{v} \in \mathbf{H}^1(\Omega);\, \gamma_\Gamma \vec{v} = \vec{g}} \|\vec{v}\|_1\,,$$

where γ_Γ denotes the trace mapping $\gamma_\Gamma : \mathbf{H}^1(\Omega) \to \mathbf{H}^{1/2}(\Gamma)$. We let $(\mathbf{H}^{1/2}(\Gamma))^*$ denote the dual space of $\mathbf{H}^{1/2}(\Gamma)$ and $< \cdot, \cdot >_\Gamma$ denote the duality pairing between

$(\mathbf{H}^{1/2}(\Gamma))^*$ and $\mathbf{H}^{1/2}(\Gamma)$. From the definition of the dual norm, we have

$$\|\vec{s}\|_{-1/2,\Gamma} = \sup_{\vec{g}\in\mathbf{H}^{1/2}(\Gamma);\,\vec{g}\neq\vec{0}} \frac{<\vec{s},\vec{g}>_\Gamma}{\|\vec{g}\|_{1/2}}\,.$$

Let Γ_s be a smooth subset of Γ. Then, the trace mapping $\gamma_{\Gamma_s} : \mathbf{H}^1(\Omega) \to \mathbf{H}^{1/2}(\Gamma_s)$ is well defined and $\mathbf{H}^{1/2}(\Gamma_s) = \gamma_{\Gamma_s}(\mathbf{H}^1(\Omega))$.

Since the pressure is only determined up to an additive constant by the Navier-Stokes system with velocity boundary conditions, we define the space of square integrable function having zero mean over Ω as

$$L^2_0(\Omega) = \{p \in L^2(\Omega) \mid \int_\Omega p\, d\vec{x} = 0\}\,.$$

In order to define a weak form of the Navier-Stokes equations, we introduce the continuous bilinear and trilinear forms

$$a(\vec{u},\vec{v}) = 2\nu \int_\Omega D(\vec{u}) : D(\vec{v})\, d\vec{x} \tag{8}$$

$$b(\vec{v},q) = -\int_\Omega q\,\nabla\cdot\vec{v}\, d\vec{x} \tag{9}$$

$$c(\vec{w};\vec{u},\vec{v}) = \int_\Omega \vec{w}\cdot\nabla\vec{u}\cdot\vec{v}\, d\vec{x}\,. \tag{10}$$

for all $\vec{u},\vec{v},\vec{w} \in \mathbf{H}^1(\Omega)$ and $q \in L^2_0(\Omega)$.

For details concerning the function spaces we have introduced, one may consult [1, 19] and for details about the bilinear and trilinear forms and their properties, one may consult [6, 19].

1.3. The Associated Boundary Value Problem

We consider the formulation of the direct problem for the Navier-Stokes system (1)–(3) for which the boundary and all the data functions are known. Let $\Gamma(\alpha)$ be the boundary which includes the segment Γ_α (see Figure 1) defined for a given $\alpha \in H^2(I)$. Given α, we can compute q by using (5).

A weak formulation of the Navier-Stokes system is given as follows: *given* $\vec{f} \in \mathbf{H}^{-1}(\Omega(\alpha))$ *and* $\vec{g} \in \mathbf{H}^{1/2}(\Gamma(\alpha))$, *find* $(\vec{u},p) \in \mathbf{H}^1(\Omega(\alpha)) \times L^2_0(\Omega(\alpha))$ *satisfying*

$$\begin{cases} a(\vec{u},\vec{v}) + c(\vec{u};\vec{u},\vec{v}) + b(\vec{v},p) = <\vec{f},\vec{v}> \\ b(\vec{u},q) = 0 \\ <\vec{u},\vec{s}>_{\Gamma(\alpha)} = <\vec{g},\vec{s}>_{\Gamma(\alpha)} \end{cases} \tag{11}$$

$\forall\,(\vec{v},q,\vec{s}) \in \mathbf{H}^1_0(\Omega(\alpha)) \times L^2_0(\Omega(\alpha)) \times \mathbf{H}^{-1/2}(\Gamma(\alpha))$. Existence, uniqueness, and regularity results for solutions of the system (11) are contained in the following theorem; see, e.g., [6, 16, 19].

Theorem 1.1. *Let $\Omega(\alpha)$ be an open, bounded set of $\Re^2$ with Lipschitz-continuous boundary $\Gamma(\alpha)$. Let $\vec{f} \in \mathbf{H}^{-1}(\Omega(\alpha))$ and $\vec{g} \in \mathbf{H}^{1/2}(\Gamma(\alpha))$ and let $\vec{g}$ satisfy the compatibility condition* (4). *Then, there exists at least one solution $(\vec{u}, p) \in \mathbf{H}^1(\Omega(\alpha)) \times L^2(\Omega(\alpha))$ of* (11)*; the set of velocity fields that are solutions of* (11) *is closed in $\mathbf{H}^1(\Omega(\alpha))$ and is compact in $\mathbf{L}^2(\Omega(\alpha))$; and if $\nu > \nu_0(\Omega(\alpha), \vec{f}, \vec{g})$ for some positive ν_0 whose value is determined by the given data, then the set of solutions of* (11) *consists of a single element. Now, let $\Gamma(\alpha)$ be piecewise $C^{1,1}$ with convex corners, $\vec{g} \in \mathbf{H}^{3/2}(\Gamma(\alpha))$, and $\vec{f} \in \mathbf{L}^2(\Omega(\alpha))$. Let $(\vec{u}, p)$ denote a solution of* (11). *Then, $(\vec{u}, p) \in \mathbf{H}^2(\Omega(\alpha)) \times H^1(\Omega(\alpha)) \cap L^2_0(\Omega(\alpha))$ and the set of solutions of* (11) *is closed in $\mathbf{H}^2(\Omega(\alpha))$ and compact in $\mathbf{H}^1(\Omega(\alpha))$.*

2. The Shape Design Problem

We now formulate the model shape design problem. We define the closed convex set

$$\mathcal{Q}_{ad} = \big\{ \alpha \in H^2(I) \mid 0 < c_0 \le \alpha \le d_0 \\ \alpha(a) = z_1,\ \alpha(b) = z_2, \text{ and } \alpha'(a) = \alpha'(b) = 0 \big\}$$

and introduce the variable q belonging to the set

$$\mathcal{B}_{ad} = \big\{ q \in L^2(I) \mid c_2 \le q \le d_2 \text{ almost everywhere} \big\}$$

defined by

$$\int_I qv\,dx = \int_I \frac{d^2\alpha}{dx^2} v\,dx \qquad \forall v \in L^2(I)\,. \tag{12}$$

The constants c_2 and d_2 are such that the set $\mathcal{Q}_{ad}$ is not empty. From the Sobolev imbedding theorem, we have that $H^2(I) \subset \mathcal{C}^1(\bar{I}) \subset \mathcal{C}^{0,1}(\bar{I})$ and therefore, if $\alpha \in \mathcal{Q}_{ad}$ and $q \in \mathcal{B}_{ad}$, then $\alpha \in C^{1,1}$ at least.

The shape design problem can then be stated in the following way: *given $\vec{u}_0 \in \mathbf{H}^1(\Omega_0)$ and $\vec{f} \in \mathbf{L}^2(\Omega(\alpha))$ and $\vec{g} \in \mathbf{H}^{3/2}(\Gamma(\alpha))$ satisfying the compatibility condition* (4), *find $(\vec{u}, p, q, \alpha)$ such that*

$$\mathcal{J}(\vec{u}, q, \alpha) \le \mathcal{J}(\tilde{u}, \tilde{q}, \tilde{\alpha}) \tag{13}$$

for all $(\tilde{u}, \tilde{p}, \tilde{q}, \tilde{\alpha}) \in \mathbf{H}^2(\Omega(\alpha)) \times \mathbf{H}^1(\Omega(\alpha)) \cap L^2_0(\Omega(\alpha)) \times \mathcal{B}_{ad} \times \mathcal{Q}_{ad}$ satisfying (11) *and* (12).

The admissible set of states and controls is given by

$$\mathcal{A}_{ad} = \{(\vec{u}, p, q, \alpha) \in \mathbf{H}^2(\Omega(\alpha)) \cap \mathbf{V}(\Omega(\alpha)) \times H^1(\Omega(\alpha)) \cap L^2_0(\Omega(\alpha)) \times \mathcal{B}_{ad} \times \mathcal{Q}_{ad} \\ \text{such that } \mathcal{J}(\vec{u}, q, \alpha) < \infty \text{ and } (\vec{u}, p, q, \alpha) \text{ satisfies (11) and (12)}\}\,.$$

The existence of optimal solutions for shape design problem is given by the following result.

Theorem 2.1. *There exists at least one optimal solution $(\vec{u}, p, q, \alpha) \in \mathcal{A}_{ad}$ of the optimal shape design problem* (13).

3. The Lagrange Multiplier Method

3.1. Preliminaries

We introduce auxiliary variables that allow us to transform the inequality constraints into equalities and then invoke well-known derivations for equality constrained minimization problems; see, e.g., [2] or [20]. We begin by replacing

$$c_0 \le \alpha \le d_0 \quad \text{and} \quad c_2 \le q \le d_2 \quad \forall x \in I \tag{14}$$

by

$$|\alpha - \alpha_0|^2 - \alpha_m^2 + s_0^2 = 0 \qquad \forall x \in I \tag{15}$$

$$|q - q_0|^2 - q_m^2 + s_2^2 = 0 \qquad \forall x \in I \tag{16}$$

for some $s_2 \in L^2(I)$ and $s_0 \in H^2(I)$, where $\alpha_0 = (c_0 + d_0)/2$, $q_0 = (c_2 + d_2)/2$, $\alpha_m = (d_0 - c_0)/2$, and $q_m = (d_2 - c_2)/2$. Clearly, if (15)–(16) are satisfied, then so are (14). Also, note that if $(\vec{u}, p, q, \alpha)$ is a solution of the shape design problem, then there exist s_0, s_2 such that α, q and s_0, s_2 satisfy (15)–(16).

We let $\Gamma(\alpha)$ be piecewise $C^{1,1}$ in agreement with the proposed model problem and $\vec{g} \in \mathbf{H}^{3/2}(\Gamma(\alpha))$ where $\int_{\Gamma(\alpha)} \vec{g} \cdot \vec{n} ds = 0$ with $\vec{g} = \vec{0}$ on $\Gamma_2 \cap \Gamma_4$, $\vec{g} = \vec{g}_1 \in \mathbf{H}^{3/2}(\Gamma_1)$ on Γ_1 and $\vec{g} = \vec{g}_3 \in \mathbf{H}^{3/2}(\Gamma_3)$ on Γ_3.

Let $\mathbf{B}_1 = (\mathbf{H}^2(\Omega) \cap \mathbf{H}_0^1(\Omega)) \times (L_0^2(\Omega) \cap H^1(\Omega)) \times \mathcal{B}_{ad} \times \mathcal{Q}_{ad} \times H^2(I) \times L^2(I)$, $\mathbf{B}_2 = \mathbf{H}^{-1}(\Omega) \times L_0^2(\Omega) \times \mathbf{H}^{1/2}(I) \times \mathbf{H}^{1/2}(\Gamma(\alpha) - \Gamma_\alpha) \times L^2(I) \times H^2(I) \times L^1(I)$ and $\mathbf{B}_3 = \mathbf{H}^{-1}(\Omega) \times L_0^2(\Omega) \times \mathbf{H}^{1/2}(I) \times \mathbf{H}^{1/2}(\Gamma(\alpha) - \Gamma_\alpha) \times L^2(I) \times W^{2,1}(I) \times L^1(I)$. We equip $\mathbf{B}_1$, $\mathbf{B}_2$ and $\mathbf{B}_3$ with the usual graph norms for the product spaces involved. We define the nonlinear mapping $M : \mathbf{B}_1 \to \mathbf{B}_3$ by $M(\vec{U}) = \vec{b}$ for $\vec{U} = (\vec{u}, p, q, \alpha, s_0, s_2) \in \mathbf{B}_1$ and $\vec{b} = (\vec{l}_1, l_2, \vec{l}_3, \vec{l}_4, l_5, l_6, l_7) \in \mathbf{B}_3$ if and only if

$$\left\{\begin{array}{l}
\nu a(\vec{u},\vec{v}) + c(\vec{u};\vec{u},\vec{v}) + b(\vec{v},p) - \displaystyle\int_\Omega \vec{f}\cdot\vec{v}\,d\vec{x} = \int_\Omega \vec{l}_1\cdot\vec{v}\,d\vec{x} \quad \forall\,\vec{v}\in\mathbf{H}_0^1(\Omega) \\
b(\vec{u},z) = \displaystyle\int_\Omega l_2\, z\, d\vec{x} \quad \forall\, z \in L_0^2(\Omega) \\
\displaystyle\int_{\Gamma_\alpha} \vec{u}\cdot\vec{s}\,ds = \int_I \vec{l}_3\cdot\vec{s}\,ds \quad \forall\vec{s}\in\mathbf{H}^{-1/2}(I) \\
\displaystyle\int_{\Gamma(\alpha)-\Gamma_\alpha} (\vec{u}-\vec{g})\cdot\vec{s}\,ds = \int_{\Gamma(\alpha)-\Gamma_\alpha} \vec{l}_4\cdot\vec{s}\,ds \quad \forall\vec{s}\in\mathbf{H}^{-1/2}(\Gamma(\alpha)-\Gamma_\alpha) \\
\displaystyle\int_I v\,q\,dx + \int_I \frac{d\alpha}{dx}\frac{dv}{dx}\,dx = \int_I l_5\, v\, dx \forall\, v\in H_0^1(I) \\
(\alpha-\alpha_0)^2 - \alpha_m^2 + s_0^2 = l_6 \quad \forall\, x\in I \\
(q-q_0)^2 - q_m^2 + s_2^2 = l_7 \quad \forall\, x\in I
\end{array}\right. \tag{17}$$

with $\alpha(a) = z_1$, $\alpha(b) = z_2$, and $\alpha'(a) = \alpha'(b) = 0$. The set of constraint equations in the optimal shape design problem can be expressed as $M(\vec{u}, p, q, \alpha, s_0, s_2) = (\vec{0}, 0, \vec{0}, \vec{0}, 0, 0, 0)$.

Given $(\vec{u}_1, p_1, q_1, \alpha_1) \in \mathcal{A}_{ad}$, we define another nonlinear mapping $Q : \mathbf{B}_1 \to \Re \times \mathbf{B}_3$ by $Q(u, p, q, \alpha, s_0, s_2) = \vec{b} = (a, \vec{l}_1, l_2, \vec{l}_3, \vec{l}_4, l_5, l_6, l_7)$ if and only if

$$\begin{pmatrix} \mathcal{J}(\vec{u}, q, \alpha) - \mathcal{J}(\vec{u}_1, q_1, \alpha_1) \\ M(\vec{u}, p, q, \alpha, s_0, s_2) \end{pmatrix} = \begin{pmatrix} a \\ \vec{b} \end{pmatrix}. \tag{18}$$

3.2. Differentiability

These mappings are strictly differentiable, as is shown in the following lemma. We recall the notion of strict differentiability (see [20]). Let X and Y denote Banach spaces, then the mapping $\varphi : X \to Y$ is strictly differentiable at $x \in X$ if there exists a bounded, linear mapping D from X to Y such that for any $\epsilon > 0$ there exists a $\delta > 0$ such that whenever $\|x - x_1\|_X < \delta$ and $\|x - x_2\|_X < \delta$ for $x_1, x_2 \in X$, then

$$\|\varphi(x_1) - \varphi(x_2) - D(x_1 - x_2)\|_Y \le \epsilon \|x_1 - x_2\|_X \,.$$

The strict derivative D at the point $x \in X$, if it exists, will often be denoted by $D = \varphi'(x)$. The value of this mapping on an element $\widetilde{x} \in X$ will often be denoted by $\varphi'(x) \cdot \widetilde{x}$. In the next theorem we can identify $X = \mathbf{B}_1$ and $Y = \mathbf{B}_2$.

Lemma 3.1. *Let the nonlinear mappings $M : \mathbf{B}_1 \to \mathbf{B}_3$ and $Q : \mathbf{B}_1 \to \Re \times \mathbf{B}_3$ be defined by* (17) *and* (18), *respectively. Then, these mappings are strictly differentiable at a point $(\vec{u}, p, q, \alpha, s_0, s_2) \in \mathbf{B}_1$ and its strict derivative is given by the bounded linear operator $M'(\vec{u}, p, q, \alpha, s_0, s_2) : \mathbf{B}_1 \to \mathbf{B}_2$, where $M'(\vec{u}, p, q, \alpha, s_0, s_2) \cdot (\widetilde{u}, \widetilde{p}, \widetilde{q}, \widetilde{\alpha}, \widetilde{s}_0, \widetilde{s}_2) = \bar{b}$ for $(\widetilde{u}, \widetilde{p}, \widetilde{q}, \widetilde{\alpha}, \widetilde{s}_0, \widetilde{s}_2) \in \mathbf{B}_1$ and $\bar{b} = (\bar{l}_1, \bar{l}_2, \bar{l}_3, \bar{l}_4, \bar{l}_5, \bar{l}_6, \bar{l}_7) \in \mathbf{B}_2$ if and only if*

$$\begin{cases} \nu a(\widetilde{u}, \vec{v}) + c(\widetilde{u}; \vec{u}, \vec{v}) + c(\vec{u}; \widetilde{u}, \vec{v}) + b(\vec{v}, \widetilde{p}) = \displaystyle\int_\Omega \bar{l}_1 \cdot \vec{v} \, d\vec{x} \quad \forall \vec{v} \in \mathbf{H}_0^1(\Omega) \\ b(\widetilde{u}, z) = \displaystyle\int_\Omega \bar{l}_2 \, z \, d\vec{x} \quad \forall z \in L_0^2(\Omega) \\ \displaystyle\int_{\Gamma_\alpha} \widetilde{u} \cdot \vec{s} \, ds + \int_{\Gamma_\alpha} (\vec{V}(\widetilde{\alpha}) \cdot \vec{n}) \, (k + \frac{\partial}{\partial n}) \vec{u} \cdot \vec{s} \, ds \\ \qquad\qquad = \displaystyle\int_I \bar{l}_3 \cdot \vec{s} \, ds \quad \forall \vec{s} \in \mathbf{H}^{-1/2}(I) \\ \displaystyle\int_{\Gamma(\alpha) - \Gamma_\alpha} \widetilde{u} \cdot \vec{s} \, ds = \int_{\Gamma(\alpha) - \Gamma_\alpha} \bar{l}_4 \cdot \vec{s} \, ds \quad \forall \vec{s} \in \mathbf{H}^{-1/2}(\Gamma(\alpha) - \Gamma_\alpha) \\ \displaystyle\int_I \widetilde{q} \, v \, dx + \int_I \frac{d\widetilde{\alpha}}{dx} \frac{dv}{dx} \, dx = \int_I \bar{l}_5 \, v \, dx \quad \forall v \in H_0^1(I) \\ 2\widetilde{\alpha}(\alpha - \alpha_0) + 2\widetilde{s}_0 \, s_0 = \bar{l}_6 \quad \forall x \in I \\ 2\,\widetilde{q}\,(q - q_0) + 2\widetilde{s}_2 \, s_2 = \bar{l}_7 \quad \forall x \in I \\ \widetilde{\alpha}(a) = \widetilde{\alpha}(b) = \widetilde{\alpha}'(a) = \widetilde{\alpha}'(b) = 0 \,, \end{cases} \tag{19}$$

where $\vec{V}(\widetilde{\alpha}) = (0, \widetilde{\alpha})$, κ denotes the curvature, and $\vec{n}$ is the normal vector to Γ_α. Moreover, the strict derivative of Q at a point $\vec{U} = (\vec{u}, p, q, \alpha, s_0, s_2) \in \mathbf{B}_1$ is given *by the bounded linear operator* $Q'(\vec{U}) : \mathbf{B}_1 \to \Re \times \mathbf{B}_2$, *where* $Q'(\vec{U}) \cdot \widetilde{U} = (\overline{a}, \overline{b})$, *for* $\widetilde{U} = (\widetilde{u}, \widetilde{p}, \widetilde{q}, \widetilde{\alpha}, \widetilde{s}_0, \widetilde{s}_2) \in \mathbf{B}_1$ *and* $(\overline{a}, \overline{b}) \in \Re \times \mathbf{B}_2$ *if and only if*

$$\begin{pmatrix} \mathcal{J}'(\vec{u}, q, \alpha) \cdot \widetilde{U} \\ M'(\vec{U}) \cdot \widetilde{U} \end{pmatrix} = \begin{pmatrix} \overline{a} \\ \overline{b} \end{pmatrix}, \tag{20}$$

where

$$\mathcal{J}'(\vec{u}, q, \alpha) \cdot \widetilde{U} = \beta \int_I q \widetilde{q} \, dx + \int_{\Omega(\alpha)} \nabla(\vec{u} - \vec{u}_0) \cdot \nabla \widetilde{u} \, d\vec{x}$$
$$+ \frac{1}{2} \int_{\Gamma_\alpha} (\vec{V}(\widetilde{\alpha}) \cdot \vec{n}) \nabla(\vec{u} - \vec{u}_0) : \nabla(\vec{u} - \vec{u}_0) \, ds \, .$$

From (19), we note that the regularity of the Gateaux derivative cannot be the same as the solution of the Navier-Stokes system. In fact, the boundary conditions for the Gateaux derivative implies a different degree of regularity. Note that since $\vec{u} = \vec{0}$ on Γ_α we have that $\nabla \vec{u} = \partial \vec{u} / \partial n$ along Γ_α.

Next, we prove some further properties of the derivatives of the mappings M and Q.

Lemma 3.2. *Let* $(\vec{u}, p, q, \alpha, s_0, s_2) \in \mathbf{B}_1$ *denote a solution of the optimal control problem. Then we have*

- *the operator* $M'(\vec{u}, p, q, \alpha, s_0, s_2)$ *has closed range in* $\mathbf{B}_2$*;*
- *the operator* $Q'(\vec{u}, p, q, \alpha, s_0, s_2)$ *has closed range in* $\Re \times \mathbf{B}_2$*;*
- *the operator* $Q'(\vec{u}, p, q, \alpha, s_0, s_2)$ *is not onto* $\Re \times \mathbf{B}_2$.

The first-order necessary condition follows easily from the fact that the operator $Q'(\vec{u}, p, q, \alpha, s_0, s_2)$ is not onto $\Re \times \mathbf{B}_2$; see, e.g., [8, 11, 12].

Theorem 3.3. *Given* $(\vec{u}, p, q, \alpha) \in \mathcal{A}_{ad}$. *If* $\vec{U} = (\vec{u}, p, q, \alpha, s_0, s_2) \in \mathbf{B}_1$ *is a solution of the optimal shape design problem, then there exists a nonzero Lagrange multiplier* $(\lambda, \vec{W}) \in \Re \times \mathbf{B}_2^*$, *where* $\vec{W} = (\vec{w}, r, \vec{\theta}, \vec{\eta}, \mu, \tau_0, \tau_2)$, *satisfying the Euler equations*

$$\lambda \mathcal{J}'(\vec{u}, q, \alpha) \cdot \widetilde{U} + \left\langle \vec{W}, M'(\vec{U}) \cdot \widetilde{U} \right\rangle = 0 \quad \forall \widetilde{U} = (\widetilde{u}, \widetilde{r}, \widetilde{q}, \widetilde{\alpha}, \widetilde{s}_0, \widetilde{s}_2) \in \mathbf{B}_1 \, , \tag{21}$$

where $\langle \cdot, \cdot \rangle$ *denotes the duality pairing between* $\mathbf{B}_2$ *and* $\mathbf{B}_2^*$.

3.3. The Optimality System

We now examine the first-order necessary condition (21) to derive an optimality system from which optimal states and controls may be determined.

Theorem 3.4. *Let* $(\vec{u}, p, q, \alpha, s_0, s_2) \in \mathbf{B}_1$ *denote a solution of the optimal design problem. Then, if* $s_0 \neq 0$ *and* $s_2 \neq 0$, μ *is the solution of*

$$\int_I \frac{d\mu}{dx} \frac{d\zeta}{dx} dx + \int_{\Gamma_\alpha} \Big(\nabla(\vec{u} - \vec{u}_0) : \nabla(\vec{u} - \vec{u}_0) - \nu \frac{\partial \vec{u}}{\partial n} \cdot \frac{\partial \vec{w}}{\partial n} \Big) \cdot$$
$$\cdot (\vec{V}(\zeta) \cdot \vec{n}) \, ds = 0 \quad \forall \zeta \in H_0^1(I) \tag{22}$$

$$\int_I (\mu + \beta q) \, v \, dx = 0 \quad \forall v \in L^2(I) \tag{23}$$

for all $\alpha \in H^2(I)$ *with* $\alpha(a) = z_1$, $\alpha(b) = z_2$, *and* $\alpha'(a) = \alpha'(b) = 0$, *where* $(\vec{w}, r) \in \mathbf{H}_0^1(\Omega(\alpha)) \times L_0^2(\Omega(\alpha))$ *satisfies the adjoint problem*

$$\begin{cases} \nu a(\vec{w},\vec{v}) + c(\vec{v};\vec{u},\vec{w}) + c(\vec{u};\vec{v},\vec{w}) + b(\vec{v},r) = \\ \qquad\qquad -\displaystyle\int_{\Omega(\alpha)} \nabla(\vec{u}-\vec{u}_0)\cdot\nabla\vec{v}\,d\vec{x} \quad \forall\,\vec{v} \in \mathbf{H}_0^1(\Omega(\alpha)) \\ b(\vec{w},q) = 0 \quad \forall\, q \in L_0^2(\Omega(\alpha)) \\ \vec{w} \in \mathbf{H}_0^1(\Omega(\alpha))\,. \end{cases} \tag{24}$$

If $s_0 = 0$ *we have* $\alpha = c_0$ *or* $\alpha = d_0$. *If* $s_2 = 0$ *we have* $q = c_2$ *or* $q = d_2$ *which gives* α *through* (12) *and the appropriate boundary conditions.*

As a consequence of the optimality system we have to solve

$$\begin{cases} \nu a(\vec{u},\vec{v}) + c(\vec{u};\vec{u},\vec{v}) + b(\vec{v},p) = <\vec{f},\vec{v}> \quad \forall\,\vec{v} \in \mathbf{H}_0^1(\Omega(\alpha)) \\ b(\vec{u},r) = 0 \quad \forall\, r \in L_0^2(\Omega(\alpha)) \\ <\vec{u},\vec{s}>_{\Gamma(\alpha)} = <\vec{g},\vec{s}>_{\Gamma(\alpha)} \quad \forall \vec{s} \in \mathbf{H}^{1/2}(\Gamma(\alpha)) \end{cases} \tag{25}$$

$$\begin{cases} \nu a(\vec{w},\vec{v}) + c(\vec{v};\vec{u},\vec{w}) + c(\vec{u};\vec{v},\vec{w}) + b(\vec{v},r) = \\ \qquad\qquad -\displaystyle\int_{\Omega(\alpha)} \nabla(\vec{u}-\vec{u}_0)\cdot\nabla\vec{v}\,d\vec{x} \quad \forall\,\vec{v} \in \mathbf{H}_0^1(\Omega(\alpha)) \\ b(\vec{w},r) = 0 \quad \forall\, r \in L_0^2(\Omega(\alpha)) \\ \vec{w} = \vec{0} \quad on \quad \Gamma(\alpha) \end{cases} \tag{26}$$

$$\int_I \frac{d\mu}{dx}\frac{d\zeta}{dx}\,dx + \int_{\Gamma_\alpha}\Big(\nabla(\vec{u}-\vec{u}_0) : \nabla(\vec{u}-\vec{u}_0) - \nu\frac{\partial\vec{u}}{\partial n}\cdot\frac{\partial\vec{w}}{\partial n}\Big)\cdot \\ \cdot(\vec{V}(\zeta)\cdot\vec{n})\,ds = 0 \quad \forall \zeta \in H_0^1(I) \tag{27}$$

$$-\int_I \frac{d\alpha}{dx}\frac{dv}{dx}\,dx = \int_I q\,v\,dx \quad \forall v \in H_0^1(I) \tag{28}$$

$$\int_I (\beta q + \mu)(\widetilde{q} - q) \geq 0 \tag{29}$$

with $\alpha(a) = z_0$ and $\alpha(a) = z_1$ and for all $\widetilde{\alpha} \in \mathcal{Q}_{ad}$ and for all $\widetilde{q} \in \mathcal{B}_{ad}$.

3.4. The Shape Gradient

The numerical solution of the coupled system (25)–(29) of variational equations and inequalities is formidable. In practice, one does not solve the system simultaneously, instead, one invokes an iteration such that at each step the method requires the sequential solution of the Navier-Stokes system (25) and the adjoint system in (26). The solution of the two Poisson equations (28) and (29) is not very expensive but care should be taken to check that the shape function α is in $\mathcal{C}_{ad}$. The iteration is often implemented so that at each step a guess for the boundary shape Γ_α is made, after which the state system (25) can be solved and the design functional evaluated. To determine the new iterate for the boundary shape, a gradient or quasi-Newton method can be used. Efficient implementations of such methods usually require knowledge of the shape gradient, i.e., the gradient of the functional with respect to changes in the boundary shape.

For any candidate optimizer $(\vec{u}, p, q, \alpha)$ satisfying the state system (25), the shape gradient is given in terms of the adjoint variables by

$$\begin{aligned} \mathcal{J}'(\vec{u}, q, \alpha)(\vec{w}, \mu, \zeta) &= \int_I \frac{d\mu}{dx}\frac{d\zeta}{dx} dx \\ &+ \int_{\Gamma_\alpha} \left(\nabla(\vec{u} - \vec{u}_0)\, \nabla(\vec{u} - \vec{u}_0) - \nu \frac{\partial \vec{u}}{\partial n} \cdot \frac{\partial \vec{w}}{\partial n} \right) (\vec{V}(\zeta) \cdot \vec{n})\, ds\,. \end{aligned} \tag{30}$$

The vector $\vec{V}(\zeta)$ is simply $(0, \zeta)$, $\vec{n}$ is the unit normal to Γ_α and the adjoint variables $\vec{w}$ and μ are determined from (26) and $\mu + \beta q = 0$, respectively. The shape gradient can be computed for multiple directions ζ with a single linear adjoint system solution. On the other hand, if sensitivity equations are used, one must solve a linear sensitivity system for each distinct direction ζ.

References

[1] R. Adams, *Sobolev Spaces,* Academic, New York (1975).

[2] V. Alekseev, V. Tikhomirov, and S. Fomin, *Optimal Control,* Consultants Bureau, New York (1987).

[3] G. Armugan and O. Pironneau, *On the problem of riblets as a drag reduction device,* Optim. Control Appl. Meth., **10** (1989), 93–112.

[4] E. Dean, Q. Dinh, R.Glowinski, J.He, T. Pan and J. Periaux, *Least squares domain embedding methods for Neumann problems: applications to fluid dynamics,* in Domain Decomposition Methods for Partial Differential Equations, D. Keyes, et al., Eds., SIAM, Philadelphia, (1992).

[5] N. Di Cesare, O. Pironneau, and E. Polak, *Consistent approximations for an optimal design problem,* Report 98005 Labotatoire d'Analyse Numérique, Paris, (1998).

[6] V. Girault and P. Raviart, *The Finite Element Method for Navier-Stokes Equations: Theory and Algorithms*, Springer, New York, (1986).

[7] R. Glowinski and O. Pironneau, *Toward the computation of minimum drag profile in viscous laminar flow,* Appl. Math. Model., **1** (1976), 58–66.

[8] M. Gunzburger, L. Hou and T. Svobodny, *Analysis and finite element approximations of optimal control problems for the stationary Navier-Stokes equations with Dirichlet controls,* Math. Model. Numer. Anal., **25** (1991), 711–748.

[9] M. Gunzburger and H. Kim, *Existence of a shape control problem for the stationary Navier-Stokes equations,* SIAM J. Cont. Optim., **36** (1998), 895–909.

[10] M. Gunzburger, H. Kim, and S. Manservisi, *On a shape control problem for the stationary Navier-Stokes equations,* to appear in Math.Model. Numer. Anal.

[11] M. Gunzburger and S. Manservisi, *The velocity tracking problem for Navier-Stokes flows with bounded distributed control,* SIAM J. Cont. Optim., **37** (1999), 1913–1945.

[12] M. Gunzburger and S. Manservisi, *A variational inequality formulation of an inverse elasticity problem,* to appear in Comp. Meth. Appl. Mech. Engrg.

[13] O. Pironneau, *Optimal Shape Design in Fluid Mechanics,* Thesis, University of Paris, Paris, (1976).

[14] O. Pironneau, *On optimal design in fluid mechanics,* J. Fluid. Mech., **64** (1974), 97–110.

[15] O. Pironneau, *Optimal Shape Design for Elliptic Systems,* Springer, Berlin, (1984).

[16] J. Simon, *Domain variation for drag Stokes flows,* in Lecture notes in Control and Information Sciences **114**, A. Bermudez, Ed., Springer, Berlin, (1987), 277–283.

[17] T. Slawig, *Domain Optimization for the Stationary Stokes and Navier-Stokes Equations by Embedding Domain Technique,* Thesis, TU Berlin, Berlin, (1998).

[18] S. Stojanovic, *Non-smooth analysis and shape optimization in flow problems,* IMA Preprint Series **1046**, IMA, Minneapolis, 1992.

[19] R.Temam, *Navier-Stokes Equations,* North-Holland, Amsterdam, (1979).

[20] V. Tikhomirov, *Fundamental Principles of the Theory of Extremal Problems,* Wiley, Chichester, (1986).

Department of Mathematics
Iowa State University
Ames IA 50011-2064, USA
E-mail address: gunzburg@iastate.edu

LIN, DIENCA
University of Bologna
Via dei colli 16
40136 Bologna, Italy
E-mail address: manser@lamu.ing.unibo.it